T0276113

FRAC WITHOUT A K

JOHN W. ELY

AuthorHouse™
1663 Liberty Drive
Bloomington, IN 47403
www.authorhouse.com
Phone: 833-262-8899

This book is printed on acid-free paper.

ISBN: 978-1-6655-5236-3 (sc)
ISBN: 978-1-6655-5237-0 (e)

Library of Congress Control Number: 2022903206

Print information available on the last page.

Published by AuthorHouse 04/06/2022

authorHOUSE®

CONTENTS

PREFACE

This is going to be the third edition of the initial book titled "Stimulation Treatment Handbook". The first book came out in 1985 and was dedicated primarily to quality control, with a lot of discussion relating to larcenous situations and the lack of real-time monitoring on fracturing treatments.

The second book, titled "Stimulation Engineering Handbook" introduced in 1994, also contains a great deal of quality control, but also has much more updating of the QC process and implementation of fracturing treatments including post frac recommendations. I am going to do something, with this book, that I do not believe has been done before. I am going to include the chapters from the 1994 book as leadoff into the actual 2022 book chapters. I believe that this will be something unique in that I will be, many times critical of what I said before, as well as giving new technology, techniques, equipment etc., comparing what was done in 1994 to the present day. We will delve into what is going on for all phases of fracturing including fracture design, job optimization and re-fracturing.

One of the center points of my belief is that we must learn from the past and be capable of altering our perception of what is happening to be able to do the very best job possible.

One of the people who I admired most in this industry, Dr. Steve Holditch, had a saying which he always restated most of the time in meetings. Basically, he stated, "I reserve the right to get smarter". Some of the saddest times of my career was when well-intentioned Engineers and or researchers re-invented past failures.

We will have a new chapter titled *"Reinvention of past failures and massive overuse of intuitive reasoning"*. This chapter may irritate some, but further illustrates the saying "If we don't study the past, we are doomed to repeat it".

I should also state at the onset of the book that my biggest pet peeves are when journalists or others spell frac with a K. They will be confined to the 7th level of Dante's description of the opposite of paradise.

DEDICATION

One of the major errors of my career, as stated in the second book, was dedicating my first book to my under-paid and dedicated secretary, instead of my wife. I tried to make up for that in the second book, dedicating it to her. At the writing of this book, we have passed 56 years of marriage. Although we are still together, to this day I suffer for the mistake I made with the dedication in the first book. I therefore dedicate this book to her for all her sacrifices made by my traveling and to show her my upmost appreciation. I also would like to dedicate this to my three children, their three spouses, after 12 grandchildren, with two spouses, and 5 Great Grand Children. Without their support, over the years, I would not have been able to enjoy life so well in our great country.

I have recently been honored with the John Franklin Carll award from the Society of Petroleum Engineers. I am deeply honored and humbled by the award and wanted to thank Brad Wolters and Mike Gatens for their work on allowing me to receive this honor. I also want to thank Gerald Coulter, John Lee, Brad Robinson, Mark Semmelbeck, Jon Harper, Esteban Nieto, David Lancaster, Pat Carlson, Mark Pearson, and George Voneiff for their letters written. Additionally, I have been named a Legend of Hydraulic Fracturing by the SPE 2021 Hydraulic Fracturing Conference in February of the same year. I know Martin Rylance had a great deal to do with the legend award and I appreciate his efforts.

Summaries of Chapters

As will be the case for all the chapters in this book, I am going to publish, in its original form, the original chapters and then critique or simply exhibit the changes that have occurred as time has passed. It has been amazing that some of the technology has literally expanded exponentially such as the type and handling of fracturing proppant and the huge changes in understanding very low permeability reservoirs. At the same time, there has been little or no development in the area of conventional fluids, i.e., for usage in conventional fracturing treatments for moderately permeable reservoirs.

I will attempt to bring to the readers of this book a very objective, sometimes subjective, comparison of everything from proppants, chemicals, equipment, fracture design, fracture optimization, and refracturing techniques. There may be, as was the case in the other books, some diversion to important oilfield war stories.

I have gone through the original chapters and will summarize changes below. This will basically give the reader some idea of what is covered throughout the book in case you do not want to read everything. The largest changes that have occurred in hydraulic fracturing relate to very high-volume slick water treatments with small size, lower quality sand. This will be a subject we will spend some time on, enlarging the chapter on fluids to contain more than conventional fracturing fluids. We will discuss topics such as why high-strength proppant is not required, why small or low-crush proppant, many times, exceeds the performance of more expensive higher-quality proppant.

Chapter Summaries

1. The eye of the beholder new chapter will be a brief summary of technology learned since the original book was published in 1994.
2. In the chapter on equipment, we will illustrate dramatic changes particularly in sand handling and transport. Although conventional pumping equipment in 1994 has increased from around 2000 HHP to greater than 2500 HHP, the rapidly increasing use of electric equipment with very high horsepower silent turbine power is, I believe a harbinger of times to come. The vast majority of pumping equipment available today is the standard diesel power 2500 HHP equipment. With the huge downturn of the combined low oil prices and Covid 19, there is a huge amount of the standard diesel horsepower stacked upon hundreds of acres of land. It will require years to either to use up all of this equipment through scavenging or refitting and putting back in service. We will also discuss the rebirth and decline of turbine only powered fleets. Although some time will be spent of important history, our major emphasis will be on the future of all equipment, electric powered blenders and pumps.
3. The chapter on safety meetings will be enhanced to discuss more emphasis on more protection for individuals on the wellsite. Many of the safety enhancement has come through digital control of multiple pieces of equipment by a single person and safety parameters wrote into the control software. The chapter on safety will show improvements on fireproof apparel as well as what is termed comfortable stylish garments which are also fireproof.
4. The chapter on compatibility was written with conventional crosslink fluids in mind. We will delve into the early and ongoing problems with the smaller number of chemicals used in slick water fracturing. Understanding the problems relating to salinity and ion character of the minimal chemicals used in slick water fracturing is extremely important in obtaining a successful treatment. The increase in recycling of both flowback and produced water creates unique challenges for formulators of friction reducers, scale inhibitors, biocides and surfactants.
5. The old chapter 5 from the 2nd edition described the complexity of keeping up with multiple chemicals and assuring what is shown on computer screens compared with actual pumped volumes. The massive volumes, high rates and continuous pumping of zipper frac jobs where 16 to 20 million gallons of fluid and more than 20 million pounds of proppant are continuously pumped can create nightmares where stringent controls are not present. The lack of calibration of flow meters, densitometers, and chemical metering pumps can yield a nightmare for those who want to optimize treatments based on actual volumes. We are increasingly dependent upon remote monitoring, even on the wellsite, of virtually everything pumped. There are techniques which are simple to utilize which will, if conducted on a regular basis, give some hope of the data being recorded is factual. If you haven't realized where I am going, A.I. will be totally worthless in optimizing frac treatments if we don't have correct data on what we have pumped.
6. In the chapter on Monitoring fracturing fluids, we will discuss in length why conventional cross-link gels do not perform well in ultra-low permeability reservoirs and at the same time illustrate that these slick water treatments have been conducted since the early days of fracturing, albeit

smaller and poorly designed, mostly using the wrong size proppant. As part of this new chapter we will discuss what we have learned in relationship to conventional frac fluid design and taking some of the knowledge learned there in the design of hydraulic fracturing treatments utilizing only slick water and small sand. We will, utilizing our experience, give general guidelines where conventional fracturing designs start and ultra-low permeability designs end. The major emphasis of this chapter will be to put forward all the revolutionary changes that have occurred with slick water and small low conductivity proppant.

7. Quality control of fracturing fluids

   The 1994 chapter on quality control for water-based fluids was built primarily around conventional crosslinked fracturing fluids. There will be some corrections and additions but for the most part, the fluids we were using in the late 80's are virtually identical to what is used today in conventional fracs. We will discuss the meaning of the word crosslink fluid and dispel some of the misuse of low concentration crosslinked guar and the implications on evaluation of so called "hybrid fluids".

   We are continuing to do conventional fracs on low permeability rock. The reservoirs I am referring to are typically .05-10 millidarcies in permeability, not the typical 200-700 nano Darcy perm of the typical "shale play". The major problem today in successfully completing this type of reservoir is, due to the low volume of treatments conducted, there is very few service companies who have the equipment or personnel to properly execute a perfect proppant transport gel and the utilize forced closure techniques on vertical wells.

   The 1994 chapters concerning QC on oil based fluids, foamed fluids and intense QC have been combined in the new addition. There will be a few updates on techniques to enhance quality control and some time spent discussing why you should not utilize hydrocarbon-based fluids or in many cases, where logistically possible, energized fluids. The basis of on-site QC came directly from Gas Research work done in the early 80's stretching into the mid 90" s. It was discovered that there was a major disconnect between laboratory formulated fluids and field execution. We will go into detail how this disconnect can drastically affect the data gathering process and, ultimately artificial intelligence use enabling optimum stimulation treatments.

8. The Original Chapter 11, on forced closure, will be slightly modified to clarify for those not having conducted single stage conventional treatments. The process can be used in single stage slick water treatments but is not practical in multiple stage horizontals.

   The 2022 version also included in chapter will simply illustrate some information gathered in the 28 years following 1994.

9. The original chapter termed Job Implementation Guidelines will have a title change for 2022 illustrating a better description of what the chapter describes. Although guidelines are present it is better titled "Available processes which have achieved success with conventional fluids." There have been tremendous successes achieved using conventional crosslinked which were given little fanfare and these systems can still be utilized to enhance production of oil and gas. The wildly successful use of slick water has created a dearth of understanding of multiple conventional processes which are still available. The chapter will be updated and made more relative to the 21st century.

10. Chapter 10 is new to the book discussing proppant selection.

    We did not do a chapter on proppant selection in either of the previous books because there was a plethora of information available relating to brown sand, white sand, ceramic, bauxite and resin coated versions of all of these. With the treatment of ultralow permeability reservoirs and the lack of noted proppant crushing, even at depths more than 15,000 feet, the old guidelines do not hold. I will endeavor to address locally available proppant and why these lesser quality proppants are working. This phenomenon has forever changed our industry. We will also discuss a very serious problem relating to lack of oversight of sand sieve size with hundreds and perhaps thousands of fracture treatments conducted with particle size wildly out of API specification. There are problems with the actual API guidelines which we will discuss at length. The wide use of wildly out oof size proppant occurring now, could turn into a major embarrassment for our industry. We will show 100 mesh which is an API 70/140 mesh delivered as 50/100 or sometimes larger and 40/70 delivered consistently as 30/50 or larger. Unless we believe, which I certainly do not, that sand size or distribution does not matter, we will have no way of evaluating the size and distribution on well performance.

11. The Sand water revolution for low permeability reservoirs is a new chapter

    This is the most important chapter in the book. Approximately 20 years ago, led perhaps by George Mitchell, virtually everything changed in hydraulic fracturing. We have taught and published that fractures are single planar cracks that are packed with proppant. We were taught that at great depths and high closures that strength, high conductivity proppants were required.

    Today, we have found that when one gets into lower permeability matrix that little or no crushing occurs and locally available sand with crush strength of 6,000 pounds works well at great depths and high frac gradients. The fluid of choice is typically water only and most of these jobs are at low sand concentrations of the weaker sand. In this chapter I will give my personal opinion of how this works and what this has done to our industry. The days of high viscosity high proppant concentrations has not totally ended but encompass less that 5 percent of the jobs pumped.

12. The original chapter 13 relating to acidizing will be little changed. Very few people know that acidizing, particularly fracture acidizing, was equally as popular as proppant fracturing well into the late 1980's There was significant research and testing relating, not only to various types of acid, but also reaction times. There was a large industry relating to surfactants, non-emulsifiers, acid retarders, acid inhibitors, acid fluid loss agents, diverting agents etc.

    Although proppant fracturing is typically less costly and more successful, matrix and acid fracturing have numerous cost-effective solutions for enhancement of hydrocarbon production.

    As was the case for chapter 12 the title will be changed to Acidizing systems. If the reader wants to go into depth in acidizing calculations consult the SPE monograph on acidizing by John Gidley. I do believe that there are thousands of applications where acidizing should be the preferred technique in optimization of production in oil and gas wells.

13. The original chapter 14 on cementing quality control remains largely unchanged. There is simply nothing more important to successful stimulation than complete zonal isolation. The lack of success of non-cemented packer and sleeve type completions has further illustrated the extreme importance of treating exactly where you perforate. Much of the confusion, even with sophisticated diagnostic tools, on where one has treated, or where production exists, relates to no cement behind pipe. My company has had thousands of success stories relating to pseudo limited entry perforating, but this technique has no possibility of success if excellent cement isolation is not achieved.

    A very large research project co-funded by Exxon and Halliburton stated that the best means of achieving mud cake removal and achievement of a cement bond is to utilize both rotation and reciprocation of pipe during cement operations. This research was unique in that it involved physical measurement of bonding. Pipe movement and solid centralization combined with cleaning and abrasive spacers will almost certainly result in a good cement job. Pipe movement in long laterals is very challenging but has been utilized by some of our customers. In most cases solid centralizers with cleaning and abrasive sweeps are all that can be utilized in long laterals. As I have stated before, there are times when I think God is punishing the oil industry. We have found reservoirs where sleeve type completions have worked well as well as single point entry instead of spaced clusters, but the vast majority of successful stimulation in ultralow permeability reservoirs is accomplished with cluster perforations in cemented pipe. It should be noted that the exceptions mentioned were in the far North.

14. The original chapter 15 on monitoring and readout control is perhaps the single most changed area in fracturing outside of total electric pumping equipment. The changes that have occurred and are still occurring are extremely exciting to those of us who ran frac jobs in open air and depended on hand signals or at best sound power communication headphones. When I entered the fracturing industry Dowell, now Schlumberger, had moved from World War II Allison airplane engines to turbine pumpers with up to 1000 HHP. Halliburton had twin pumpers which would yield 1100 HHP at best. Halliburton had men standing on the trucks and Dowell had remote single truck operators. When the initial frac vans hit the industry, they were said to be a passing fad as real men did their work outside and would have better control of the equipment. As one who has lost a significant amount of hearing capability and has shielded fracometers and densitometers, "Open air strip chart recorders", with my body during storms and pressure leaks, the frac vans were a blessing. I will go into detail in the chapter on the revolutionary changes in metering, monitoring and control of frac equipment.
15. This totally new chapter titled " Re-inventions of past failures and massive overuse of "Intuitive Reasoning". This chapter will probably be somewhat controversial to re-inventors and die- hard believers in failed philosophies. I will be nice.
16. The chapter on philosophy of Q.C. is short and will simply summarize the changes that have occurred not only with quality control for conventional viscous frac treatments, but I well discuss the Q.C. needed for slick water fracs and the huge cost to the industry if competent oversight does not take place.
17. I have decided to add a chapter on directional control of blowouts which took up a significant part of my career. Because of my direct involvement in a Dubai Petroleum Blowout and a subsequent patent and SPE paper I attended multiple directional kill blowouts around the world. The technology is amazing and has nothing to do with the top kill blowouts that occurred in Kuwait. It will also allow for some fine true war stories on locations where money doesn't matter.

CHAPTER 1

THE EYE OF THE BEHOLDER (1994)

In early 1973, I made a commitment to change my occupation from being a researcher which involved field travel, to being a district engineer. Accordingly, I spent a great deal of time broadening my work experience, which had been in fracturing and acidizing, to include other treatments such as cementing, tools, and testing. Then I traveled with my family to south Iran. Upon arriving in Ahwaz, I couldn't wait to attend my first cementing treatment. I was told a large multi truck, 5,000-sack cement job was under way just west of the city. After receiving directions to the location, I departed in a small vehicle. (Note: On the way, I learned not to get off the "beaten path," and I was almost killed by some soldiers who were guarding a production station. But that's another story.)

Poor Expectations = poor service

I arrived on location about 2:00 a.m. The cement job had just begun. By stateside standards, it was a large treatment. It had more than 4,000 sacks of cement with three high-horsepower cement pumping units going downhole. However, what I first saw made me panic. Instead of a smooth operation with the pumps alternately sucking and mixing cement and pumping it downhole with little vibration, I saw the three 2 in. lines leading from the trucks jumping more than 3 ft. into the air. Iranian workers were mixing cement that had the consistency of dry oatmeal, and they were trying to suck up this mixture with a downhole pump. Coolies beating on hammer bins tried to feed the cement into hoppers that led into high-pressure mixers pumping at 2,000 psi. The pumps vibrated because of the cement's high viscosity. The lines jumped into the air ultimately destroying the manifold, which was connected to a 3 in. line going to the cement head on the floor.

I watched an Arab coolie with his head wrapped in cloth shout profanities toward my company as he hit the bins with a sledgehammer. It was the ultimate nightmare for someone used to smooth pumping operations. Not knowing what to do, I walked over to the person in charge. Fearing that he would verbally abuse me or tell me to take my trucks and leave the location, I didn't know what to say. Suddenly, the man put his hand out, introduced himself, and casually commented that it looked like a normal job and that everything seemed to be progressing well. I couldn't believe it! I had not seen treatments such as these; and this, in fact, *was* standard operating procedure.

Overall, this treatment had not been properly designed or properly conducted. Instead, individuals had accepted poor service and poor quality because they were the norm-not because they were right. In short, they had accepted them not because they were good, but because extremely poor service was an accepted standard.

The pumps' extreme vibrations were not the service company's fault. They occurred because the oil company would not use dispersants and chemical additives to lower the cement viscosity. The oil company had also insisted on using saturated salt water, which tends to make cement more viscous. Additionally, they were mixing cement approaching 22 lbs./ gal, which would be difficult under the best of circumstances. Primarily, many problems were caused by using improper equipment, yet these people did not want to change. They thought this approach was the cheapest and the best. As time passed, however, I attempted to educate the oil company to use thinners, dispersants, and modern cementing equipment to accomplish a cementing treatment. However, when I left two years later, the same type of treatment and the same kind of service were still being conducted.

I have seen similar scenarios in other parts of the world, including the United States. However, in some areas of the United States, cementing and stimulation treatments are being conducted professionally with excellent service from the particular service company and good quality control measures taken by the oil company engineers. To accomplish this, I believe two criteria go hand in hand:

1. Quality control by the oil company.
2. Good service from the service company.

Quality service from competent individuals

Over the years, I have seen numerous jobs where company representatives were inept or didn't even show up for the treatment. However, on the whole, when there are competent, knowledgeable company personnel on the location-who check and recheck and who understand what's going on, both on the surface and -downhole-the service results are always good. This relates to acidizing, fracturing, cementing, and tools for any treatment. A much higher percentage of success is achieved under these circumstances.

It has not been uncommon over the last few years for service companies to gear up for a large, high-horsepower, massive frac treatment. They come onto the location with 20-40 frac tanks, 10-15 pump trucks, and as many as 5-7 blenders, assorted iron, frac vans, and other equipment. The stimulation treatment costs can approach $1 million.

All of this hardware is laid out and prepared. On the day of the treatment, an inexperienced engineer arrives to take charge. It is obvious to the service company personnel that this individual is frightened and dazzled by the experience. For those people who have attended large treatments-and even for those of us who have seen hundreds-the setup is always awe-inspiring. There is certainly reason for fear and trepidation.

Great care needs to be taken for the safety of the individuals on the location. It boggles my mind that a company will spend hundreds of thousands of dollars and not take the time to control the quality of what it is buying. For instance, just a company engineer who climbs on the tanks to inspect the gel or who gets up on a blender, walks the lines, and asks about the design will ensure that more care is taken with a treatment. An incident that occurred during the boom years of 1979-1981 in the Austin Chalk area illustrates this point.

I had known a particular gentleman for many years. He was well known for his lack of intelligence. He had trouble keeping a job even during the boom years of the industry. He was found on some occasions sleeping on the side of the road when he was supposed to be on location. He arrived at one of the locations where I was conducting a fracture treatment. He came on location, climbed on top of the tanks, opened the hatches, looked in all of them, climbed off of the frac tanks, and then proceeded over to the sand storage unit. He opened all the hatches there, inspected the sand, walked the lines, and looked at the well head. Needless to say, I was totally surprised and, quite frankly, impressed. I couldn't wait to rush over to the individual, who at that time was working for a small oil company, and congratulate him on his diligence. When I asked him where he had learned to do some of the primary checks concerning quality control, he stated very frankly that he had no idea what he was doing or what he was seeing, but he had observed another individual who had done this and was told that he got a better fracture treatment. Indeed, this is true, and it is proven by the fact that individuals who act like they care about what is going on and make some effort to observe it, do indeed get better fracture treatments.

I am sure that service company representatives are not intentionally dishonest. Yet, service companies are plagued by two basic problems; namely, they employ human beings who are subject to error, and they use mechanical and electrical equipment that is also subject to error. The more people involved who are interested in the quality of the treatment, the better that treatment will be.

Over the past few years, a new buzz word has evolved; that is, *TQM or Total Quality Management.* Since I wrote my first book about quality control before 1985, I can honestly say that I was involved in quality control before Mr. Deming's approach reached the United States. The advent of so much emphasis on quality control, combined with the very traumatic effects of some very large industry damaging lawsuits, such as the Parker and Parsley lawsuit in West Texas, has changed the face of our industry for the better. Because of Quality Control, there is no cheating, stealing, shorting of product or service anywhere in our oilfields today. Obviously, this was not the case in years past, as demonstrated by the lawsuits that occurred in west Texas and other areas. Since the writing of the first book, quality control and "intense quality control" and some of the new procedures and chemicals have greatly enhanced the productive capacity of wells. I did not know about the extent of problems with fluids and systems and their effects on potential production. Since the writing of this book, hundreds of wells have been documented where more than 60% of them would have been plugged without proper quality control procedures. It is my contention that thousands of United States wells exist where the majority of the propped fractures are plugged with unbroken gel.

Not long ago, I was giving a presentation in Oklahoma City on quality control and job execution. Following my presentation, another gentleman gave a presentation on horizontal drilling. In his presentation, he noted that he had, in fact, drilled a horizontal well intersecting a vertical fracture. This vertical fracture had been created nine years earlier with a delayed cross-link, titanate gel system. The bottom hole temperature was 240°. When they intersected the vertical fracture, they cored through it and brought the core to the surface. They found the fracture stuck together with polymer that was still stable nine years after the fracture treatment. If there is stable fracturing gel at 240°, then there are hundreds of wells where broken gel problems exist. In fact, later the book discusses the contention that the tremendous concern about "water sensitivity" is nothing more than poor quality control in addition to breakers. In many instances, it is the ability of the service companies to run the breakers (particularly in hotter wells) and to execute the treatment. With the advent of delayed breakers (encapsulated breakers, etc.), companies now have the ability to accomplish a fracturing treatment in good order and still degrade the fluids back to water.

The war has not been won in having total quality management, intense quality control, or whatever terminology is used in all fracturing, stimulations, cementing, and other treatments. However, we have made progress in these areas, and we have found the tremendous advantages of following quality control and having qualified, capable, and well-trained individuals overseeing hydraulic fracturing, cementing, or other well service treatments.

Frankly, I became spoiled by the first work that I did as a chemist (touting a new fracturing fluid) in an area where the service from local personnel was excellent. The engineering was high quality, and strict control was exercised by the major oil company involved. Therefore, superb quality control and good job performance resulted because the company had first-line people who were informed about every phase of the operation. They asked questions, observed, and made sure that things were done properly and on time.

Later, my initial experience in quality control at my first fracturing treatment was a traumatic one. I was on a job near Stinnet, Texas, in 1968. On this treatment, several new things were attempted. First, a continuous-mix cross-linked fluid was run; second, 1/4 in. glass beads were pumped into the well using a downstream injector. Thinking back, I guess I just assumed this was standard procedure, and I was not particularly awed by the activities. Looking back, I learned a great deal on that particular treatment about the oilfield, about common sense, and about the reality of field operations.

Stinnet is in the Texas Panhandle, north of Borger and northeast of Amarillo. The treatment was conducted in early December. My company was faced with 12°F and 40-mi/hr winds. In addition, I came down with a severe case of the flu, which greatly hampered my efforts.

The downstream injection of glass beads was accomplished by suspending the glass beads in a 200 lb. guar gum gel. The gel was prepared in a P tank by bubbling air through water, agitating the solution, and adding the beads to it. The mixture was then transferred into a tube trailer some 50 ft. long, consisting of four large tubes approximately 10 3/4 in. in diameter. These tubes contained the gel and beads. The concentration of the beads was not high, and it was going to be pumped at the end of the treatment. Most of the treatment used river sand, and the treated formation was the Brown dolomite, a fairly shallow zone in north Texas.

As mentioned, the treatment was going to be a continuous-mix one. The cross-linked gel consisted of a guar gum, which was dispersible, a slowly dissolving buffer, and a powdered cross-linking agent. The chemistry of this system was very complex. If the guar gum hydrated too slowly because of a low-water temperature, the viscosity increase would be slow and cross-linking poor. If the guar gum hydrated too quickly, lumping would occur and result in a poor cross-link. This system worked well in distilled water in the laboratory at 70°F. However, in field operations, it was an absolute nightmare.

At that time, dry metering of materials was not advanced. To get the gelling agent to the blender tub, I ran it up one of the sand screws. One blender was used to meter the gelling agent; this blender pumped into a holding tank. A second blender sucked out of the holding tank with only sand added. The system was extremely critical in relationship to temperature and pH.

Further problems were the low temperatures and high winds. Hot oilers were on location to heat the fluid, but continuous heating was necessary. Otherwise, the fluids would cool, and the chemistry of the system would be destroyed. I had a van at the location and a technician, who caught samples and evaluated the fluid as the company prepared to pump it downhole.

Several times we checked samples and were ready to pump, but mechanical problems delayed the job. When personnel on the mechanical side were ready to pump, the fluids had cooled, and we needed to further adjust the pH to perform the treatment.

My first "baptism under fire" in the oilfield came when I announced to the local manager that I was going to add 1 qt. of hydrochloric acid to the 500 bbl. tank to adjust the pH so we could run the job. This was, in fact, the required amount, and it did subsequently work. But the manager cursed, ranted, and raved about such a critical system coming out of research. After adding the acid, getting precisely the pH, and having the temperature correct, I initiated the job.

On the same job, the company representatives also were quite nervous and looked over my shoulder as I caught samples throughout the treatment. During the pad of the treatment, everything worked well. My company metered the material properly, basically by having a lot of conscientious individuals who cut the sacks and counted them as time passed. Yet, about the time I thought everything was lined out, catastrophe occurred-river sand was added to the fluid.

This river sand must have contained contaminants that drastically altered the pH. To this point, I had been catching samples and passing them on to the customer. I continued to catch samples, but there was no cross-linking. In fact, there was little hydration of the gel because the pH was raised to an alkaline condition and the gelling agent would not viscosify. Anticipating disaster, I notified the treater, who told me to keep out of his way and not to jeopardize the treatment.

The "frac gods" were good that day because all the river sand and glass beads were put away without any problem. Also, the incident may have further proved that high concentrations of very large proppants can be placed into the Brown dolomite formation without any viscosity at all. This particular hypothesis was reinforced when I later talked to someone experienced in the vicinity, who said some area farmers actually fracture wells with a water hose and cut sacks of sand into the water as the hydrostatic head of the water breaks down the formation. Then they flush the well with the same water hose. Stimulation is accomplished by breaking through near the wellbore damage. Other service companies fracture this formation with blenders, rather than high-pressure pumps.

The lessons of this treatment have served me well. From then on, I was extremely negative toward any complex continuous-mix treatment. The fluid used on this treatment was later taken apart and developed as a batch-mix system, although research personnel tried for two or three years to make the continuous system work. I discovered that, rather than strong acids or strong bases, buffered systems-particularly strong buffers-are much better to use than strong caustic or acid solutions for control in the field. I also learned to consider the environment (in this case, ambient temperature conditions) when developing a fluid system.

I was continually plagued with strange, abnormally buffered waters. When developing a fracture fluid or system, be aware of waters near freezing or treating fluids exceeding 130°F. If a fracturing fluid is to work, it must withstand the rigorous tests of low and high temperatures and of widely buffered water systems. Also, I learned to work and cooperate with field operations. No fluid system ever succeeded in the field if the field personnel could not understand how to use it.

One particular area I continually confront is the batch-mix versus continuous mix treatments. I (and everyone in my company) insist upon batch mixing on all treatments that we design and supervise. I do attend some treatments and QC where continuous-mix operations are under way; but if I have a choice, I still insist on batch-mixing fracturing fluids.

Service companies have come a long way in developing continuous-mix fracturing fluids and equipment. It is my experience, however, that the advantages gained through continuous-mix operations are greatly negated by the disadvantages. Recently, I was very impressed with a continuous-mix operation that occurred in an area near Sonora, Texas. However, this particular treatment was the exception. Continuous-mix operations are very advantageous to the service companies. They eliminate a great deal of service company preparation when preparing fluids prior to the treatment, and they eliminate a great deal of work required to batch mix all of the linear gels required in any fracturing treatment. The problem with any continuous-mix operation is that unless sufficient holding tanks are available where complete hydration of the gel occurs and the ability to monitor very closely the viscosity of these fluids, there are variations in gel concentrations that affect the ultimate cross-link viscosity downhole. When a cross-linker is added to a fracturing fluid, all hydration {viscosity increase) ceases at that point. Cross-linking negates any further base gel viscosity development. As I have also noted on continuous-mix treatments, even linear gel treatments {addition of breakers prior to full hydration also negates hydration of the gel and in the case of enzyme breakers), allow for much more rapid degradation of the system.

There are enough problems on a fracturing treatment that occur when going through standard quality control and intense quality control as is discussed in later chapters. Therefore, to further complicate the operation by running everything continuously is simply not worth it. Environmental situations and logistical considerations, such as offshore wells, may require continuous-mix procedures, but in no way do I see any reason other than for very, very large massive fracture treatments to be conducted other than batch mix. For very large treatments (i.e., 20-30 frac tanks), we allow continuous- or semi-continuous mix operations, but insist upon having holding tanks on location for not only complete hydration of the gel, but also for allowing sufficient time to do quality control procedures if problems occur in the gel. To conduct small treatments in a continuous-mix procedure, such as 20-40,000 gal treatments, is ludicrous. By the time you are lined out on cross-linking (addition of gel, etc.), the treatment is over. One of the interesting things that will be discussed in the intense quality control procedure is the effect ofsmall changes in cross-linker, gelling agent, concentration, and so forth, and its effect upon downhole viscosities. When there is a failure rate of more than 70% for batch-mix operations, that rate is even higher when trying to run multiple systems continuously.

Also, a great effort is being made by service companies to build better microprocessor controls in mixing and blending equipment for continuous-mix operations; and perhaps in the future, these will allow for quality-fracturing treatments in a continuous-mix manner. Now, the customers receive less for their dollars through insufficient hydration and lower viscosity downhole through continuous mix operations, and the only real advantages of continuous mix are to the service company and not to the operator.

Returning to historical events in the development of fracturing fluids, perhaps the biggest mistake made by research personnel is the development of fracturing fluids that, although functional under laboratory conditions, will not work in field conditions. Many fluids have been tried that were impossible to run under "real" conditions with standard oilfield equipment. Countless efforts have been made by both mechanical and chemical research people to develop complex systems or equipment. However, the harsh environment of the oilfield typically negates complex systems.

One company developed a very complicated blending system that contained a great deal of computer electronics. This system failed on its first job because a driving rainstorm hit the location. The company was embarrassed when most of the fracturing fluid drained onto the ground because the electronically controlled valve could not be closed. I have also seen fracturing fluids that contained a high concentration of polyacrylamide. This system could be mixed in a Waring blender by a conscientious technician but could not be run in the field. Because of persistent management, systems like these are tried again and again. Unfortunately, they plug wells and cause job failures.

Later in this book, systems and processes are pinpointed that were doomed to fail and are not particularly applicable to stimulation treatments. A successful fluid must be developed around available equipment and ease of addition. Virtually no system can be made trouble-free, but with a concerted effort and an understanding of basic equipment, practical, feasible fracturing fluids can be developed for the oilfield.

Common sense is a must when quality controlling fracturing operations. Customers have asked me during a large high-rate treatment specifically how much sand had been used at a particular time. I answered "approximately" 1,500 sacks. These people became upset when I could not give them a more precise number (i.e., within a few sacks). Presently, no service company's metering capabilities can monitor more accurately than about 100 sacks. The key is to know the relative rate at which chemicals are added and how much with which to start and finish. If these factors are known on all fracturing treatments and how much fluid is pumped, the success rate is better.

A totally different experience I had spoiled me in many ways and gave me another impression about the oilfield. Fortunately, the development of a new high-temperature fracturing fluid coincided with the development of high pressure intensifier equipment and high-pressure iron. Neither the high-temperature fluid nor the pumps and iron would have succeeded without the other. But the coming together of this iron and fluid was not altogether without problems and mishaps.

One of the earliest treatments in south Texas using this fluid and iron was perhaps my most exciting fracturing treatment. It was a large treatment for that time; it was late 1969 or early 1970. The treatment involved pumping about 300,000 gals of fluid at approximately 13,000 psi. The pump rate was around 13 bbls/min. The treatment was unusual and was termed a "40-car job" because of the large number of management personnel on the location, both for the oil company and the service company. As is often the case, too many managers tend to cause confusion, and some problems can ensue.

On this job the problems were not due to the presence of upper-level management. Early in the treatment, while pumping a prepad of slick water, some new chicksan joints on the low-pressure side of the intensifiers developed a severe leak. In fact, a man was looking at one chicksan that had been leaking, and the next minute he disappeared. Water gushed all over everything. Later, we ascertained that the chicksans had been sent from the factory with the seal rings in backward.

After replacing these chicksans, the engineer in charge asked the company man if he was ready to proceed with the treatment. The company man said "yes." However, he apparently did not understand that the engineer was ready to pump because he had shut-in the well. The engineer turned to the personnel on the pumping units, gave them the high sign, and 5,000 hydraulic horsepower was brought to bear almost instantly. The intensifier pumps used at that time were brought on instantly with all of the pump operators putting their pumps in gear and applying full throttle.

With the well shut-in, of course, there was a very rapid increase in pressure. In fact, some 100 ft. of the 100 lbs./ft. iron going from the pumps to the wellhead actually moved the wellhead. The wellhead moved before a pop-off valve on the low-pressure side blew at about 5,000 lbs. Because of the nature of the pumps, we achieved a pressure of at least 25,000 lbs. on the wellhead.

For a long time, this job proceeded rather smoothly. This was going to be a three-stage treatment diverted with ball sealers. Following the first stage, the company engineer evaluated the differential friction pressure after the first balls had hit on the first stage and decided he did not want to drop the number of balls that had already been loaded. This job was before the advent of positive-feed ball injectors. The old corn-planter ball injectors would not function well in viscous fluids because you never knew for sure if the balls were dropping. The oil company specified a positive-type ball injector. The positive feeding was going to be accomplished using a hydraulic pump (i.e., a grease injector system would pump balls through a series of plug-type valves).

The system was loaded by taking off the hose leading to the hydraulic pump and by dropping balls through the ball valves into spacers. The balls were dropped into a closed ball valve with spacers between two (a three-stage system was set up prior to the job with the balls in pipes between three valves). After dropping the first set of balls, the company man determined that not enough holes were opened, so he wanted to drop fewer balls in the second stage. This posed quite a problem because there was really no safe way to remove the balls and reload them while continuing to pump.

However, the on-site engineer indicated that this would be no problem for him; he would simply take the hose off and open all three valves. At that time, we were pumping 5 lbs. of sand per gallon of fluid and treating at approximately 13,000 lbs. He unscrewed the hydraulic hose and quickly opened all three valves.

A long, long string of gel shot straight up into the air, and the ball sealers went out of sight. Everyone stared in amazement, and the engineer attempted to close the valves. What happened then was not totally unexpected. When the engineer closed the first valve, it immediately cut out, and there was no restriction in flow. He was not deterred and grabbed the second valve. He closed this valve, and it cut out immediately. By that time several people at the location saw what was going to happen, and they ran to prevent him from closing the third valve.

The attempts were to no avail; he had the third valve closed before people could reach him. The third valve did slow the flow to a slight trickle. Then he reached for the hydraulic hose, spun it up on top of the valve, and seemed very pleased with himself as he walked toward us. This success was short-lived. Suddenly, the third valve gave way, the hydraulic hose ballooned and burst, and the hose started flapping around, spraying gel and sand. Acting as if it had eyes, the hose swung around and went into an open window of a diesel truck that was on the location to keep our trucks fueled.

Before we could shut down, the cab of the truck was full of gel and sand. We shut down, repaired the broken lines, and restarted the job. The lack of staging balls resulted in a partial failure on the treatment, since we could not properly stage it, but we did get all of the sand away for the customer.

A valuable lesson from this job is that competent personnel with common sense need to be in charge. Operations for ensuring safety need to be checked again and again.

Many treatments were conducted in the early 1960s, before the term *massivefrac* appeared, where sand volumes approaching 1 million lbs. and sand concentrations above 7 lb/gal were pumped at very high pressures for periods exceeding 9-10 hours. These treatments were well documented, and I found it interesting that people invented the word *massivefrac* to describe treatments conducted post-1977. I spent a long time in south Texas thinking that fracture treatments were conducted in the following manner. All equipment was checked and pressure tested, and equipment was running at daylight before the customer arrived on the location. The treatment was almost boring, with plenty of standby equipment and standby blenders.

Any excitement on a job came from the equipment going down and standby equipment being put on. On rare occasions, a blender malfunctioned, and was switched over to a standby blender, often without the customer knowing a blender was lost. This practice was not a result of any service company's policy; it was due to local management and conscientious individuals.

Same company, different performance

A startling discovery for me was that acidizing, fracturing, and cementing jobs were not necessarily conducted in this fashion. Lack of planning and maintenance often led to very exciting treatments. One area of the country (which I will not refer to directly) is notorious for jobs not being conducted as they were designed.

One day, I was on a treatment using a gelled oil system. Many problems ensued because the base oil was contaminated. One tank was poorly gelled, and the other had very little gel in it. Some mechanical problems occurred, and the job was delayed for a while. When the customer arrived on the location, the job was started. It was apparent in the early stages of the job that the well was going to screenout. I was on the blender and ran over to the pressure recorder to see what was happening because I had noticed the pumps were slowing down. I overheard the treater telling the customer that the diverting agent used to stage the job was working very well. This news surprised me because no diverting agent had been used. We had to flush the well and restart the operation, later running some diverting agent.

This job was experimental. In fact, it was one of the first to use a particular gelled oil system. I was eager to find out how much sand had been used on the treatment. To do so, I had to almost forcibly detain a sand truck driver. After checking the sand truck, I determined that at most 30-40 sacks of sand had been used. This was very depressing because I thought the fluid, which I had spent a long time developing, would not be accepted.

The next day I was shocked to learn that the well had responded dramatically and was considered the best in the field. Many more jobs were anticipated. I didn't understand what had caused this success until I talked with a person experienced in the area. The well had been severely damaged near wellbore, either by paraffin, scale, or other material. My efforts had simply broken through this damage at the wellbore.

I now understand an important point: until you achieve about 60% of the drainage radius of a well and unless you have dramatically higher conductivity in the proppant pack, the only stimulation increase is breaking through damage near the wellbore. Most fracturing treatments before the late 1960s were not fracture stimulation treatments per se; they were damage removal treatments.

The first fracture treatment was, of course, a damage removal treatment. The amount of sand and type of sand used on that treatment was certainly insufficient to achieve any long-term stimulation, but it was certainly sufficient to remove any near wellbore damage. One of the most disconcerting things in our particular area of technology is the lack of science and technology put into design and implementation. Over the years, I have seen people using low-conductivity sand to attempt stimulation treatments in very high-conductivity formations. As a general rule, shoot for something like 50,000 times the conductivity in the proppant pack as in the formation to achieve effective conductive fracture stimulation.

In recent years, there has been a great deal of success in stimulation of high permeability formations. In my opinion, this stimulation success relates more to relative heterogeneity of the formation rather than to relative conductivity of the proppant pack. These areas of success have been in the North Slope and Offshore Louisiana in unconsolidated Miocene formations. What is most disconcerting, though, are stimulation treatments that are attempted on wells where there is no knowledge of permeability and very little knowledge of the gas in place or the extent of the reservoir? Hydraulic fracturing, acid fracturing, or other stimulation techniques, when properly applied, are excellent tools for enhancement of stimulation. Basically, what is wrong in many cases in the industry is that there is no attempt at optimization. Regrettably, companies continue to conduct the same type of treatment over and over, especially if some degree of success was achieved initially. In service company schools during 20 years in the service company industry, I always taught to utilize the biggest treatment possible, as well as the largest amount of additives if treating the first well in the field. If the treatment worked, that treatment was conducted from then on. This sort of approach sounds ludicrous, but it has been the case over and over in the industry. In fracturing schools, I term this *drawer* frac designs, where the engineer goes to his file cabinet, pulls out a design that has been conducted on an offset well, uses magic marker to mark out the well name, and repeats that treatment on the offset well. What is particularly sad about this type of operation is that if the design is wrong on the initial well, the mistakes are repeated over and over.

With fairly straightforward pressure buildup analysis or pump-in, fall-off analysis or history matching of production data, a fairly good understanding can be developed of the relative capacity of the formation. This data can then be utilized by competent engineering personnel to obtain a cost-effective stimulation treatment.

Over the years, the advertising utilized by some of the service companies has added to the confusion about the potential success with stimulation treatments. It can vary, but typically the format is as follows. A customer has a well that has dropped in production or needs assistance, so he calls his friendly expert. Let's say the well is making 4 bbl. of oil per day with just a little gas, and the customer is concerned. He and his local representative decide the well needs 400,000 gals. of "super stuff." After the treatment, the customer is amazed. The well produces 500 bbls. of oil and one-half million cubic feet of gas. The customer then decides to use "super stuff" on all his wells.

Anyone with basic knowledge of acidizing and fracturing knows that something is wrong with these results. Obviously, the well was tremendously damaged. Otherwise, no such production increase is possible. The wildest, most imaginative production increases in our industry are based on conductivity theories that relate to such things as 13-fold increases. Practical field experience shows this to be almost impossible. Practical production-increase values in the 7-fold range are the maximum. In other words, this enthused customer, whose well went from 4 to 500 b/d with 300,000 gals. of "super stuff," might have achieved equivalent production increase values with 500 gals of 15% acid.

Quality control and job supervision often start before the job is designed, from the standpoint of understanding the problems involved. A 300,000 gal treatment removes damage, but it is not the most cost-effective treatment.

"'The final solution": technical competence is required

I have many "war stories" from different times and situations. I have learned from these that the service-company performance, whether Halliburton, Western, Dowell or B.J., is not consistent across the United States. These companies have comparable equipment. What it comes down to is conscientious personnel in a particular area.

It is extremely important that oil-company personnel be informed about the situation and enforce quality control and job performance. A service company that can perform an excellent treatment in south Texas may not have the experience or capability to do it in the Rocky Mountains and vice versa. In my early days, I found that in some areas companies did not want to conduct hydraulic-fracturing. treatments. They felt comfortable with cementing and acidizing work, and they did not want to become involved with this kind of project. Obviously, poor job performance resulted.

Oil-company representatives must become more knowledgeable about the fluids and the equipment supplied by service companies. This will allow a joint effort between the service company and the oil company 'in developing first-rate service and capabilities. If the oil company depends on the service company for quality control and to oversee all problems, human error and equipment malfunction could jeopardize the treatment.

The more people involved in the quality control, the better the treatment. Much of my experience in south Texas, which combined good people, quality fluids, and strict quality control, resulted in some of the best documented and economically successful stimulation treatments ever done. When I traveled elsewhere, the great eye-opener for me was to find that the quality of service was lower. But even more surprising was the total lack of concern by the oil companies in other areas.

Ignorance and poor service = poor results

I spent 4 1/2 years in the Mideast before returning to the United States. On my last treatment there, I helped supervise an offshore acidizing treatment. It was a large treatment, to say the least, involving 220,000 gals of viscous gel, 140,000 gals of 28% acid, and sundry chemical additives. The job took two stimulation vessels and a great deal of coordination. ·

In my opinion, the reservoir did not need this type of treatment. In a way, it was a political treatment because the local government wanted the largest possible treatment.

The treatment itself was not particularly impressive. Some seven hours were required for lines to be tested because of poor quality control and poor maintenance on one stimulation vessel. After accomplishing this, the large stimulation treatment was started early one morning. It was immediately apparent that much of the equipment had not been properly maintained. Engines and pumps started to fail (many before we even got to the acid), and there were numerous problems in metering the chemicals properly. At one point, instead of having 15 units pumping downhole, we were down to one. I was working on the metering equipment and packing pumps. Several people worked very hard in 115°F heat and 95% humidity.

After completing the job, I will never forget the forebodings I felt as I had to get on the basket and go up to the floor of the jack-up. I thought I was going to be verbally (if not physically) abused by the company people present. I must have been a sight. I had been working about 15 hours with little sleep, and I had gel and dirt all over me. I was the man who had sold this job and had worked with these people for two years. I anticipated their being disappointed in me, if not in my company.

Both men representing this company ran to meet me as I got on the floor of the rig. They put their arms around me, shook my hand, and congratulated me for a good effort. They made comments like "One of the boats seemed to work harder ·than the other one" and "Everybody put forth his best." Obviously, these people had not understood what had happened. They had spent about $1 million for a stimulation treatment on the well. For that kind of money, they deserved excellent service and performance. Regardless of the efforts of the people involved, they received very poor service. But they thought that they had received an excellent effort.

Guidelines

1. Allowing poor sendee because it is the norm is unacceptable.
2. Quality control by the oil company and good performance by the service company go hand in hand.
3. Stimulation fluids and equipment must be relatively uncomplex and must adapt to field conditions.
4. Use common sense and be reasonable in evaluating the performances of the people, fluids, and equipment.
5. Success is related mainly to conscientious individuals, both from the oil company and the service company.
6. Basic understanding of stimulation theory is a prerequisite for successful treatments.

CHAPTER 1

THE EYE OF THE BEHOLDER (2022)

I will not attempt to edit chapter 1. What we will discuss in this new chapter is perspectives like those given in the final guidelines of the original 1994 chapter. This will effectively cover experiences and knowledge gained since the inception of my company. I left my favorite company "Holditch and Associates" with a heavy heart because Steve Holditch had been my friend since the early 1970's. One of the Holditch customers offered me substantial funding to be able to start my own company in May of 1991.

The early 90's well into the 2000's was, in my opinion, golden years for performance of the service companies. The guidelines set forth for quality control were, for the most part, followed by the service companies. This involved pre-job testing of the water to be used on fracs with the actual chemicals at bottom hole temperature. My company prospered during these times utilizing the great depth of technology developed through the Gas Research organization and substantial in house understanding of fracturing fluids per-se. The crosslinked fluids developed during this period are in fact virtually the same products which exist today for the small amount of conventional fracturing done in the US. I might comment that there is huge potential for conventional and unconventional fracturing throughout the world.

As already discussed, the massive change that has occurred in the past 20 years has been the move to low technology water, using only sand as the proppant, even at great depths i.e., deeper than 15,000 feet and conventional closure pressures exceeding 14,000 psi. Like many things in our industry the technology was led not by intuitive reasoning but by a search for more economical ways of stimulating extremely tight reservoirs. We had all been taught, and I personally led many schools, that we would never allow even high-quality white sand to be used at calculated closure pressures above 6,000 psi. We would go to resin coating for enhanced strength due to distribution of stress over a larger surface area. Above 8,000 psi calculated closure we would recommend some form of man-made ceramic, then above 10,000 psi closure we would only recommend Bauxite or even more dense and stronger versions of bauxite at closures above 15,000 psi.

By ignoring classical pressure selection points for proppants and yielding to dramatically lower cost, probably the most significant discovery in recent times was achieved. I personally was skewered in front of one of our largest customers for allowing white sand to be used in very deep Woodford plays. The dipping of some of the Woodford in some cases approached 15,000 feet. Early results indicated low initial I.P.'s and I was called on the carpet to explain. The reason was not even price, it related to very fast operations and following the plan of offset wells i.e., "cookie cutter wells. I was never personally told, but the deeper wells with white sand outperformed wells where stronger proppant was utilized. A similar pattern with our customers followed in the Haynesville where 1.0 frac gradients existed at 12-13,000 feet. A large customer started our using white sand and tailing in with resin coat. Based on cost they converted to sand only and eventually used a combination of 40/70 and 100 mesh. Based on our knowledge the sand wells outperformed or at least performed equally well as those treatments containing high strength proppant. As has occurred recently in other plays even lower quality sand has been successfully used and there has been a trend in many plays to all 100-mesh proppant.

CHAPTER 2

RIGGING UP AT THE LOCATION (1994)

Company personnel rarely show any interest or become involved in the layout of equipment, hoses, and iron on location. There is probably nothing more important for the treatment's success, except for the quality of the fluids, than the way the equipment is laid out. Most of this layout is governed by common sense and basic engineering practices. Over the years, I have learned some rules that make rig-up and equipment layout very simple.

Particularly in the eastern United States, equipment layout is dictated by the space available. The amount of space is not terribly significant if conducting a single-truck cement job or a small acid treatment, but space becomes very important if running a 2-4 million lb. sand frac treatment that requires large areas for tankage and sand storage.[1]

One common-sense approach to rigging up is to use a minimum of treating iron. It's upsetting to see operators rigging up a site haphazardly, laying far too much treating iron. I favor compact rig-ups. A smaller amount of treating iron reduces the danger when treating at high pressures.

Some oil companies require a minimum distance between the equipment and the wellhead. This requirement dictates how much iron is used. Obviously, if you are performing a cement or acid treatment with a drilling rig on the location, you probably need to rig-up past the catwalk or wherever else is convenient. I believe 50-100 ft. is a good, safe distance from the wellhead. I have been on many locations, particularly in the eastern United States, where we were within a few feet of the wellhead. If we had rules dictating 50-100 ft., we would have been off the side of a mountain. Use common sense and do such things as not having fuel tanks near the wellhead. I have even been on locations where we have actually rigged up the trucks down the road from the location, since there was simply not enough space to put the pumping equipment on the location. If a catastrophe occurs, perhaps the equipment could be removed and not destroyed by fire. A lot of the equipment is operated by remote control, and operators are not on the equipment, so this rule does not generally relate to personnel safety.

On numerous occasions, I have been at locations where the equipment was 100ft from the wellhead, and the operator stations for that particular equipment were within 15-20 ft. of the wellhead. The safety of the operators is far more important than the equipment.

---

1 Two interesting papers on laying out massivefrac treatments are "Logistical and Operational Considerations for Hydraulic Fracturing" (Cecil D. Parker, *JPT,* July 1981) and "A Case History for Massive Hydraulic Fracturing the Cotton Valley Lime Matrix, Fallow and Personville Fields" (H. G. Kozik and S. A. Holditch, *JPT,* February 1981).

Iron is critical

On the line from the manifold to the wellhead, the most important criterion is that the inner diameter of the treating iron does not limit the pump rate. Table 2-1 lists maximum treating rates for various iron sizes. These rates are not "set in cement;" you can pump faster than 25 bbls/min. down 3 in. pipe. What you suffer is increased abrasion on the pipe and higher friction pressure. A major amount of abrasion problems that occur are in chicksans and swings that. are in line. As the abrasive fluid turns a corner, you see tremendous amounts of abrasion in these turning joints. It is not uncommon to see chicksans and swings wash out during a treatment, particularly where exceeding maximum pump rates, as shown in Table 2-1. I have had it stated to me that it doesn't really matter if the service company is wearing out the iron, its their cost. More importantly, based on Murphy's Law, this will probably occur on my job and the disaster that occurs because of the wash out of the iron will be a disaster with my well. One of the most common things to do on a fracturing treatment is to insist upon sufficient size frac iron to be sure that the maximum pump rate is not exceeded through swivels and chicksan joints.

Table 2-1 Maximum Recommended Pump Rate Through Discharge Lines

Line (inches)	ID (Inches)	Working Pressure (psi)	Pump rate (bbls/min)
2	1.775	15,000	8
3	2.750	15,000	20
4	3.750	6,000	37

Note: This table does not give set values for the absolute maximum rates for the treating iron. Exceeding these rates, utilizing high-sand concentrations, yields a short life with severe abrasion for the treating iron. When excessive rates are used without abrasive materials, high-friction pressures result.

Place pressure monitor close to the wellhead

The first thing a company person should do before a rig-up is to find the location in the line where the treating pressure will be measured. You pay for hydraulic horsepower. On high-rate treatments where the treating rate to the wellhead may approach or exceed the maximum velocity for the treating iron, quite possibly excessive back pressures will be seen between the pumps and the wellhead. This can create large horsepower costs, which are billed to the customer. To alleviate this expense, the pressure transducer used by the service company must be placed at or near the wellhead.

I have been on locations where the treating pressure shown on the fracometer exceeded 7,000 lbs., and the treating pressure at the wellhead was around 5,000 lbs. Table 2-2 gives two methods for calculating hydraulic horsepower. A company engineer should always mentally calculate the hydraulic horsepower used and determine if anticipated pressures and rates fall within the realm of the treatment design.

Are the valves open?

Another anecdote from my life in the oilfield concerns treating pressures. We were treating a well down 7 in. casing. This fairly large acidizing treatment was conducted on a very hot day. We were all depressed because, when we started pumping into the well, we only achieved 1-2 bbl/min. at 7,000 lbs.

We had hoped when the acid reached the perforations that we would see a break and be able to increase our rate. The pressure gauge was at the wellhead, and the proper horsepower was used. We saw no pressure break; I treated for 7 1/2 hours to conduct the treatment, but the rate only increased to about 2 1/2 bbls/min. after pumping large quantities of 28% acid.

On this location and for the particular company, one procedure was indeed carved in stone; namely, service companies were not allowed at any time to touch any of the valves on the tree. We rigged up to the well and waited for company

Table 2-2 Hydraulic Horsepower
$$\text{Hydraulic horsepower } (hhp) = \frac{\text{pressure (psi)} x\ rate\ (bbl/min)}{40.8}$$ hhp = pressure x rate x 0.02456

Note: Hydraulic horsepower (hhp}is differentiated from brake horsepower [bhp). The service company typically brings units to your location that may have 1000, 1200, 1600, 1800, or 2000 bhp. Never expect these units to produce more than 80% of brake. The company engineer should frequently calculate the hydraulic horsepower being used based on rate and pressure. He can compare that amount to the number of pumping units used and their capacities and ensure that there are no substantial errors in the flow meters.

representatives to open or close the tree. This day, after we had rigged up, the company man had arrived, gone to the wellhead, opened up the well, and reported to us that everything was go. After completing this long, tiring treatment, I did not care about the rules and regulations; I just wanted to shut-in the tree, rig down, and go home jumped on the very large Christmas tree and was only able to make a half turn. In my attempt to close off the wellhead, I discovered the company engineer had not opened the tree.

All we had accomplished during those seven hours was to seriously erode a Cameron valve! This brings up a point that I try to make: There is nothing magical about cementing, acidizing, nitrogen, or fracturing treatments. If pressures are not what you expect, there is generally a malfunction or a problem. Regardless of rules and regulations, make sure all valves are open when treating a well. This can save your life. Much of the equipment is capable of exerting pressures exceeding 20,000 psi, which can result in disaster if iron explodes.

Is the check valve properly installed?

Most service companies, whether on a cementing, acidizing, or fracturing treatment, run back-pressure or check valves in the line. These valves should be as near the wellhead as possible and, of course, are used if a catastrophe (such as a rupture in the well, in the line, or at some point downstream) occurs, The check. valves keep the well from becoming live at that point. A check valve that is a great distance from the wellhead is essentially useless.

Keep several things in mind concerning check valves. Most important, the check valve should be installed properly. Most check valves have an arrow showing the direction of flow (see Figure 2-1).

Another embarrassing moment in my travels in the oilfield relates to a check valve. While I was preparing emulsified acid, some experienced operators rigged up the lines and placed a very large, very heavy 4 in. check valve upside down. This particular valve had an arrow on it that was at least 2 ft long. The arrow was obviously pointing in the wrong direction. We noticed the problem when we opened the wellhead to start the treatment and found the well was flowing.

These things sound petty; they sound like common sense checks that anyone with experience should make to avoid preventable problems. A short time spent checking, asking questions, and making observations may save your life. Large quantities of frac iron have ruptured and exploded because of pumping against closed stops or backward check valves. Look first! (You may save the most important life there is-your own.)

Basically, there are two types of check valves: a dart and a flapper. The dart type check valve cannot be used with ball sealers. Ball sealers will not pass through the check valve. The dart-type check valve is often used with trailer manifolds with a check at the manifold before it leads off to the truck. This gives further control in the system near the pumping unit, but it is not uncommon to have dart-type check valves in the line near the wellhead.

Figure 2-1 3 in. in-line check valve. Always study check valves to be sure that the arrow on the side of the check valve is pointed in the proper flow direction. See that flapper-type valves are situated so they will be functional should reverse flow occur.

Asking a few questions can save a lot of grief on a treatment. One precaution to take with flapper-type check valves is to be sure that they are pointed in the right direction. Also, many of these check valves must be set up so the flapper will operate automatically when flow toward the wellhead ceases. That is, the flapper should fall into the flow line if pumping ceases or reverse flow starts to occur.

Figure 2-2 shows both dart- and flapper-type check valves. The flapper-type valves must be in the proper position before they will function. If the flapper-type check valve is not set up in a proper position, the flapper may not close, and the basic function of restricting flow in the opposite direction will not work. Experienced operators position this valve so it will automatically function when flow ceases.

Another consideration when positioning these systems is a means of releasing the pressure between the wellhead and the check valve. If you have ever been unable to release pressure or have tried to break down unions with pressure on them, you will understand this situation. It is very difficult to do and is extremely dangerous.

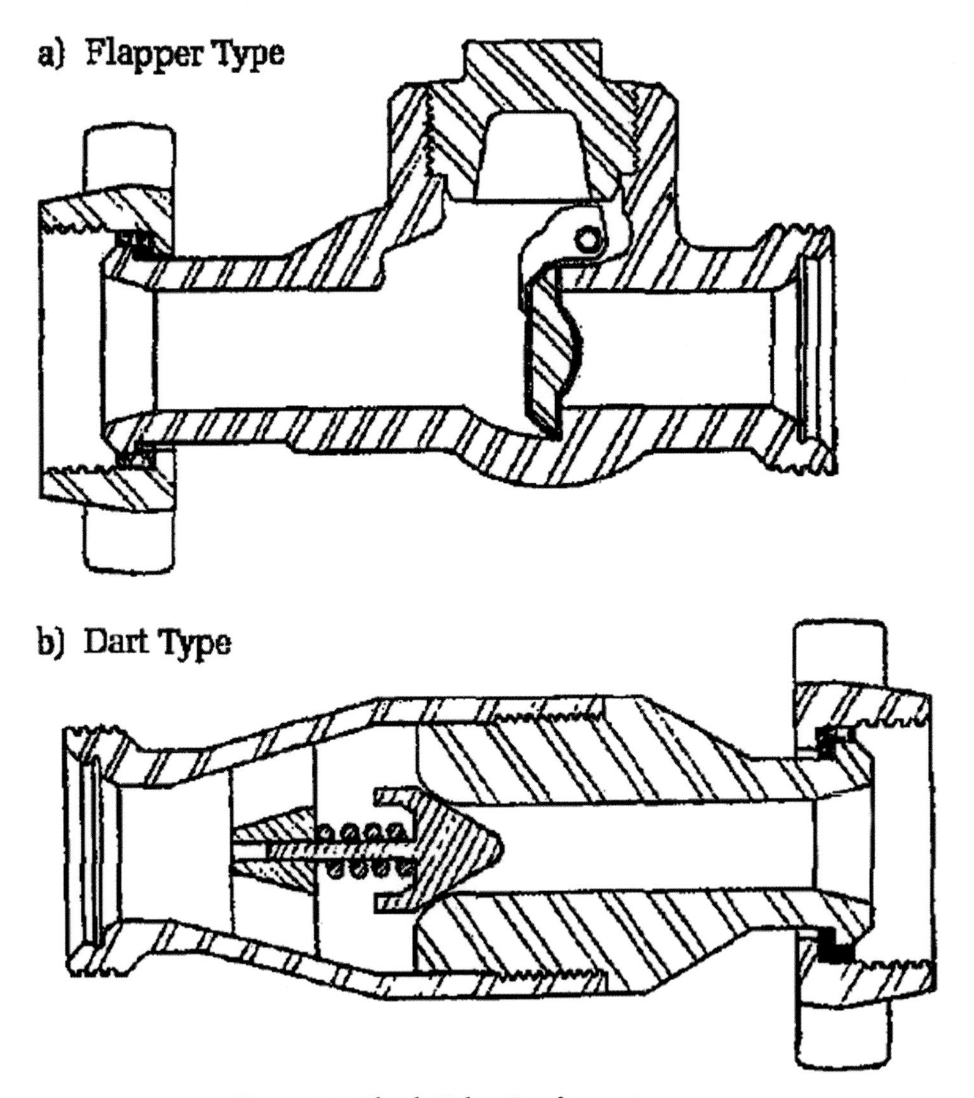

Figure 2-2 Check Valve Configurations

Be sure that during the rig-up, pressure can be released once the well is shut-in between the well and the check valve. This typically can be accomplished through the line used to prime the pumps, or it can be done through a release valve on the wellhead (i.e., a valve above the master valve that can be shut-in). Then, the pressure can be released above the master valve by the dipping valve, and so on.

The most common factor relating to check valves is that, in most cases, these checks do, in fact, leak. Many check valves are installed in service company manifold trailers and are very difficult to check and maintain on an orderly basis. Recently, I was on a job that I considered to be a very safe rig-up where the check valves were actually tied into the valve that was on to the wellhead. As happens in many situations, though, a problem had occurred where the service company had inadvertently cross-linked the entire flush volume of the well. Because of this, I was unable to run in a packer because of the viscosity of the fluid. I was very discouraged because I had no way to flow the well with the checks on the wellhead. The service company hand on location laughed and said, "No problem," and he opened the well and it flowed back unrestricted. The check hadn't held for some time, according to the service company representative. Therefore, it is extremely important for the checks to hold and then these check valves should be tested prior to the job. In another recent instance, we had a gauge leak on a pump truck during a fracture treatment. The entire frac job had to be shut down for a period of time to bull plug off the hose back to the gauge. The service company representatives did not mention that quite obviously the check between the frac line and the truck was leaking. As soon as the truck was shut down and there was no flow in the direction of the wellhead, the leak should have stopped and it did not. It is a good procedure to pretest check valves on equipment at a regular interval. Questioning your service company about the last time the check valves were analyzed is the prudent and proper thing to do to assure safe operating procedures. The check valves that are used at the wellhead are your last fail-safe options, if a line breaks on one of the pump trucks during the treatment. If the check doesn't hold, then you have a live well on your hands or at least one that you will have to move into the wellhead area to shut a valve with fluid, gas, and so forth, spraying in the general vicinity. This is particularly detrimental in a high-pressure gas well where you are pumping propping agent. Ongoing maintenance of checks and blocking valves is always important, and this will be discussed more later in the chapter.

Isolation tool installation is critical

Before progressing to the pumping equipment, let's discuss tree savers, well-head isolation tools, tree guards, or whatever the service company calls them. Figure 2-3 gives a basic description of the tree guard. This tool is utilized so the treating pressure is not placed on the Christmas tree.

In the early days of high-pressure fracturing, the industry was limited on pres-sure to the maximum allowable pressure of the Christmas tree. If you needed to pump at 12,000-13,000 lbs., as was the case in early treatments in the McAllen Ranch in south Texas, then you had to go to the expense of putting a 15,000 lb. Christmas tree on the well. This was particularly dysfunctional because the flowing pressure of these wells was well below 10,000 psi.

To by~ pass this situation or alleviate the problem, service companies developed wellhead by-pass tools that would stab into the tubular goods with a packer arrangement (Figure 2-3), effectively isolating the Christmas tree from the treating pressure. Those people experienced in the industry know that tubular goods can usually withstand treating pressures exceeding Christmas trees. It is not uncommon to have 5,000 lb. Christmas trees and treat in excess of 10,000 lb. on a fracturing treatment. Isolation tools, tree guards, and similar equipment are often used on acid jobs and fracturing treatments, but they are seldom used on cementing treatments.

Several factors must be considered when using this tool. Be absolutely certain of the ID of the tubular goods, so the packer on the bottom mandrel of the isola-tion tool will fit. Also, be sure of the measurements of the Christmas tree, so the mandrel can be spaced to allow stabbing into the tubular goods when the tool is being set. This requires precise calipering and measuring of all distances between the tubing where the tool is set and the top of the tree where the isolating tool is placed. There are several types of this tool in the industry; however, I will not give all the nuances of this device. Some guidelines for maximum rates for varying tubular sizes are listed in Table 2-3.

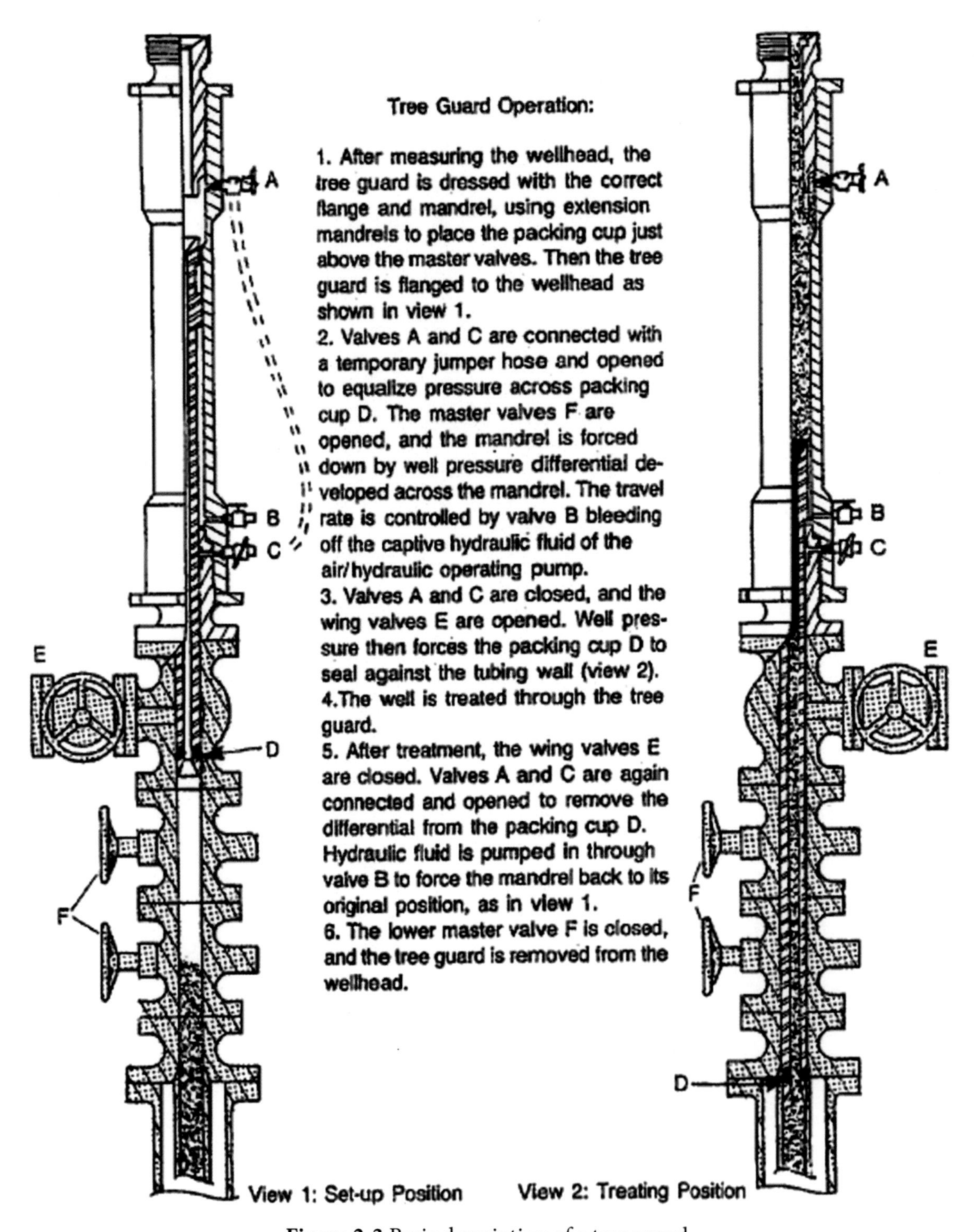

Figure 2-3 Basic description of a tree guard

These rates should not be exceeded. This ensures that you do not excessively erode the tubing or cut it while pumping abrasive proppants through the tool. A particular problem with an isolating tool is that, even when used improperly, it can cut the tubing right below the packer and the mandrel. This can be caused by too high of a rate or poor configurations used by the service company on the outlet port of the isolation tool. Much research has been done to optimize the configuration of the bottom of the mandrel so maximum rates can be achieved without abrading tubing below the packer and the mandrel.

Table 2-3 Dimensions and Maximum Pump Rates for an Isolation Tool

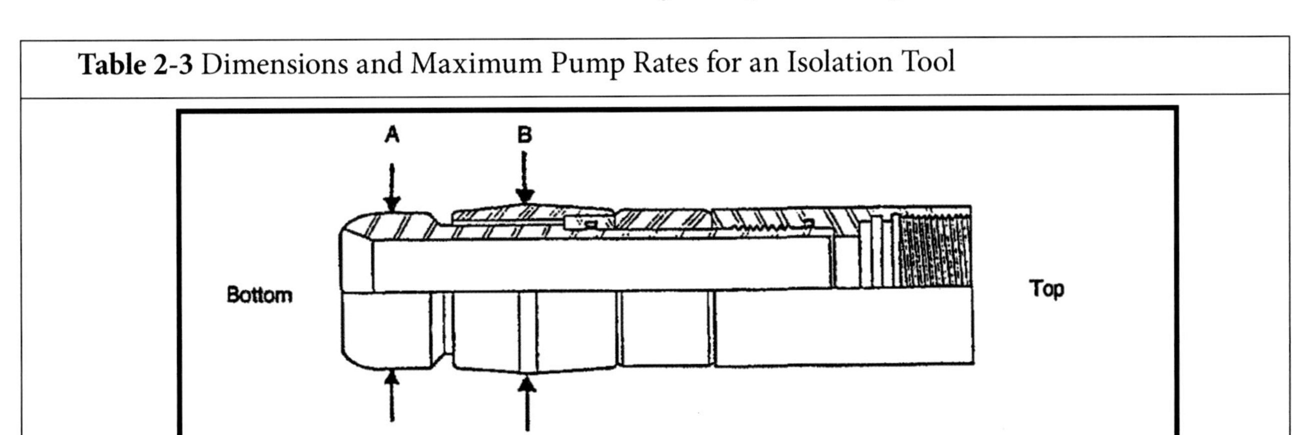

Tool/Stroke Length	Tubing/ Casing Size	Weight Range (lbs)	Working Pressure (psi)	Mandrel ID (inches)	Max. Rate 128	Flow (bpm) 200
#1-A/36"	2 1/16	3.25	15000	1.06	8.2	13.1
	2 3/8	4.7	15000	1.06	8.2	13.1
	2 7/8	6.5 – 8.7	15000	1.06	8.2	13.1
#1-B/35.8"	2 3/8	4.7	15000	1.31	12.5	20.0
	2 7/8	6.5-10.7	15000	1.31	12.5	20.0
#1-D/48"	2 3/8	6.5-8.7	15000	1.31	12.5	20.0
	2 7/8	6.5-10.7	15000	1.31	12.5	20.0
#4-A/44"	2 7/8	6.5-8.7	15000	1.57	18.0	28.7
	3 ½	9.2-15.8	15000	1.57	18.0	28.7
#4-A/44"	2 7/8	4.7	15000	1.75	22.3	35.7
	3 ½	9.2-15.8	15000	1.75	22.3	35.7
	5 ½	15.5-20	5500	1.75	22.3	35.7
#2-A/60"	2 7/8	6.5-8.7	15000	1.57	18.0	28.7
	3 ½	9.2-15.8	15000	1.57	18.0	28.7
#2-C/60"	2 7/8	6.5-10.7	20000	1.31	12.5	20.0
	3 ½	9.2-15.8	20000	1.31	12.5	20.0
#2-B/60"	3 ½	12.95-17	15000	1.57	18.0	28.7
#2-E/60"	3 ½	9.2	15000	1.57	18.0	40.8
	4	9.5-13.4	15000	1.87	25.5	40.8
	4 ½	9.5-15.1	15000	1.87	25.5	59.0
	5	18-20	15000	2.25	36.9	59.0
	5 ½	15.5-23	15000	2.25	36.9	59.0
	7	23-32	11000	2.25	36.9	59.0
	7 5/8	39	10000	2.25	36.9	59.0

#2-G/60" 80"	4 ½	9.5-15.1	15000	2.87	36.9	96.0
	5	18-20	12000	2.87	60.0	96.0
	5 ½	15.5-20	12000	2.87	60.0	96.0
	7	17-32	7500	2.87	60.0	96.0
#2-F/80"	5 ½	15.5-26	15000	3.00	65.6	105.0
	7	17-32	10000	3.00	65.6	105.0
	7 5/8	29-7-33.7	9000	3.00	65.6	105.0
	9 5/8	43-61	6000	3.00	65.6	105.0

Courtesy Halliburton Services

When setting these tools, see that the Christmas tree is truly vertical and that the mandrel going down toward the tubing is not coming into the setting position at an angle. Setting the tool is very important, and the supervising company engineer should be present during this activity. Some tools are designed to visually monitor the progress of the mandrel proceeding downward. Other tools are hydraulically pumped down; the hydraulic fluid volume is measured as the tool progresses downward. The tools can be pumped down with hydraulic pumps or by using the pressure in the well. Most tools are rigged up to do either.

Therefore, be certain the tool is set and completely down before initiating the treatment. I recommend that a wing valve be open from the annulus during the treatment to indicate any problems with the isolating tool. If the tool is unseated, the pressure will be released and not placed on the annulus, which typically can-not withstand the high pressures of a stimulation treatment.

I have been on treatments where the open wing valve was, in fact, directed toward the frac van. On one particular treatment, when the tree saver gave way, the van was sprayed down with propping agent and severe abrasion occurred before we were able to shut down the pumps. More commonly, the wing valve is opened into a line that goes into a pit. It is very important that someone observe this to be sure that there is no leakage going on during the treatment. Commonly in the industry today, there are separate contractors from the service company supplying many of the wellhead isolation tools or treesavers, and their personnel will be observing for any flow during the treatment. If it is a treesaver or isolation tool from a specific service company, be sure that someone is observing for flow throughout the treatment. One should never close the wing valve and apply the pressure to the well-head if that wellhead is not capable of holding pressures during a treatment.

It is interesting to note that I have had isolation tools give way during a pad on a treatment and have lowered the rate to stay within my maximum pressure and continued the treatment. Basically, and this is not necessarily a specific recommendation, the treatment is run at maximum pressure and increase rate back to the design rate as the hydrostatic head helps in surface treating pressures. It would be, perhaps, a good rule of thumb to terminate a fracturing treatment if the treesavers fail early in the treatment, but if it fails late, you may be able to continue the treatment due to lower surface treating pressures because of hydrostatic head of the proppant. This, of course, does not relieve the problem of a screen-out or increasing net pressures due to various phenomenon going on in the fracture. It has been fairly rare to have a failure of a wellhead isolation tool during an actual treatment. Most of the failures occur with being unable to obtain a seal just prior to the fracture.

Always find out the maximum rate your service company will use on the tool they have in the hole. Do not exceed this rate during the treatment. Another problem occurs when removing this type of tool from the hole. For some reason, the tool may become jammed. When you try to remove it, shear sleeves may break, leaving the tool in the hole. This is typically caused by improper tool alignment, packer swelling, or poor quality control on the shear sleeve of the tool.

One suggestion for removing the tool is to allow the pressure to equalize above and below the packer of the isolation tool. Let the rubbers in the tool relax for 15-20 minutes after the treatment. This often allows you to remove the tool with-out parting the shear sleeve.

If you part the shear sleeve or pull the mandrel in two, remember that the tool itself can be used to remove the lower portion. A smaller mandrel can be run through the packer mandrel with dogs to recover that lower portion. This technique has been used successfully many times when, for one reason or another, the lower part of the isolation tool has been left in the hole.

One area that is an absolute necessity when utilizing treesavers is that, unless you are dealing with very low bottom hole pressures, you should always insist that the service company, or the treesaver company, supply a remote actuated hydraulic valve on top of the treesaver. Because you have effectively isolated the wellhead valves if the lines part above the top of the trees aver prior to the check or block valve, you will have an uncontrolled situation and no way to shut-in the well. Immediately on top of the treesaver, you should have a hydraulically actuated blocking valve. You should never do a job on a high-pressure well without a remote actuated block valve on top of the treesaver.

In summary, the wellhead isolation tool, tree guard, or treesaver used by your service company is an excellent tool. However, with this tool, you must consider the maximum rates and properly measure the ID of tubular goods and the distance from the top of the tree to the tubing in the well. Additionally, consider any possible tree angles other than vertical when placing the tool in the hole. There is a tremendous dollar saving with this type of device, but great care must be taken in its placement and removal from the wellhead. Be sure your service company representatives leave the annulus open with a flow line that is staked down to ensure that the isolation tool is functioning properly during the treatment. Next, I will address the basic rig-up of iron and connections from small trucks or small treatments and work into larger treatments. There are some major considerations for smaller treatments. Basically, you must have treating iron large enough to accommodate the pumping rate you plan to use. Also, be sure the iron is staked down correctly (see Figure 2-4 and Figure 2-5). In conjunction with the service company rigging up the location, make certain that treating lines from the truck to the ground are not in a rigid position (i.e., any vibration that occurs during the treatment cannot break them).

Figure 2-4 Properly staked treatment iron. Properly staked treatment iron is (essential to prevent dangerous situations, in case vibrations or other reasons cause (the iron to part. Should the iron split at any point, it must be staked down to prevent its flying around and striking equipment or personnel. Stake down iron at 10 to 15 ft. intervals and secure it before pumping.

Figure 2-5 This is perhaps the ultimate staking using a dead man where energized fluids are used. This approach may be the very best.

Flexible connections are necessary

In the early days of hydraulic fracturing, I have attended fracture treatments where dangerous rig-ups were noted. These involved personnel working very hard to achieve rigid hookups of iron with conventional treating iron. If there is no flexibility in the iron, any vibration of the iron and any movement will cause the iron to part. Quite frankly, it has been more than 20 years since I have seen a rigid hookup on a fracturing location.

The exceptions to rigid hookups are what Halliburton terms, "Big Inch Iron." Halliburton uses treating iron, which is indeed rigid in nature, for high pressure work. The highest pressure version, 20,000 lb. iron, is 6 in. OD and 3 in. ID. Although Halliburton does have some Big Inch swivels, the majority of the iron hookups simply use "sweeps," "elbows," and so forth, to configure the iron from the wellhead to the ground. There are no knock-up connections on this iron, and it is tied together with metal-to-metal seals and external clamps. There is a 6 x 4 in. version, and indeed there are larger versions for very high rate treatments. The heavy nature and extreme strength of this iron makes it an exception of the afore mentioned rigid hookups. To date no other service company utilizes this type of iron and typically use Weco type knock-up iron with integral union connections.

For all other hookups, including Halliburton's standard iron, it is always advisable to be sure that sufficient chicksans are' present to negate any potential rigidity in the treating iron.

Test all valves before the treatment

Whether it is a small rig-up (one or two trucks with T's and Y's) or a very large location with multiple trucks, each truck should have an individual plug-type valve or check valve, in addition to a check valve or plug valve at the wellhead so each truck can be isolated from the treatment. This may seem to be an unnecessary safety factor, but it is important.

Many times, my life depended on a properly functioning check valve. Obviously, the check valve worked on these occasions. However, the check valve often failed and a plug valve was not in the line; subsequently, it caused fires and deaths. Check valves tested prior to the treatment can malfunction during the treatment due to erosion or mechanical failure. On high-pressure treatments, those exceeding 10,000 psi, I strongly recommend the use of both the plug valve and check valve between each pump truck and the manifold. On the majority of treatments done in the United States, the service companies either use the check or a blocking valve between each truck and the manifold trailer or a ground manifold. Dowell Schlumberger typically uses a block valve to isolate each truck. Halliburton Services, BJ Services, and The Western Company typically use the check. You must insist on having a check and a block in the treating lines as close as possible to the wellhead. I have noted on many occasions in many areas of the United States that service companies do not, until requested, supply isolation for the trucks and/or specific checks and block valves at the wellhead. Do not allow a treatment to be conducted without sufficient checks and blocking valves. The life you save may be your own.

This brings up another consideration: is it the oil company person's responsibility to ensure that the service company tests check valves and plug valves before the treatment? Particularly in boom times, people become overworked, and things that need to be checked are not. If a treatment fails, it is too late to determine that check valves and plug valves are leaking, and you have no control over your well. You can easily ascertain if a check valve or plug valve is leaking simply by placing pressure against one side and knocking off a union. On larger treatments, where many trucks are manifold together with a trailer or ground manifold, testing plug valves and check valves becomes even more important. With ground or trailer manifolds, it is quite common for one or more of the check valves or plug valves to leak. You can easily and quickly check for leaks and make repairs before a treatment for everyone's safety.

I am always interested in seeing the approach by many service companies in rigging up to trailer or ground manifolds. These manifolds are designed to minimize the amount of iron used on a rig-up and to alleviate much of the work in putting T's and Y's together. The manifold allows several trucks to be tied into one blender, and all lines can lead into one or two lines going to a wellhead.

I have spent a lot of time in the field trying to emphasize the point of using minimum flexible iron in the rig-up. If there is enough space on the location, the rig-up to a trailer or ground manifold should be symmetrical because these manifolds are designed to accommodate 4-12 pumps. You should be able to stand at the end of one pumping unit and see only one exhaust. The pump trucks should be in a row, and the same amount of iron should be used for each rig-up with typically one suction hose to each pump. Figure 2-6 illustrates proper rig-up from a high-pressure pump to a frac manifold.

On some locations, if space is limited, the rig-up may differ considerably. But a little planning lessens the danger of maximum quantities of iron that can leak or rupture. Figure 2-7 shows a symmetrical rig-up using a trailer manifold.

Some basic factors must be taken into account when using ground or trailer manifolds. The things mentioned earlier also hold true for the trailer manifold. The maximum rate/size of the iron must be considered with trailer manifolds. Small ground manifolds do not cause problems because a maximum of four trucks tie into one manifold, which ties into a single line. However, remember the maximum rates that were discussed earlier; a manifold with a 4 in. ID should only accommodate about 40 bbl/min/side. This figure can be stretched somewhat, but the life of the manifold is lessened and you can see excessive friction pressures when you get into higher rates.

A typical trailer manifold (see Figure 2-7) has two high-pressure 4 in. ID sections. On each of the two sections, there are six 3 in. ID Weco connections. The two 4 in. sides are manifold together, so multiple lines can come off the trailer manifold to the wellhead. The maximum rates previously discussed help reduce friction pressure and excessive abrasion when pumping proppants.

Another consideration, to which little thought is given, relates to the suction side of the trailer manifold. Frequently, there is insufficient diameter on the manifold trailer to accommodate high rates, or sometimes a velocity on the suction side of the manifold is too low, allowing proppants to settle and ultimately causing serious problems with the treatment. Figure 2-7 shows a high-and low-rate suction manifold set up to accommodate two very different situations. The U-tube, 4 in. manifold shown above the 8 in. manifold, which can be isolated, allows the same trailer to be utilized for 100 bbl/min. treatments as well as treatments at less than 10 bbl/min. with thin fluids carrying high-density proppants like bauxite. The oil company person in charge on a high-or low-rate treatment should make observations. He should question the service company people about their ability to supply pressurized fluids from the blender to the truck farthest from the blender to make sure the treatment is controlled. If you starve the last pump in line for fluid, the truck will run roughly and very inefficiently. Other problems appeared with new technological developments.

Figure 2-6 Treating iron. These pictures illustrate 3-in. treating iron coming off the discharge side of the pump to the ground and up to a frac manifold. Note (top) that there is a single-wing chicksan coming off the pump to a pup joint followed by another single-wing tied to a double-wing chicksan connected to the manifold trailer (bottom). This allows movement in any direction with no immobile iron that could cause breaking or rupturing of iron, a potential catastrophe.

Figure 2-7 Symmetrical rig-up using a trailer manifold showing low-and high-rate suction manifold on a trailer manifold. (Note the 4 in. ID suction placed above the 8 in. ID manifold.) The two manifolds allow versatility for the use of the trail-er on low-and high-rate treatments.

To illustrate what can happen with too large of a suction manifold, I will relate what happened to Nowsco Services in treating and utilizing delayed cross-link fluids while pumping bauxite at low rates. On the initial treatment while using a new delayed cross-link fluid, we started losing discharge pressure on our blender and were barely able to finish the treatment. On completion of the treat-ment, we found that our discharge manifold on the blender and our trailer manifold were completely packed off with bauxite. The high-density bauxite had settled out because of the low velocity in the manifold. We immediately had to build a dual low-pressure manifold as previously mentioned for the lower rate treatments. By using the 4 in. section, the proppant settling was negated.

Another example of this kind of problem relates to foam frac treatments where high concentrations of proppant are pumped at very low rates. In these situations, you need to run 3 in. hoses, on many occasions, directly from the discharge manifold of the blender directly to the pump truck. It comes down to evaluating the required velocity to carry proppantor, on the other hand, having sufficient size to be able to pump at very high rates. Recently, I was on several high-rate treatments in the Rocky Mountains and found a trailer manifold that had been built with a single 8 in. suction manifold decreasing to 4 in. at the far end of the manifold. I had severe problems with staving of the pumps at the end of the manifold, and eventually I had to replace that manifold to achieve our required 60 bpm rate with high-viscosity fracturing fluids. The discussion of rate per pump needs to address suction hoses. One common error in rigging up pumping units, on acidizing, cementing, or fracturing treatments, is not applying the correct amount of suction hose for adequate velocity or adequate fluid flow for the pump rate. As a general rule, use one 4 in. suction hose for each 10-12 bbl/min. of fluid to be pumped. If the rates fall below 5 bbl/min., drop to a 3 in. suction hose to keep proppant, if utilized, from settling out in the hose and plugging off.

If hose diameter is too large on a low-rate treatment, you can get proppant slugging, which can damage the pumps if the proper valves are not used. The most common error tends to be on the opposite end; that is, not allowing enough suction hose between the blender or pressurizing pumps and the pump trucks. If you are using high-rate pumps with pumping rates up to 25 bbl/min., have at least two suction hoses per pump. This is particularly true if you are pumping fluids with much viscosity. When there is insufficient booster pressure or fluid volume to the pumps, the pumps run roughly, and there is a great deal of vibration.

To reiterate guidelines for suction hoses, between the frac tanks and blender and between the blenders and pump trucks, the following rules hold. Frac tanks to blenders have one 20ft section of hose per 10 bbl/min. if40# gel or less is utilized. For higher gel concentrations, run two 20 ft 4 in. hoses per 10 bbl/min. Between the blender and the discharge manifold and pump trucks, a good rule is to use one 20 ft hose per 10 bbl/min. Therefore, for a 40# gel treatment to be con-ducted at 40 bbl/min., a minimum of four suction hoses between the blender and frac tank should be used and a minimum of four hoses should be run between the blender and the manifold trailer or pump trucks on the fracture treatment. For a high-viscosity treatment such as one using 60# gel, you should have, for a 40 bbl. treatment, a minimum of eight suction hoses between the tanks and the blender and four hoses between the discharge of the blender and the manifold trailer or pump trucks.

With the advent of continuous mix treatments and holding tanks and pressurizing units, some of the service companies are using 8 in. or larger suction hoses between the frac tank manifolds and their continuous mix unit, as well as 8 in. discharge hoses between their continuous mix units and blenders. Usually an 8 in. single hose is sufficient to maintain rates up to and including 60 bbl./min.

The reason for being very specific about hose requirements on fracture treat-ments and specifying them in this book is that the item most altered or changed on every fracture treatment is the number of suction and discharge hoses utilized on the treatment. For whatever reason, there is always a reluctance to lay the proper number of hoses in order to be assured a proper suction and discharge rate. To further explain this problem, fewer hoses will suffice if there is sufficient hydrostatic head, but when sucking out multiple tanks and the volumes in the tanks get quite low, there is no help from hydrostatic head. Therefore, it is quite easy to suck air and to loose suction with blenders coming out of frac tanks. Conversely, if there is a restriction in rate coming from the blender to the pump trucks, then the pumps with cavitate. You will lose efficiency, and the operation can become quite dangerous from lines jumping. If you are on a fracturing treatment and see a great deal of line vibration, many times this is due to lack of booster pressure or a restriction due to too few hoses between the blender and the pump trucks. Obviously, another rea-son for this is poorly maintained pump trucks with bad valves, seats, and so forth.

Adequate horsepower and pressure capability?

Table 2-4a-m lists several types of pumps used in the oil field and gives their rates and pressure capabilities with varying horsepower and plunger sizes. It is extremely important that company personnel understand the pump rate, pressure capability, and hydraulic horsepower of pump units to be used. This is helpful when selecting suction hose, and it is very important when making a quick evaluation, for instance, to determine if enough available horsepower and standby horsepower is on the location to accomplish the treatment.

Unfortunately, I have arrived on locations to find insufficient horsepower or improper fluid-end sizes to accomplish a treatment. This occurred because personnel were inadequately trained and did not understand the basics of pump mechanics. By reviewing Table 2-4, you can easily see that it is improper to use a 6 1/2 in. pump-to-pump at 8,000-10,000 lbs. The life of the power unit, trans-mission, etc., would be extremely short if you did. On the other hand, using a 4 in. pump-to-pump at maximum rate and low pressure would be inefficient.

On many large treatments, the service company is required to have 100% excess horsepower on the location. The oil company person must query the service company about horsepower and fluid-end sizes of the units that are pumping and those on standby. Often times, due to logistics and timing errors, the wrong fluid ends arrive at the location. Jobs are jeopardized because of it. Frequently, such errors result from human mistakes, but certainly it is unwise to attempt to treat at 10,000 lbs. for a long. time with large-size ends. Along the same line, it is unwise to run small-size fluid ends at maximum rpm for a long time. For the benefit of the operating company and the service company, choose the most efficient operating range for the units on the location. Upon request, the service company will gladly supply optimum working ranges for pumps.

One interesting story relates to some of the first massive hydraulic treatments conducted in east Texas in the Cotton Valley lime and sand formations. These early treatments were massive with sand volumes exceeding 3 million pounds. Pump times, because of treating down tubular goods, sometimes approached 12

Table 2-4a OPI 1800 AWS Pump

Pump Specifications

Rated brake horsepower	1,800
Stroke length	8 inch
Maximum rod load	225,000 lb s.
Overall length	89 ¾ inch
Overall width	59 ¾ inch
Overall height	44 5/16 lbs
Weight	11,900
Gear Ratio	6,353:1

Plunger, Diameter, inch	**Gallons per Revolution**	**Pump Room**									
		50		**100**		**200**		**300**		**330**	
		gpm	**psi**	**gpm**	**psi**	**gpm**	**psi**	**gpm**	**psi**	**gpm**	**psi**
7 ½	4.95	229	5,093	459	5,093	918	2,857	1,377	1,905	**1,515	1,732
6 ¾	3.72	186	6,288	372	6,288	744	3,527	1,115	2,352	1,227	2,138
6 ½	3.45	172	6,781	345	6,781	690	3,804	1,304	2,536	1,138	2,305
6	2.94	147	7,958	294	7,958	588	4,464	881	2,976	969	2,706
5 ¾	2.70	135	8,665	270	8,665	540	4,861	809	3,241	890	2,946
5 ½	2.47	123	9,470	247	9,470	494	5,313	741	3,542	815	3,220
5	2.04	102	11,459	204	11,459	408	6,429	612	4,286	673	3,896
4 ½	1.65	83	14,147	165	14,147	330	7,937	496	5,291	545	4,810
4	1.31	65	**17,905	131	**17,905	261	10,045	392	6,696	431	6,088
3 ¾	1.15	57	**20,372	115	**20,372	229	11,429	344	7,619	379	6,926
Brake horsepower		802		1,604		1,800		1,800		1,800	
Pinion rpm		318		635		1,271		1,906		2,096	

* Based on 85% mechanical efficiency (ME] and 100% volumetric efficiency (VEl-intermittent service only.

* Application to be approved by OPI Engineering. Reprinted courtesy OPI Engineering

Reprinted courtesy OPI Engineering

Table 2-4b OPI 1800 AWS Pump

Pump Specifications

Rated brake horsepower	1,300
Stroke length	8 inch
Maximum rod load	200,000 lb s.
Overall length	89 ¾ inch
Overall width	59 ¾ inch
Overall height	44 5/16 lbs
Weight	11,700 lbs
Gear Ratio	7.0:1-7.5:1

Plunger, Diameter, inch	**Gallons per Revolution**	**Pump Room**									
		50		**100**		**200**		**300**		**330**	
		gpm	**psi**	**gpm**	**psi**	**gpm**	**psi**	**gpm**	**psi**	**gpm**	**psi**
7 ½	4.95	200	4,527	459	4,127	918	2,064	1,147	1,651	1,377	1,376
6 ¾	3.72	186	5,589	372	5,095	744	2,548	929	2,038	1,115	1,698
6 ½	3.45	172	6,027	345	5,495	690	2,747	862	2,198	1,034	1,832
6	2.94	147	7,074	294	6,448	588	3,224	734	2,579	881	2,149
5 ¾	2.70	135	7,702	270	7,021	540	3,511	674	2,809	809	2,340
5 ½	2.47	123	8,418	247	7,674	494	3,837	617	3,070	741	2,558
5	2.04	102	10,186	204	9,286	408	4,643	510	3,714	612	3,095
4 ½	1.65	83	12,575	165	11,464	330	5,732	413	4,586	496	3,821
4	1.31	65	**15,915	131	**14,509	261	7,254	326	5,804	392	4,836
3 ¾	1.15	57	**18,108	115	**16,508	229	8,254	287	6,603	344	5,503
Brake horsepower		713		1,300		1,300		1,300		1,300	
Pinion rpm (7.0:1)		350		700		1,400		1,750		2,100	
Pinion rpm (7.5:1)		375		500		1,500		1,875		2,250	

* Based on 85% mechanical efficiency (ME) and 100% volumetric efficiency (VE]-intermittent service only. **Application to be approved by OPI Engineering.
Reprinted courtesy OPI Engineering

Table 2-4c OPI 1800 AWS Pump

Pump Specifications

Rated brake horsepower	1,000
Stroke length	8 inch
Maximum rod load	200,000 lbs.
Overall length	89 ¾ inch
Overall width	59 ¾ inch
Overall height	44 5/16 lbs
Weight	11,700 lbs
Gear Ratio	4.95:1, 7.0:1, 7.5:1

Plunger, Diameter, inch	Gallons per Revolution	Pump Room									
		50		100		200		300		330	
		gpm	psi	gpm	psi	gpm	psi	gpm	psi	gpm	psi
7 ½	4.95	229	4,527	459	3,175	918	1,587	1,147	1,270	1,377	1,058
6 ¾	3.72	186	5,589	372	3,919	744	1,960	929	1,568	1,115	1,306
6 ½	3.45	172	6,027	345	4,227	690	2,113	862	1,691	1,034	1,409
6	2.94	147	7,074	294	4,960	588	2,480	734	1,984	881	1,653
5 ¾	2.70	135	7,702	270	5,401	540	2,701	674	2,160	809	1,800
5 ½	2.47	123	8,418	247	5,903	494	2,952	617	2,361	741	1,968
5	2.04	102	10,186	204	7,143	408	3,571	510	2,857	612	2,381
4 ½	1.65	83	12,575	165	8,818	330	4,409	413	3,527	496	2,939
4	1.31	65	**15,915	131	11,161	261	5,580	326	4,464	392	3,720
3 ¾	1.15	57	**18,108	115	12,698	229	6,349	287	5,079	344	4,233
Brake horsepower		713		1,300		1,300		1,300		1,300	
Pinion rpm (4.95:1)		248		495		900		1,238		1,485	
Pinion rpm (7.0:1)		350		700		1,400		1,750		2,100	
Pinion rpm (7.5:1)		375		500		1,500		1,875		2,250	

* 4.95:1 power end.

**Based on 85% mechanical efficiency (ME) and 100% volumetric efficiency (VEl-intermittent service only. Application to be approved by OPI Engineering.

Reprinted courtesy OPI Engineerin

Table 2–4d

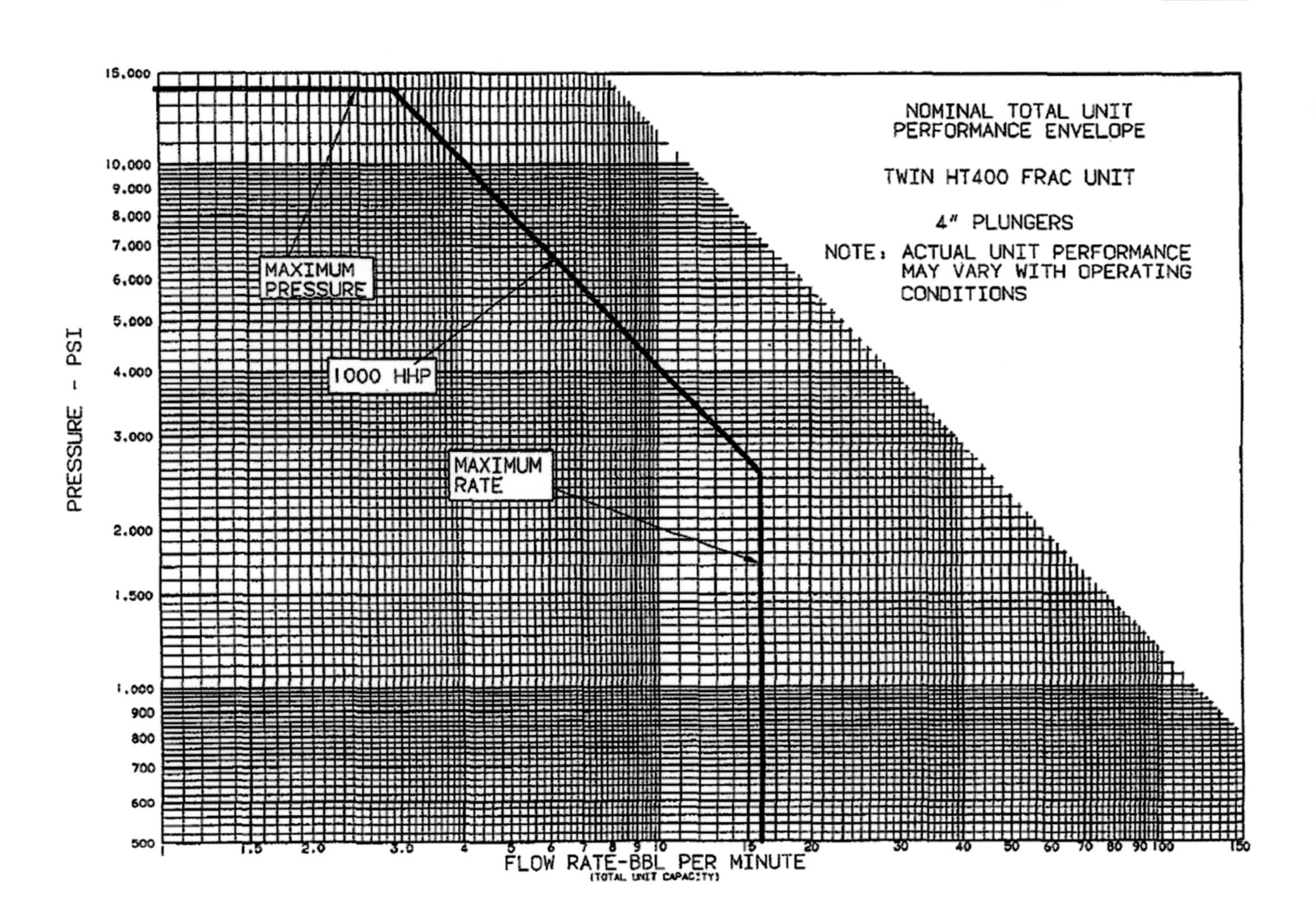
NOMINAL TOTAL UNIT
PERFORMANCE ENVELOPE
TWIN HT400 FRAC UNIT
4" PLUNGERS
NOTE: ACTUAL UNIT PERFORMANCE
MAY VARY WITH OPERATING
CONDITIONS
MAXIMUM
PRESSURE
1000 HHP
MAXIMUM
RATE
PRESSURE - PSI
15,000
10,000
9,000
8,000
7,000
6,000
5,000
4,000
3,000
2,000
1,500
1,000
900
800
700
600
500
1
1.5
2.0
3.0
4
5
6
7
8
9
10
15
20
30
40
50
60
70
80
90
100
150
FLOW RATE-BBL PER MINUTE
(TOTAL UNIT CAPACITY)

Table 2–4e

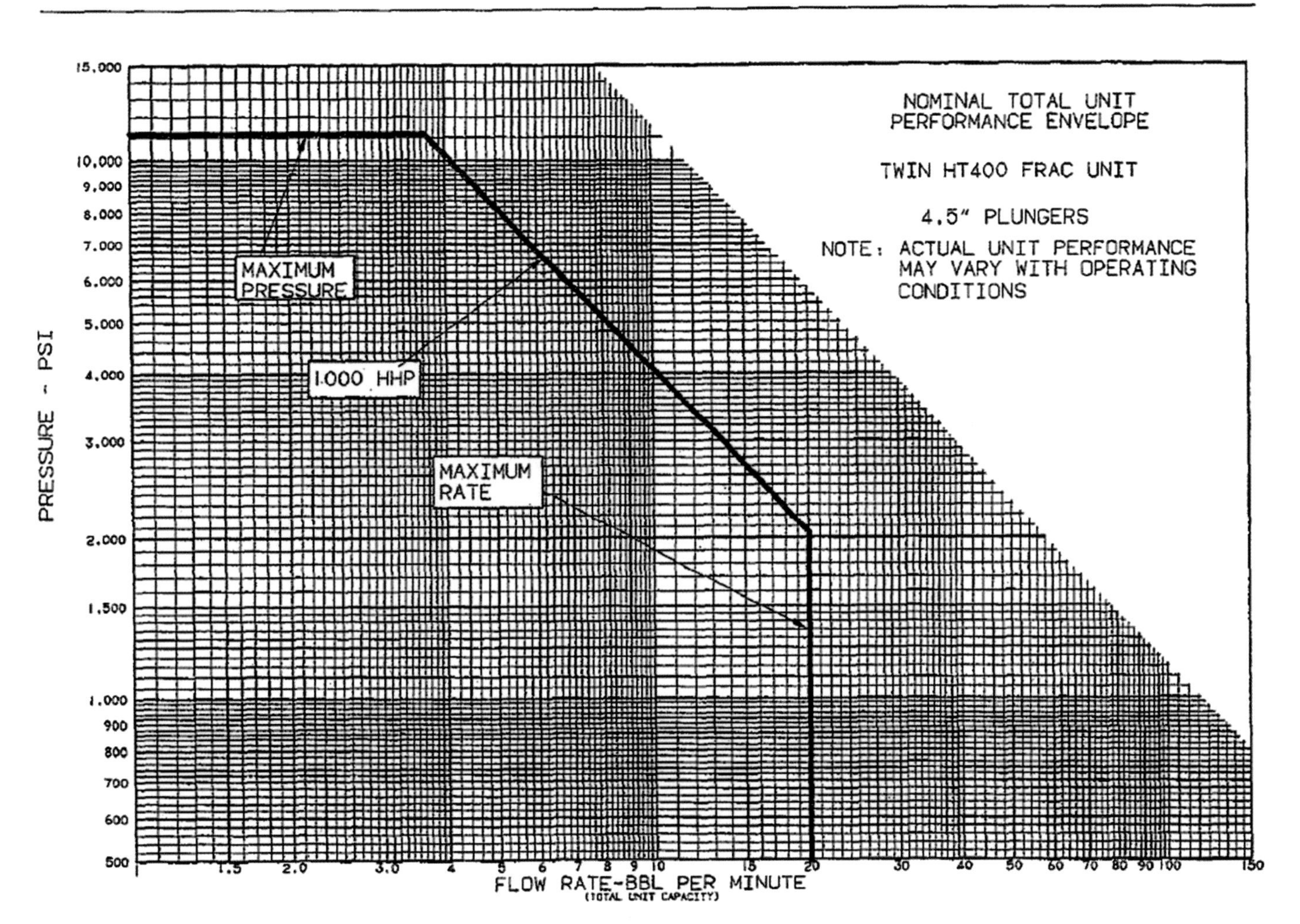
NOMINAL TOTAL UNIT
PERFORMANCE ENVELOPE
TWIN HT400 FRAC UNIT
4.5" PLUNGERS
NOTE: ACTUAL UNIT PERFORMANCE
MAY VARY WITH OPERATING
CONDITIONS
MAXIMUM
PRESSURE
1.000 HHP
MAXIMUM
RATE
PRESSURE - PSI
15,000
10,000
9,000
8,000
7,000
6,000
5,000
4,000
3,000
2,000
1,500
1,000
900
800
700
600
500
1.5
2.0
3.0
4
5
6
7
8
9
10
15
20
30
40
50
60
70
80
90
100
150
FLOW RATE-BBL PER MINUTE
(TOTAL UNIT CAPACITY)

Table 2–4f

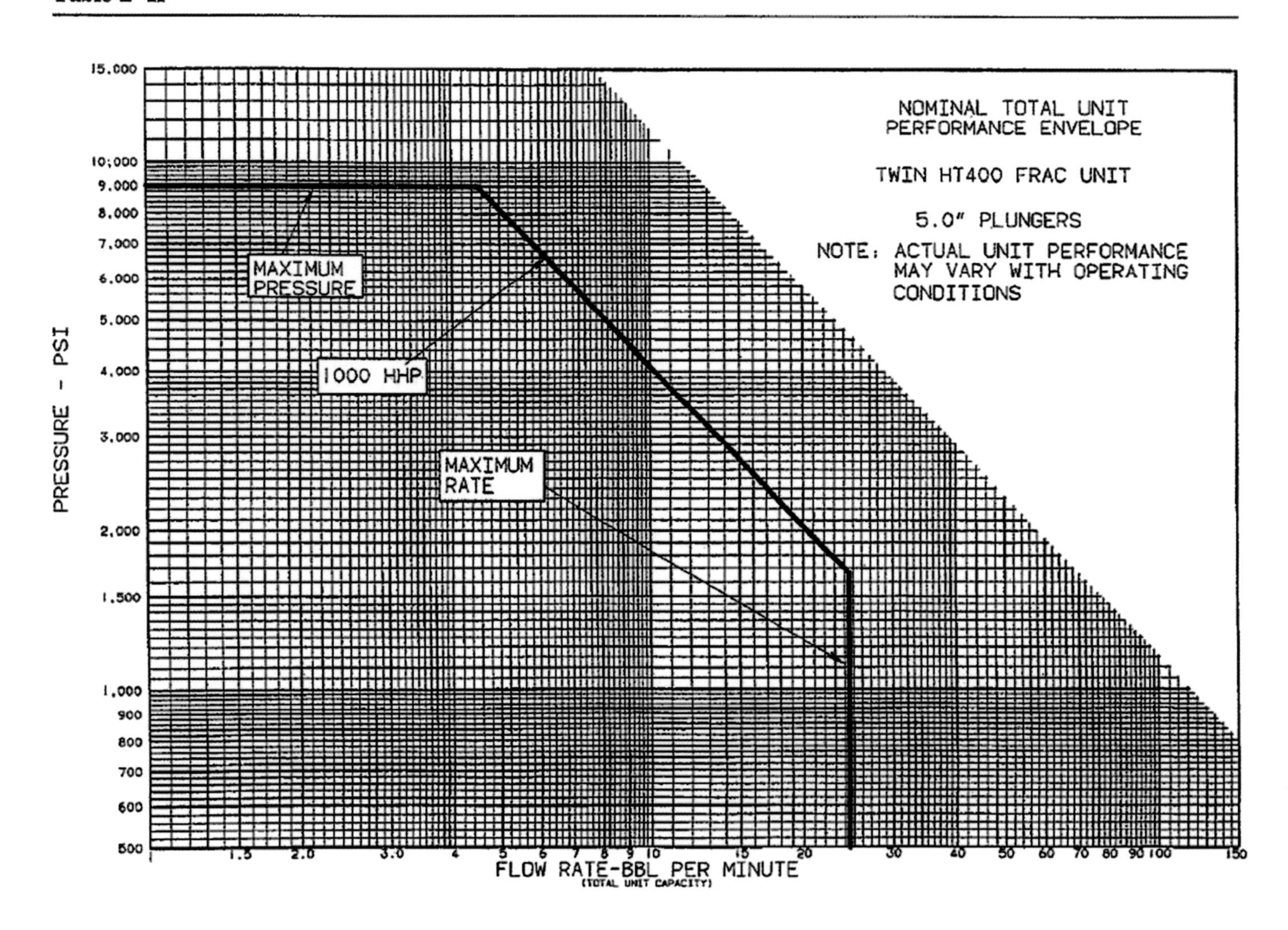
NOMINAL TOTAL UNIT
PERFORMANCE ENVELOPE
TWIN HT400 FRAC UNIT
5.0" PLUNGERS
NOTE: ACTUAL UNIT PERFORMANCE
MAY VARY WITH OPERATING
CONDITIONS
MAXIMUM
PRESSURE
1000 HHP
MAXIMUM
RATE
PRESSURE - PSI
15,000
10,000
9,000
8,000
7,000
6,000
5,000
4,000
3,000
2,000
1,500
1,000
900
800
700
600
500
1.5
2.0
3.0
4
5
6
7
8
9
10
15
20
30
40
50
60
70
80
90
100
150
FLOW RATE-BBL PER MINUTE
(TOTAL UNIT CAPACITY)

Table 2–4g

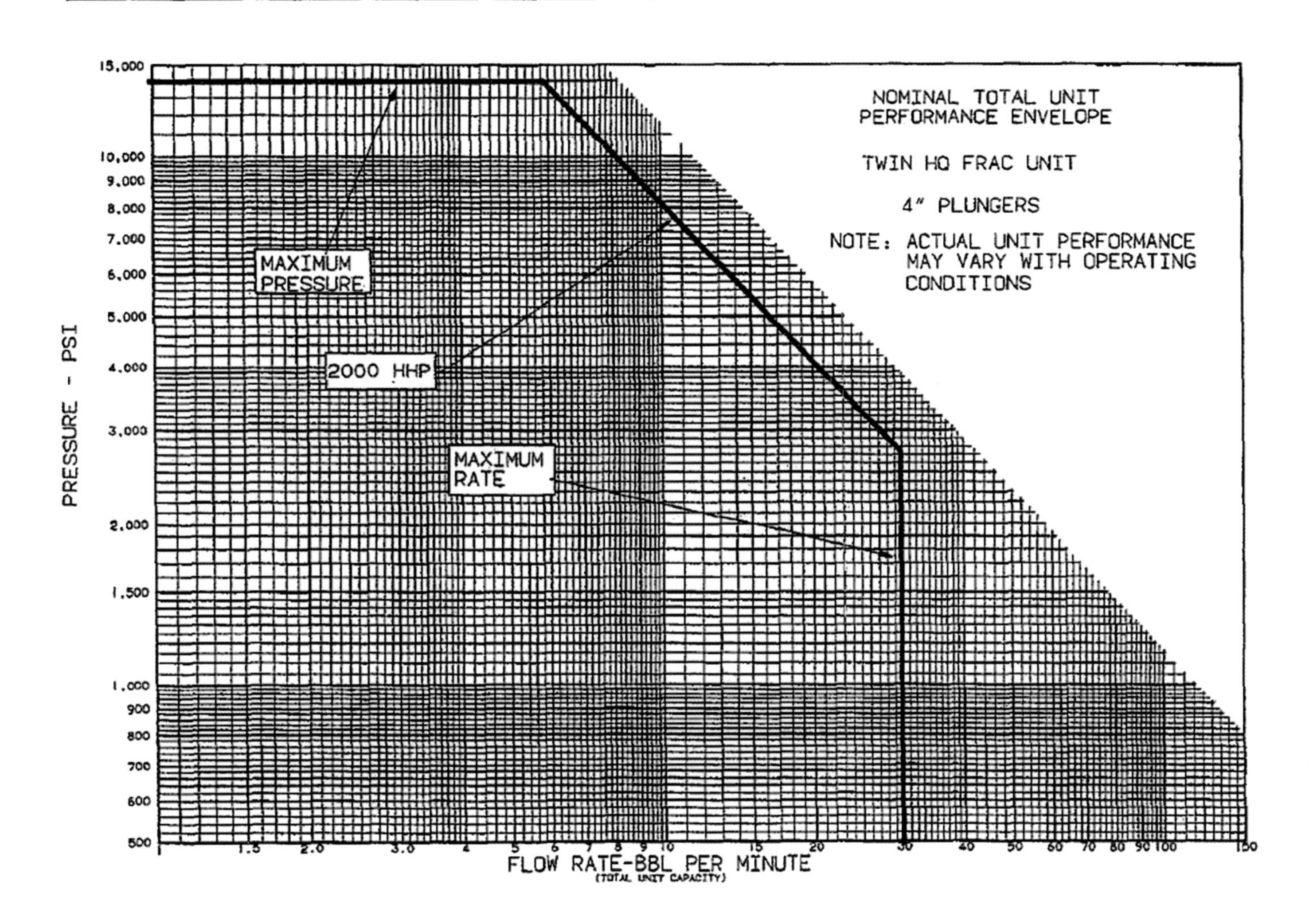
NOMINAL TOTAL UNIT
PERFORMANCE ENVELOPE
TWIN HQ FRAC UNIT
4" PLUNGERS
NOTE: ACTUAL UNIT PERFORMANCE
MAY VARY WITH OPERATING
CONDITIONS
MAXIMUM
PRESSURE
2000 HHP
MAXIMUM
RATE
PRESSURE - PSI
15,000
10,000
9,000
8,000
7,000
6,000
5,000
4,000
3,000
2,000
1,500
1,000
900
800
700
600
500
1
1.5
2.0
3.0
4
5
6
7
8
9
10
15
20
30
40
50
60
70
80
90
100
150
FLOW RATE-BBL PER MINUTE
(TOTAL UNIT CAPACITY)

Table 2–4h

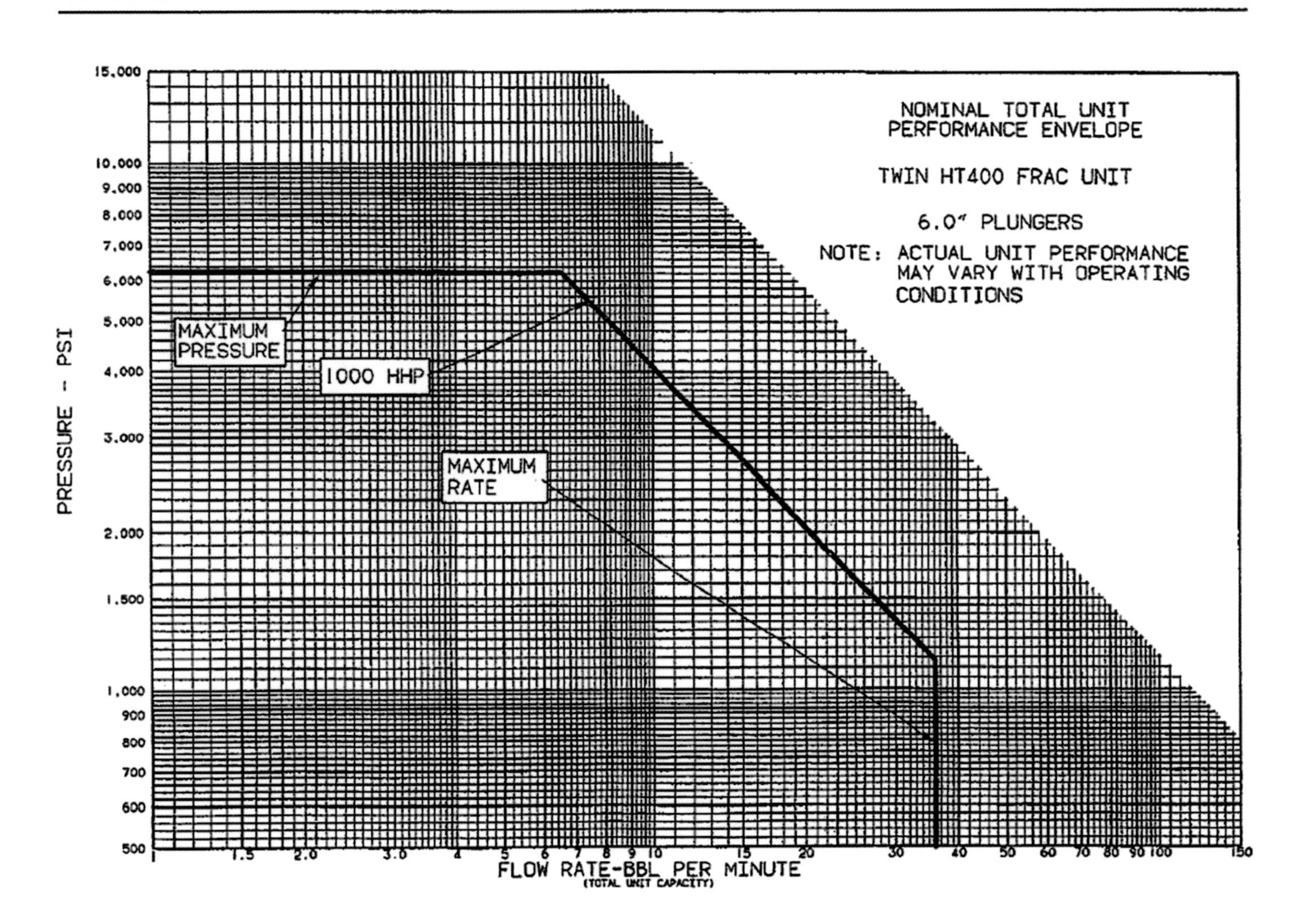
NOMINAL TOTAL UNIT
PERFORMANCE ENVELOPE
TWIN HT400 FRAC UNIT
6.0" PLUNGERS
NOTE: ACTUAL UNIT PERFORMANCE
MAY VARY WITH OPERATING
CONDITIONS
MAXIMUM
PRESSURE
1000 HHP
MAXIMUM
RATE
PRESSURE - PSI
15,000
10,000
9,000
8,000
7,000
6,000
5,000
4,000
3,000
2,000
1,500
1,000
900
800
700
600
500
1
1.5
2.0
3.0
4
5
6
7
8
9
10
15
20
30
40
50
60
70
80
90
100
150
FLOW RATE-BBL PER MINUTE
(TOTAL UNIT CAPACITY)

Table 2–4i

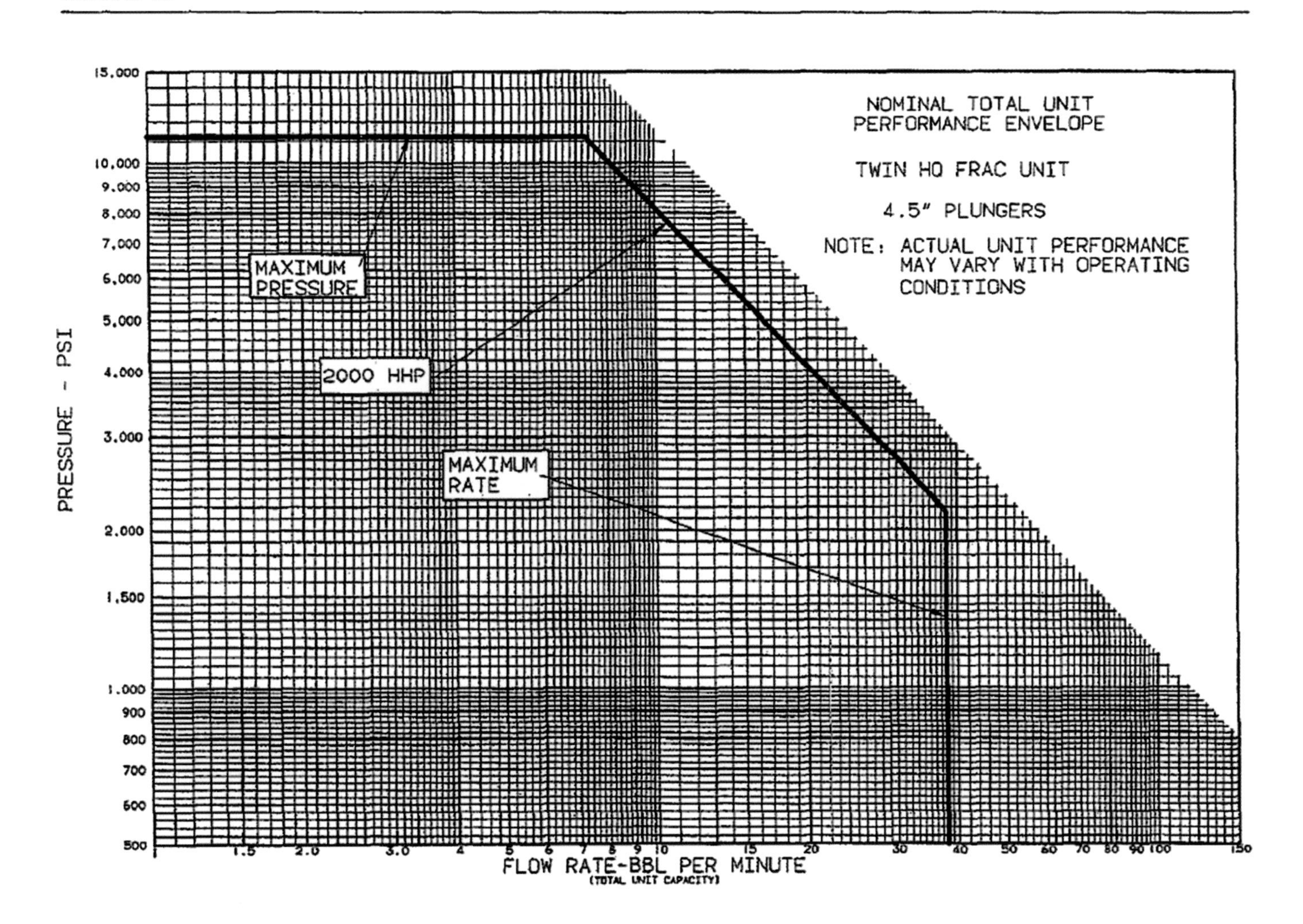
NOMINAL TOTAL UNIT
PERFORMANCE ENVELOPE
TWIN HQ FRAC UNIT
4.5" PLUNGERS
NOTE: ACTUAL UNIT PERFORMANCE
MAY VARY WITH OPERATING
CONDITIONS
MAXIMUM
PRESSURE
2000 HHP
MAXIMUM
RATE
PRESSURE - PSI
15,000
10,000
9,000
8,000
7,000
6,000
5,000
4,000
3,000
2,000
1,500
1,000
900
800
700
600
500
1.5
2.0
3.0
4
5
6
7
8
9
10
15
20
30
40
50
60
70
80
90
100
150
FLOW RATE-BBL PER MINUTE
(TOTAL UNIT CAPACITY)

Table 2–4j

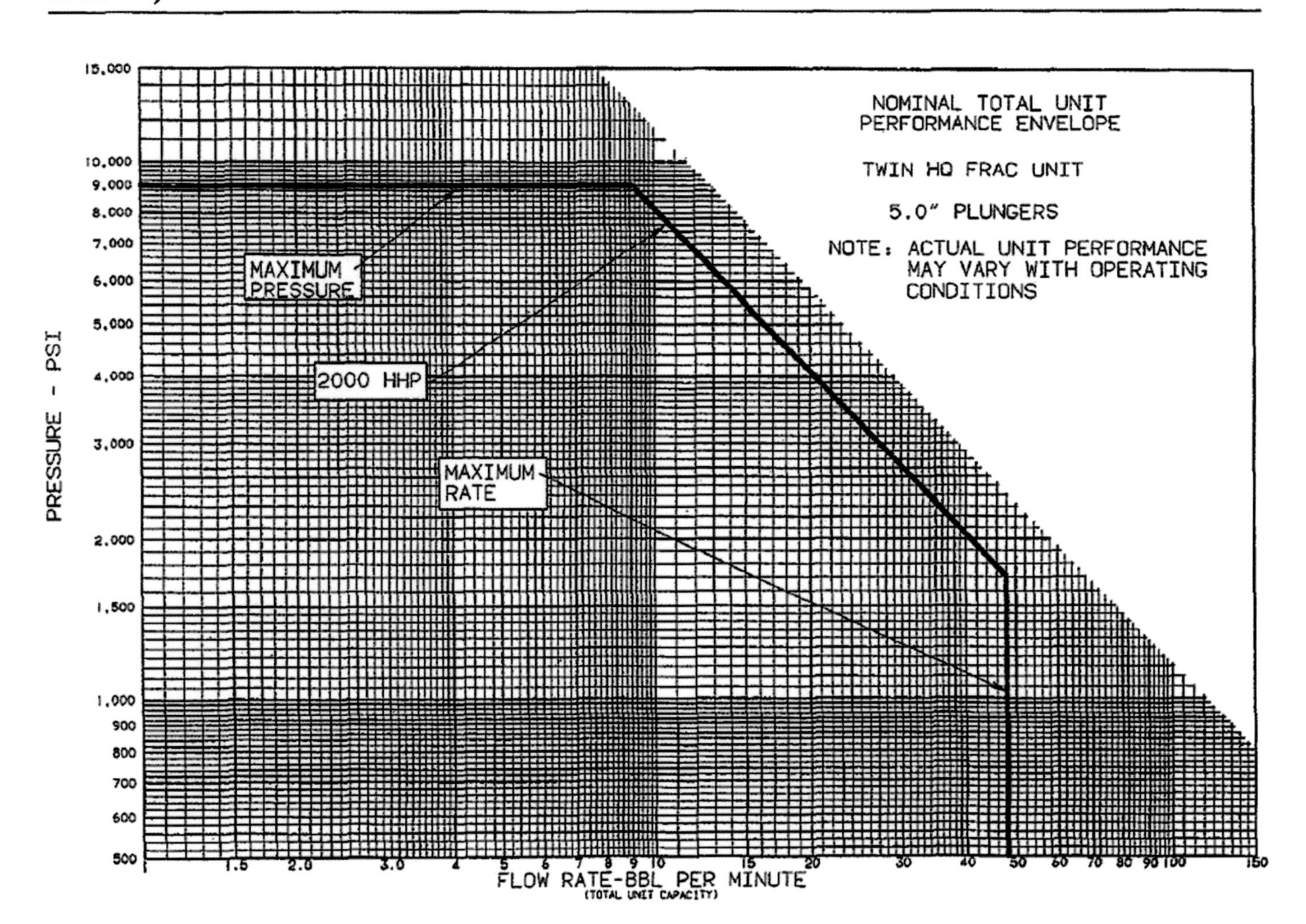
NOMINAL TOTAL UNIT
PERFORMANCE ENVELOPE
TWIN HQ FRAC UNIT
5.0" PLUNGERS
NOTE: ACTUAL UNIT PERFORMANCE
MAY VARY WITH OPERATING
CONDITIONS
MAXIMUM
PRESSURE
2000 HHP
MAXIMUM
RATE
PRESSURE - PSI
15,000
10,000
9,000
8,000
7,000
6,000
5,000
4,000
3,000
2,000
1,500
1,000
900
800
700
600
500
1.5
2.0
3.0
4
5
6
7
8
9
10
15
20
30
40
50
60
70
80
90
100
150
FLOW RATE-BBL PER MINUTE
(TOTAL UNIT CAPACITY)

Table 2–4k

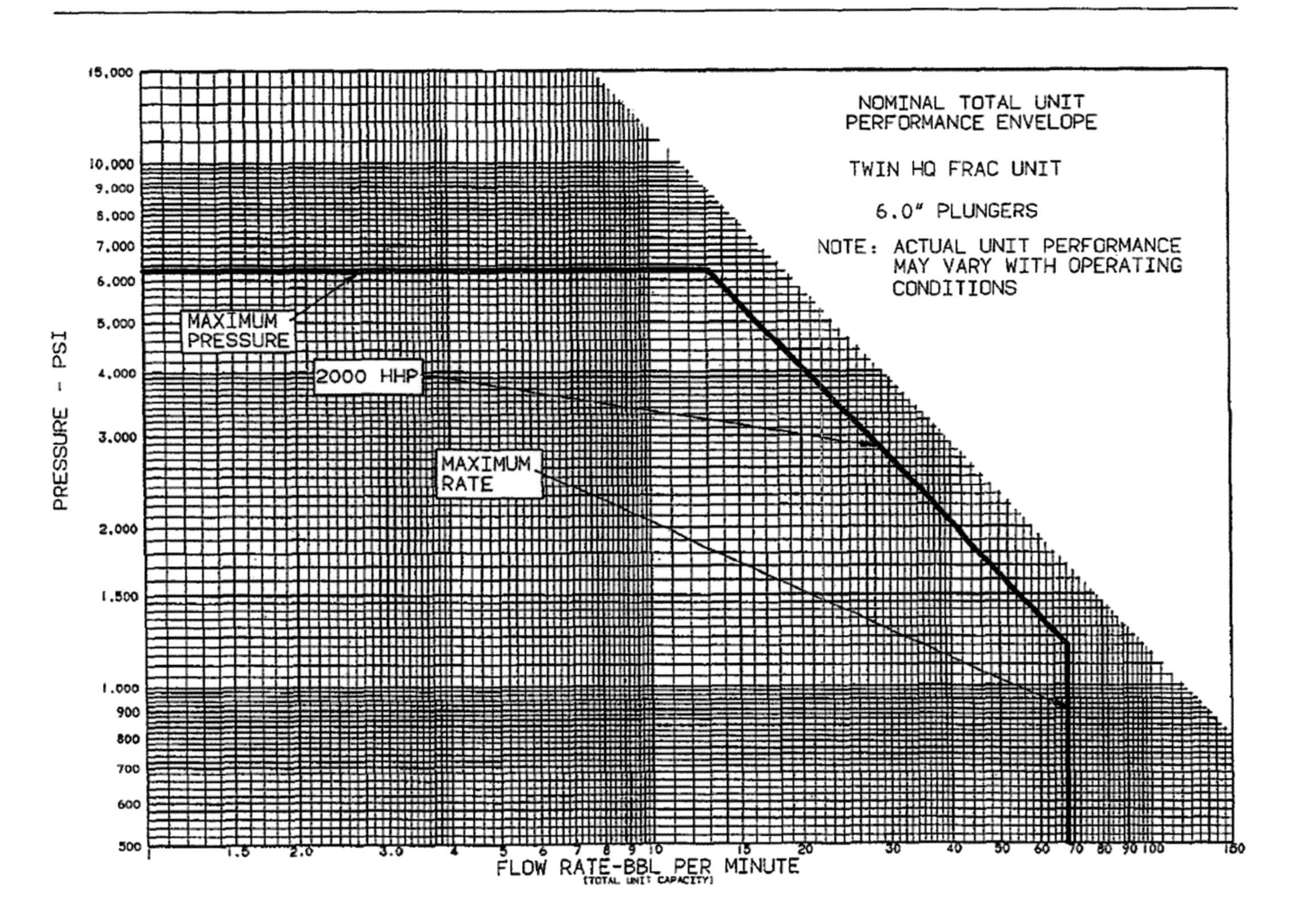
NOMINAL TOTAL UNIT
PERFORMANCE ENVELOPE
TWIN HQ FRAC UNIT
6.0" PLUNGERS
NOTE: ACTUAL UNIT PERFORMANCE
MAY VARY WITH OPERATING
CONDITIONS
MAXIMUM
PRESSURE
2000 HHP
MAXIMUM
RATE
PRESSURE - PSI
15,000
10,000
9,000
8,000
7,000
6,000
5,000
4,000
3,000
2,000
1,500
1,000
900
800
700
600
500
1
1.5
2.0
3.0
4
5
6
7
8
9
10
15
20
30
40
50
60
70
80
90
100
150
FLOW RATE-BBL PER MINUTE
(TOTAL UNIT CAPACITY)

Table 2–4l

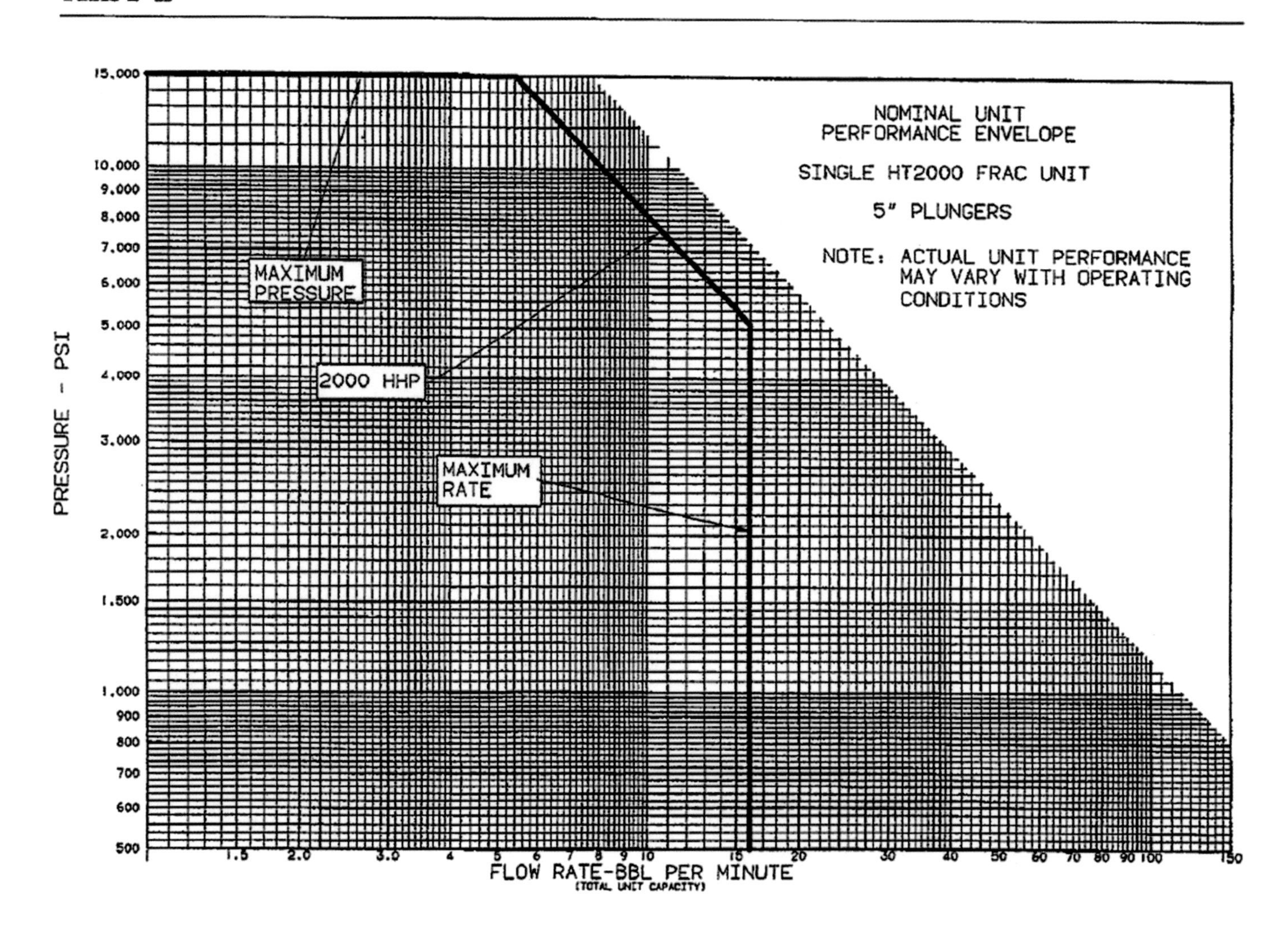
NOMINAL UNIT
PERFORMANCE ENVELOPE
SINGLE HT2000 FRAC UNIT
5" PLUNGERS
NOTE: ACTUAL UNIT PERFORMANCE
MAY VARY WITH OPERATING
CONDITIONS
MAXIMUM
PRESSURE
2000 HHP
MAXIMUM
RATE
PRESSURE - PSI
15,000
10,000
9,000
8,000
7,000
6,000
5,000
4,000
3,000
2,000
1,500
1,000
900
800
700
600
500
1.5
2.0
3.0
4
5
6
7
8
9
10
15
20
30
40
50
60
70
80
90
100
150
FLOW RATE-BBL PER MINUTE
(TOTAL UNIT CAPACITY)

Table 2–4m

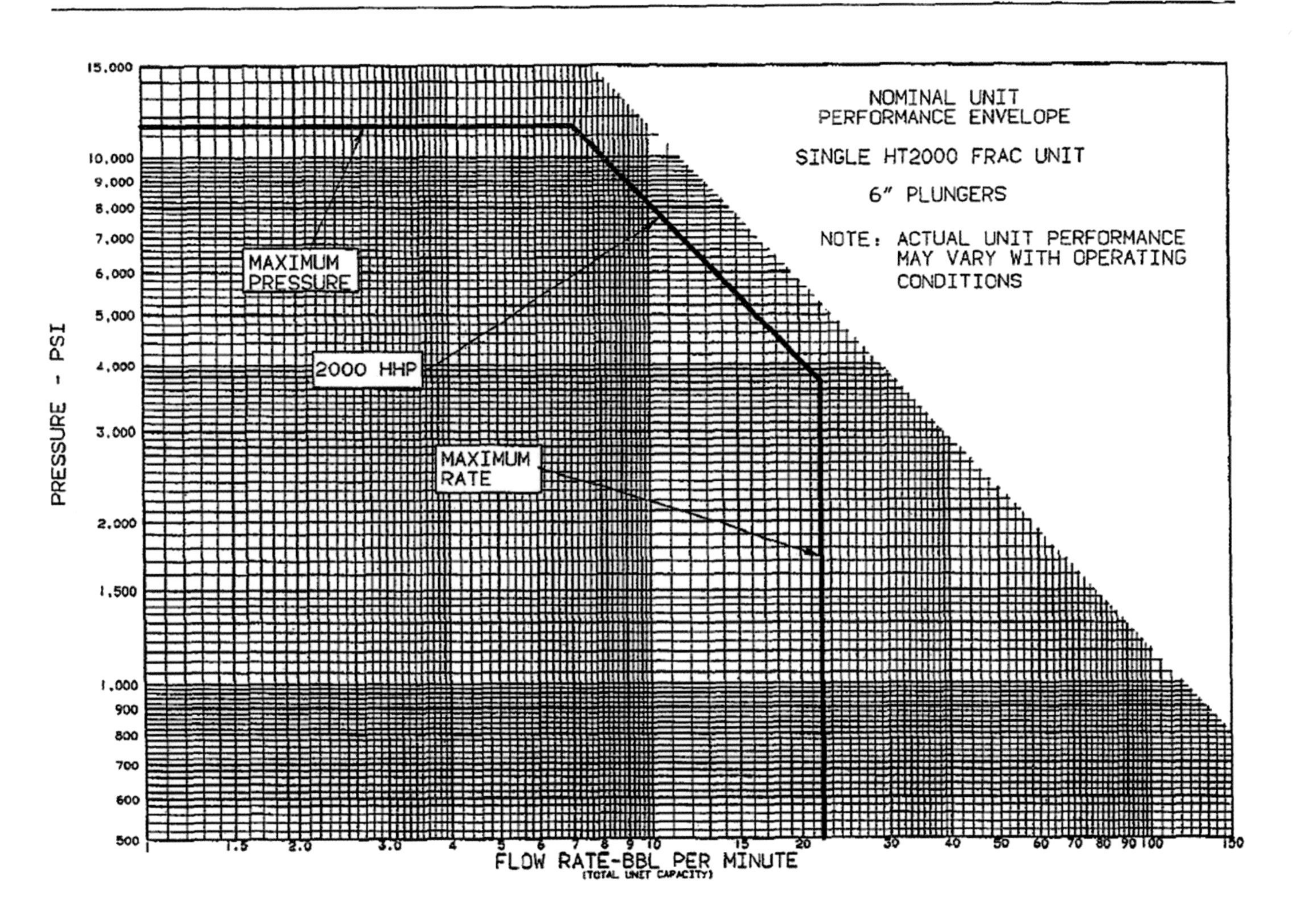
NOMINAL UNIT
PERFORMANCE ENVELOPE
SINGLE HT2000 FRAC UNIT
6" PLUNGERS
NOTE: ACTUAL UNIT PERFORMANCE
MAY VARY WITH OPERATING
CONDITIONS
MAXIMUM
PRESSURE
2000 HHP
MAXIMUM
RATE
PRESSURE - PSI
15,000
10,000
9,000
8,000
7,000
6,000
5,000
4,000
3,000
2,000
1,500
1,000
900
800
700
600
500
1
1.5
2.0
3.0
4
5
6
7
8
9
10
15
20
30
40
50
60
70
80
90
100
150
FLOW RATE-BBL PER MINUTE
(TOTAL UNIT CAPACITY)

hours, with pumping pressures from 7,000-11,000 lbs. Typically, pumping pressures were not within the range requiring intensifiers, but excessive losses of fluid ends, power ends, and transmissions were experienced by some service companies. Upon investigating, it was ascertained that when the service company placed excess equipment on the location, they were, in fact, running the required amount of equipment in a red-line situation and leaving the standby equipment on "idle." Doing this was equivalent to driving a new car at 120 mph and towing a heavy trailer. The investigation determined that running all of the equipment at 60% of capacity would enhance the units' life spans.

Sometime in your career, you will meet people who are extremely proud of their company's high-horsepower pumping units. These people point to a unit and proclaim that the unit can pull, depending on its type, from 1,100 to 2,000 hydraulic horsepower. Calculations indicate that they are getting, in some cases, 80-85% of brake horsepower from the engine, which is a remarkable performance indeed. This type of horsepower delivery should only be yielded in a test chamber at the yard to find out the unit's efficiency, or in a critical situation where no standby equipment is available. The oil company person may say, "If they want to run their equipment that way, then it is their loss." But running equipment in this manner jeopardizes the treatment. I believe in excess equipment on any job, even on a small cement or acid job. But the best situation is one where you utilize all equipment at something far less than peak capacity. I evaluate the average horsepower per unit during a treatment. At the same time, I determine if we are using a torque-type transmission, which is common today. I make certain that we are in direct drive, not torque. If your service company is running an automatic transmission and is not in direct drive or lockup, then it will lose a transmission in a very short time. Much training comes from the service company for running equipment so it will have a long life and for con-ducting lengthy treatments in a workaday fashion.

I always insist that the service company runs all of the equipment on site rather than bringing on part of the equipment and having other equipment idling. This sometimes requires a discussion with the personnel in relationship to charges, and it is something that should be settled before the start of the job. I always have the service company bring on all of the equipment in as low a gear as possible and have everything on line. Consequently, several things occur. First of all, you are minimizing stress on the equipment; and secondly, all of the equipment is on line and ready to perform if any of the equipment fails. It is always a dangerous situation to try to bring on a truck late in the treatment, particularly if you are running high-sand concentrations. If you are paying for standby horse-power on a location, the arrangement should be that all of the equipment will run. Excess horsepower will only be utilized if the primary equipment fails.

I have heard customers insist that the operators run the equipment outside recommended ranges. When training service company employees, I insist that for the good of everyone involved that equipment be run only within acceptable limits. All service companies should be able to supply fluid-end size, plunger size, operating ranges of transmissions, rates per pump, and hydraulic horsepower capabilities for all units. Operating company personnel must understand the complete capabilities of the units at the location.

I have seen situations, particularly those where flow meters are not used, that indicate flow rates with the treating pressures are greatly exceeding the horsepower at the location. If you have the basic understanding of hydraulic horsepower, then this confusion will not occur. Additionally, reducing excess horsepower charges will give you considerable savings.

Prejob flow rate check

A service company should not supply less hydraulic horsepower than expected by the operating company. People do make mistakes, and sometimes improper units arrive at the location. I recommend a little checking and preparation prior to a treatment, both by the service company engineer and by the operating company representative, to prevent problems. Nothing is more embarrassing than to progress into critical stages of the treatment, then, because of some minor mechanical failure, terminate it.

One of my important realizations occurred early in my career; namely, people make mistakes! I attended one job where more than one-half of our trucks ran out of fuel. Later, on other jobs, as I went around checking my fluids, I often checked fuel tanks. I am not suggesting that the oil· company person should check fuel tanks. However, you are responsible for questioning the service company engineer to be sure he has adequate fuel and horsepower (i.e., the essentials) before starting the treatment. He will respect you for asking, those questions, and you will feel better for asking, especially if a job is delayed because of too little fuel or horse-power. Service companies are professional organizations striving to do jobs to the best of their abilities. A cooperative oil company person can be tremendously helpful by reminding them of all the items that need to be checked, particularly ·before long treatments.

Also, you need adequate suction hose from the blender or pressurizing unit to the pump trucks. If you are going directly to the pumping units, then remember the rule previously mentioned: Use at least one 4 in. hose for every 10-12 bb/min. from the blender to each unit or an equal number of hoses from the blender to the ground or trailer manifold. Note that on acidizing and cementing trucks that the pressurizing pump is often included on the unit. Then the only consideration is to have adequate suction from the pumping unit to liquid storage.

The same basic rules hold for the suction side of the pressurizing unit or blender. You need not have smaller hoses because there are no proppants or materials to settle out on the suction side of the blender. The biggest problem on the suction side of the blender on jobs with more than 10 frac tanks is the distances that must be pumped if large numbers of tanks are on the location.

A rig-up on a massive frac treatment is shown in Figure 2-8. This location had approximately 30 frac tanks. Five blenders were used either as primary or stand-by, with three additional blenders to transfer fluids to working tanks. You do not want to suck with a blender for great distances, particularly if you are attempting to pump at high rates. A good rule is to pump out of a single blender for no more than 10-12 tanks. If you are going to have more tanks, a booster pump, transfer pump, or another blender can pump into working tanks.

Optimize fluid placement and removal

Figure 2-9 shows a technique that is used successfully in east Texas where many massive frac treatments are conducted. This technique entails elevating the rear end of the tank so that virtually all of the fluid can be sucked out of the tanks instead of leaving 2,00Q-3,000 gals on the bottom.

This technique has saved the customer considerable expense. On a large treatment, like that shown in Figure 2-8, as much as 45,000 gals of expensive fracturing fluid can be saved. This fluid is pumped downhole rather than left in the bottom of the tank. Many frac tanks are designed so when the level gets low a man hole cover can be taken off the front. Observers can move almost all the fluid from the tanks (see Figure 2-10).

Figure 2-8 Typical rig-up for massive hydraulic fracture treatment.

Figure 2-9 Elevated rear of frac tank to assist in fluid removal.

On many treatments, such as small fracturing or acidizing jobs, upright or horizontal frac tanks are not utilized. Service company transports or contract trailers are used instead. In these situations, the amount of suction hose coming from the trailers should be calculated and competent personnel should be positioned on top of the trailers. This will ensure that you do not suck air when removing the fluid from the trailers.

The major consideration (whether frac tanks, contract trailers, or service company trailer units) is that the distance from the transfer pumps and blenders is minimized to reduce the distance the fluid must be pulled. If at all possible, station the trailers so their hydrostatic heads ease problems in sucking the fluid. If the fluids transferred are extremely viscous, additional suction hoses may be required.

Make proppant accessible to the blenders

In treatments requiring sand or other types of bulk proppant, proper placement of the storage bins is important. Convenient placement allows the material to be adequately transported to the blending trucks. On large fracturing treatments, most people require standby units in case one of the blenders should go down. One of the first things to determine on large fracturing treatments is the service company's ability to get proppant to the standby blender. A standby blender should be rigged to assume all blender and metering functions immediately.

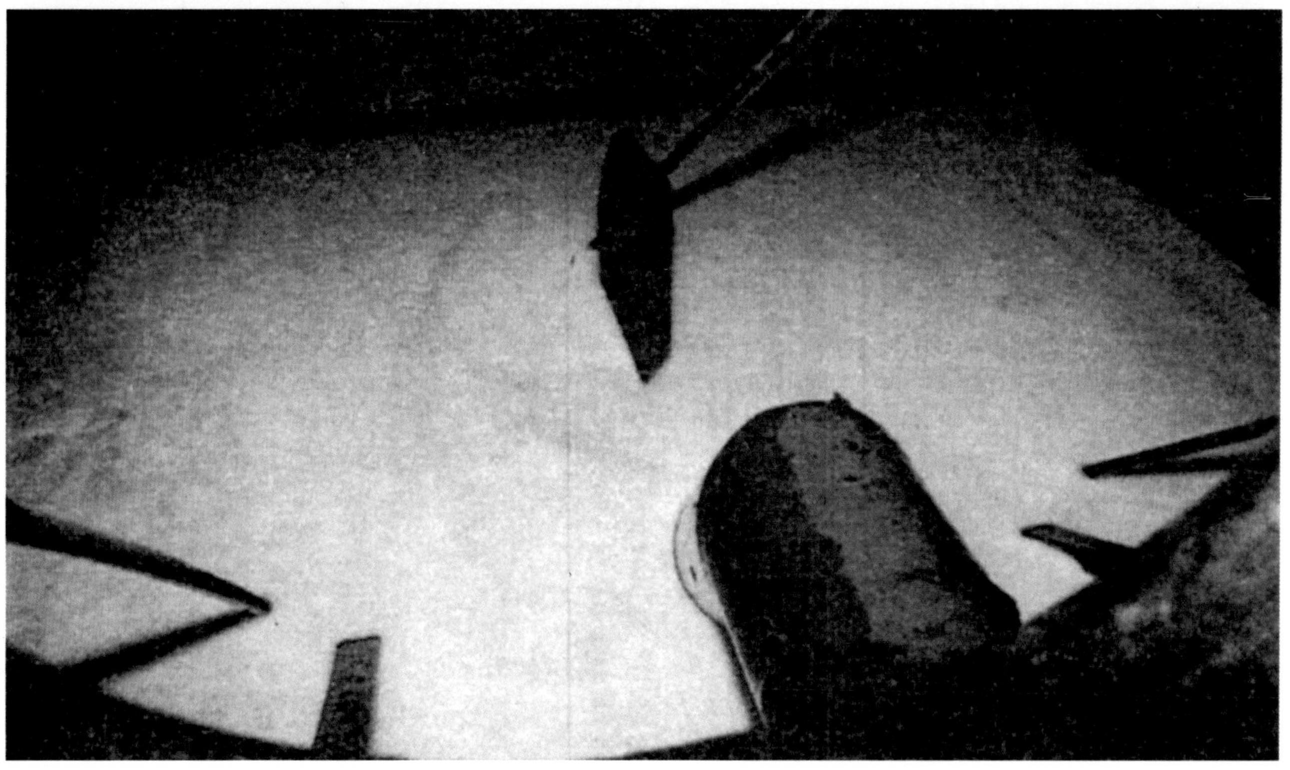

Figure 2-10 Manholes on frac tanks. Front manhole covers are removed (top) as the fluid is lowered for observation and maximum fluid removal. Observation through the manhole (bottom) allows almost all of the fluid to be removed.

Proper standby blender preparation, a difficult task on tight locations, requires prejob planning and a working relationship between the oil company engineer and the service engineer. The treatment shown in Figure 2-8 accomplished the transfer and movement of 2.7 million lbs. of sand.

It is wise for the oil company person to physically review the location with the service company engineer. Have the engineer explain in detail how the proppant is going to be added and what will occur if equipment goes down. Whether it is cement bulk equipment or sand storage equipment, some provision must be made for power unit or pressurizing unit failures on a particular system. Most service companies can maintain a job by transferring hydraulic power or air. This quality control should be available on any large treatment. Figure 2-11 shows spare hydraulic lines on sand storage equipment that can be hooked up immediately if a power unit goes down.

Figure 2-11 Sand conveyor belt connected to a sand storage unit. Shown are quick-connect hydraulic hoses that can be moved from unit to unit. Each unit has an individual hydraulic power system. Should one fail, the unit can be powered from another close by. This set-up gives unlimited standby capacity on large jobs.

Most of the things discussed for rig-up are common sense. You must have adequate space at the location for storing proppants or cement. You must efficiently transport those materials to a blending area. There must.be sufficient space for the liquids to be utilized and enough hosing or transfer equipment to move that fluid to a blending area. Downstream of the blender, ample hosing must be available to transport the fluid and possibly proppant to a trailer or ground manifold and then onward to the high-pressure pumps. You also need adequate velocity to keep the proppants in suspension and transfer hosing sized to move the fluid at low pressures.

The ID of the high-pressure iron should be large enough to allow minimum friction and a small amount of abrasion if proppants are used. Proper isolating valves and check valves need to be used to maintain a safe environment. All of these items must be double-checked to ensure that the operation can be conducted efficiently and safely.

Safety check all high pressure iron

Above all, the correct pressure-rated iron should be used for each application. Most service companies take pride in their iron and its quality control, including the various long joints, chicksans, double wings, and single wings. However, a potential problem could result because several distinct types of iron are available. One type of treating iron should not be used over 3,000 lbs., and another is rated to 6,000 lbs. There are pipes rated to 10,000lbs. and to 15,000 lbs. of working pressure, which is the most common. For any treatment exceeding 6,000 lbs., I strongly recommend that only 15,000 lbs. working pressure iron be used.

For low-pressure treatments (i.e., less than 5,000 lbs.) and high-rate treatments, 6,000 lb. iron can be used because of the larger ID (4 in.). Other than the previously mentioned Halliburton Big Inch iron, there are strings of 4 in. 10,000 · iron being built by manufacturers of fmc iron. Much of this iron has no threads and integral connections. One of my major concerns is making sure about the compatibility of iron in the field. In most areas, the most common treating iron in the oil field is 15,000 psi, 2 in. and 3 in.

After checking with your service company to see if it has the proper treating iron at the location, confirm the maximum treating pressure. Also determine what action will be taken in the form of pop-off valves on units or pump truck kickouts. Typical pop-off valves available today are hydraulically controlled devices that release line pressure when a particular preset pressure is reached. The kickout devices used by some service companies either automatically take the pumps out of gear or in some cases shut down the units. For the sake of our industry and possibly due to a lot of litigation, the use of proper pop-offs and/or pump truck kick-outs has become more common. In earlier years, many of the pop-off devices were not very reliable and, in fact, created a sense of safety when it simply did not exist. Major service companies offer hydraulically controlled pop-offs and devices to take pumps out of gear at preset pressures. I recommend that you request these on any job where there is a danger of imminent screen-out at high rates and pressures.I have recently used these type of devices very successfully on high- concentration sand treatments in coal seams where rates exceeded 60 bpm and they functioned very well. Additionally, I have used them in the south Texas area for very high-pressure treatments.

Other service companies that use intensifier pumps have pop-offs on the back side of their pumps, which allows them to maintain a constant pressure quite easily with the release of clean fluid on the back side of the pumps as maximum pressure approaches. It should be noted that hydraulic release pop-offs on the dirty side of the fluid can possibly result in a dangemus situation when abrasive fluids are flowed out of these hydraulically actuated valves. One should always have some means of shutting in the well in the event of a pop-off occurring with abrasive fluids. The best procedure is a hydraulically actuated remote blocking valve at the wellhead as used for treesavers.

With high-rate treatments where there is a potential for screen out, either failsafe pop-offs or cut-outs for the trucks need to be present because people cannot react swiftly enough to avert disaster. Often, the rupture caused by this sort of thing does not occur downhole. Underground, a rupture can become an economic disaster, but on the surface it can mean loss of life.

Pressure testing lines prior to the treatment

Hopefully, one thing that is never overlooked is the maximum treating pressure on a treatment, whether it's 6,000, 8,000, 10,000, or even 20,000 psi. Prior to pumping operations, the service company is instructed (and normally does on its own) to test their lines to a pressure in excess of the maximum treating pressure from the operating company. A good rule is to test lines at least 1,000 psi higher than will be used for maximum treating pressure on the treatment. In other words, if you're going to have a maximum operating pressure of 7,000 psi, it would be a good idea to test the lines at 8,000 psi. Obviously, one would not want to test lines above the maximum operating pressure of the frac iron. In the oil field we commonly use either 602 of 1502 or sometimes 10,000 psi operating pressure iron and never exceed the operating pressure of the iron, even though it may have been tested higher during manufacturing. It is always a mistake to exceed the maximum working pressure on iron. In other words, do not test 602 iron above 6,000 lb., do not test 10,000 lb. iron above 10,000 psi, or do not test 15,000 lb. iron above 15,000 psi, and so forth. Commonly, there has always been a great deal of consternation and argument about how a pressure test should be conducted. The following list contains very important considerations:

1. When testing lines, every pump truck should be used in the testing operation. This will ensure that there are not substantial leaks behind check or blocking valves during the treatment. If the service company is using blocking valves to isolate its trucks, and if all of these are open, then one truck could be used for the test operation. Where most of the service companies utilize checks, then every truck should be bought on during the line test operation to be sure that all iron has been tested at the maximum pressure.
2. 1. Always note the location of the pressure transducer in case it is downstream of blocking valves and checks at the wellhead. If this is the case, there could be substantial leaks upstream of the check valve and still achieve a very good test, although there are substantial leaks on chicksans, lines, and so forth from the pump trucks to the check valve. Most service companies. are very aware of these particular problems in testing lines, but it is always good to observe that all lines are tested on location to above the maximum working pressure anticipated during the treatment.
3. There will always be some small amount of leakage, either from air in the lines or simply because the seals for chicksans (etc.) do not allow for an airtight seal operation. It is feasible to allow a small amount of drop in pressure vs. time as long as it is not a visible leak, and you have achieved a successful line test. In the past, some major oil companies required straight line charts for 15 minutes to show that the lines had been tested. This is virtually impossible. In other situations, oil companies require the service companies to test the lines with hydrochloric acid. In one case, the service company attempted for 12 hours to test lines with 28% acid. After 12 hours, the operation was a failure. It is simply not feasible, with the type of iron that the service companies use, to be able to test lines and hold pressure for long periods of time with corrosive type fluids. This is not to say that these lines will not hold during a treatment, but it is simply not feasible to test lines with corrosive fluids and not expect them to leak. An example of a good pressure test is testing lines at 10,000 psi for over 2-3 minutes with the pressure dropping only 2-300 psi. I also recommend very strongly that during the pressure test that the oil company representative go out and observe the lines to be sure that leaks are not occurring.

There is another technique I recommend for proper pressure testing. If a good pressure test is achieved (whether you held it for 2-3 minutes or 15 minutes) and you got that test on the first sequence of pumping up pressure, ask the service company to drop the pressure off and redo the pressure test. If there is a potential crack or fatigue problem in any of the iron, a duplicate pressure test will catch it, and you can find a split in a fluid end or a crack in a particular chicksan piece of iron. Normally, unless the service company is extremely diligent, you will not achieve a pressure test the first time, and you will find leaks and a duplicate pressure test is not necessary.

In one of my particular situations, a service company achieved a pressure test the first time. We were in a hurry and dark was approaching. As we brought the pressure up later in the treatment and were conducting the treatment, we had a fluid end split, spraying one of the personnel with hydrocarbons. I made a commitment to myself to never again conduct a single pressure test on a treatment, and I recommend the same to you.

On a recent treatment, I was standing by a treater, and he told me that he had achieved a good line test when, in fact, the pressure had dropped off rapidly. I instructed him to pressure up the lines again and went out and observed the operation. What I saw was something resembling a sprinkler system. I was told by a gentleman with the service company that I was observing low-pressure leaks and that there would not be any leaks if we pressured up the lines again. I kept a straight face and informed the individual to pressure up his lines. What occurred, of course, were high-pressure leaks, with leaks occurring out of grease fittings on the chicksans, and so forth. That particular experience was very discouraging to me because it occurred late in 1992. I was also told by the service company that one of the reasons for the leaks was that because of excessive discounting that they were not able to repair their equipment. I informed them (and you should also inform them) that if they cannot get a good pressure test, then the operation cannot move forward. After two hours of repair, we were able to get a decent line test and conducted the treatment. *Do not ever* conduct a high-pressure fracture treatment without testing all iron on location.

Checklist on rig-up

1. Compact rig-up-minimum iron used
2. Safe, practical distance from wellhead
3. Treating iron large enough to accommodate anticipated treating rate
4. Check valve properly installed near wellhead
5. Pressure transducer at or near wellhead
6. Treesaver, if required, properly installed
7. Lines properly staked
8. All iron should be flexible
9. Check valve and plug valve on each pump pretested before job
10. Check minimum and maximum rates for trailer manifold, if used
11. Check for sufficient suction hose and evaluate velocity per hose
12. Check horsepower and plunger sizes of pumps on location
13. Check flow-monitoring equipment before job
14. Adequate hose on suction and discharge side of blender
15. Estimate available fluid-removable from tank

16. Check placement of proppant storage to assure convenient movement and access to standby blender
17. Double-check working pressure rating on frac iron
18. Always obtain a successful pressure test of all iron at 1,000 psi above anticipated working pressure prior to the treatment.

Reasons for poor job quality and performance

1. Poor job design, inadequate materials and/or equipment
2. Acceptance of poor performance
3. Human error
4. Equipment failure

CHAPTER 2

FRACTURING EQUIPMENT (2022)

The second version of the book was titled *Rigging Up on Location*. There are some valuable items in the chapter relating to velocity with abrasives, maintaining assurance that all pump-through valves are in fact open, the necessity for check valves, tree-saver assemblies, tiedown of equipment, proper trailer manifold rig-up, selecting the appropriate blending and pumping equipment with recommendations on standby usage, optimization of fluid location, and proper line testing. A lot of the equipment illustrated is considered dated, but the basic information is correct and should be followed. I have supplied numerous annotated pictures of older equipment. Note the section on historical fracturing equipment for a trip down memory lane. I do apologize for omitting some non-existent service companies, such as Acid Engineering, and even a few of the more recent pressure pumping companies such as Frac Tech or Pro-Petro. I did not have a cell phone during my early field days and most of the pictures I took were through disposable Kodak cameras. In our industry, we have undergone the loss and mergers of many service companies throughout the years and locating historical proof of their existence proved somewhat difficult. But the best of efforts was made.

I stated earlier in the book that changes in equipment control and monitoring were among the largest changes since 1994. Sand handling and storage have undergone huge changes, but in some cases like upright sand storage silos we have found ourselves back to where we first started.

Hydraulic fracturing began in 1947 in southwestern Kansas. At the end of the chapter are self-explained images of the equipment. This extremely small job with river sand and gelled gasoline was dangerous but did result in stimulation thanks to near wellbore damage. We progressed through the 1940's and into the 50's with World War II Allison airplane engines and conventional diesel driven pumps and engines with HHP in the range of 750 to more than 1,400 HHP. The next stage for Dowell, now Schlumberger, was a move that would last for many years as they migrated to turbine powered pumps in the range of 1000 HHP. Halliburton built a pump that was introduced in 1957 that, through redesign and upgrading, would yield 550 HHP. This unit was twin mounted on body loads and trailers, and became a mainstay for Halliburton well into the 1990's. This pump was also used on all cementing equipment and continues to be used in cementing equipment to this very day. Halliburton, as well as the rest of the industry, did eventually move toward single trailer units that started out in the 1000 HHP range, with units gradually transformed into today's conventional 2,500 brake horsepower units that presently dominate our industry. One company, BJ Services, developed a 2,900 HHP unit they called the *Gorilla* in the 1990's. The Gorilla was a good pumping unit but had to be permitted due to its overweight status. The plethora of 2,500 brake horsepower units stem from the weight restrictions encountered during the time. As seen, there were many different pumping units with various designs, but ultimately strict weight restrictions became the stopping point for enhancing conventional horsepower beyond 2,500 HHP using conventional, hydrocarbon-powered engines. There was a movement towards the utilization of methane gas for fuel instead of diesel and a small amount of use of available methane, but logistics was and is a constant problem for the process. As stated earlier there are massive amounts of hydrocarbon fueled equipment in existence today, and they will continue to be used for many years to come. The early 2000's saw another attempted comeback at utilizing updated turbine powered equipment. Dowell utilized turbine powered equipment for more than 30 years before moving to more conventional equipment.

The turbine units are extremely efficient and could legally yield 4-5,000 horsepower on a single trailer due to the much smaller weight. The problem with turbine powered equipment for fracturing was the lack of control of the pump equipment, particularly in emergency shut down situations, such as a screen out during sand pumping. The other problem with turbine equipment was the absolute need for skilled mechanics.

One of the more interesting developments relating to pumping was the use of pressure intensifiers for ultra-high-pressure pumping. The original development of this type of pump was initiated by Gulf Oil in the 1960's. They needed a high-pressure pump that would function at extreme pressures and handle the pumping of steel shot. Their need was for very high-pressure, abrasive drilling. A very basic explanation of how intensifier pumps function is that a hydraulic fluid is pumped across a fairly large surface area and the force is transmitted to smaller plungers thereby intensifying the hydraulic pressure typically 3-5 times. In other words, the hydraulic fluid would be circulated at 3000 psi and output would be 9-15,000 psi. The pumps typically had a 5-foot stroke and the power systems never saw abrasives. The abrasive drilling was successful as they could drill as fast as they could change connections. The drilling operations was not unlike the use of sand to cut pipe, termed Abrasi-jet, or hydro jetting. In this case, specially designed nozzles were directed downward and by pumping abrasives at high pressures they could speed their rates of drilling. By optimizing the pressure, rate, and concentration of abrasives, extremely fast drilling rates were achieved. To my knowledge the process was terminated due to the lack of control of the downhole location. The technique of drilling with high pressure abrasion has been used in other locations with continuous coil tubing. Obviously, abrasive drilling has not taken over drilling operations but does have applications in some areas.

Like many of the inventions of our time, the use of intensifier pumps although not successful initially in drilling, found application in hydraulic fracturing of deep wells completed with tubulars rather than larger I.D. casing. Unknown to those of us who have not been around for a hundred years, deep well completions were dominated by tubular completions. I believe the use of tubulars only for well completions was built around safety concerns. When the industry got a better handle on handling fracturing down casing and then running tubing when required, the use of intensifier pumps came to an end. The intensifier pumps were utilized starting in the mid 1960's and continued well into the 1990's with all major pumping companies getting into the fray. Halliburton started with 10 HT-1000's and eventually moved to the HT-3000. The HT-1000 was skid mounted and the HT-3000 were trailer units. The HT-1000 was rated at 1,000 HHP throughput and were powered by circulating water as the power fluid. The HT-3000's had a 4,000 HHP throughput and again were powered by conventional pumps circulating water as the hydraulic power units. Dowell, now Schlumberger, got into the intensifier pump competition with duplex pumps powered with hydraulic oil. They were termed pressure multipliers. BJ Services also used duplex pumps liken to the Dowell Schlumberger Units and the Western Company had hydraulic oil powered triplex intensifier pumps.

Because I developed the first high temperature fracturing fluid, I was able to attend many intensifier jobs, performing quality control on the fluids. I was a popular guy on location as I cut all the gel sacks when we batch mixed the fluid for the fracs. Also, I developed a great deal of admiration for the crews that ran the intensifiers. They were the oilfield version of Navy Seals. Typically, they were large guys, smart, and could hold their alcohol. Coincidentally, with the development of the intensifier pumps, Halliburton developed what was termed *big inch iron*. The original high pressure treating iron built for the abrasive drilling was 6-inch O.D. by 3-inch I.D. I was told the iron was designed to withstand 50,000 psi burst. I do know we tested it to 25,000 pounds by mistake one day and the iron and humans present survived. It was interesting in that the original iron did not have chiksans and was clamped together with metal seals. I was told the iron was 110 pounds per foot and could not be manhandled. In the early jobs in the late 60's we used gin-pole trucks to rig up and the use of rigid iron caused a lot of problems. On two separate occasions I witnessed something that would not occur in today's world. After spending a great deal of time with poorly designed gin pole trucks to lift and move the iron, the rig-up hands got the heavy iron to within an inch or so of the metal-to-metal seal at the wellhead which was to be clamped together and bolted. Because of the rigidity of the iron, they would have to start over. On both occasions a young engineer, watching, went to the bottom of the ladder leading to the wellhead and proclaimed that if they had started at the wellhead, they would not have the problem. On both occasions the burly hand on the top of the ladder slowly climbed down and proceeded to punch the engineer in the face. Halliburton quickly developed swings and chiksans. which enabled a much easier rig-up.

It is exciting now to see the electric powered fleets of pumps with massive turbine power. One of many impressive things is the quiet of all systems. The units yield a much smaller footprint and the only hydrocarbons utilized is the fuel to run the massive turbines powering the electric units. We are at the beginning with many improvements to be made. Presently in most locations the conventional equipment is more practical, because of ease of movement and the massive amount of equipment stacked. The HHP to weight of electric power relating to downhole pressures and rates is amazing and the greatly reduced maintenance for electric engines make them an easy choice were it not for present logistical problems. The only major thing, outside of cost, I see as a major problem will be logistics in supplying high pressure clean methane or huge volumes and rates of diesel. I am confident that the engineering of manufacturing companies like Stewart and Stevenson can solve these problems. Another major factor relates to the political environment and the rush to eliminate hydrocarbons. I will say, unequivocally, that the elimination of hydraulic fracturing in the United States will only hasten the time to achieve the latter. This is true since minuscule hydrocarbon production stems from reservoirs that do not require fracture stimulation.

Image A

1947 - First experimental fracturing treatment by Stanolind Oil in the Hugoton gas field in southwestern Kansas with river sand and gelled gasoline.

Image B

1949 - One of the first two commercial fracturing treatments ever performed, by Haliburton Oil Well Cementing Company. The first two treatments were carried out in Texas and Oklahoma.

Image C

1955 – Fracturing pump manufacturing facility; some remotely controlled pumps were powered by surplus Allison aircraft engines from P-38, P-39, P-40, and P-51 fighters, as well as from B-42 bombers.

Image D

1950's remotely controlled, truck mounted pumper, with 600 HHP Allison aircraft engine.

Image E

Image F

Western Company pumping equipment.

Image G

Dowell turbine power pumps, with horsepower range up to 1,000 HHP.

Image H

Dowell - alternate view of turbine frac pumper on location, during fracturing operations.

Image I

1957 – Halliburton HOWCO HT-400 horizontal triplex pump; often referred to as the "workhorse of the field".

Image J

Halliburton cement pump truck with mounted HT-400 triplex pump

Image K

Halliburton HOWCO HT-400 horizontal twin pumper triplex pump, Rocky Mountain operations.

Image L

Halliburton HOWCO HT-400 pumping skids mounted on pumping vessel in the Gulf of Mexico.

Image M

Halliburton HQ-2000 pump (Grizzley) quintuplex pump designed to achieve maximum reliability and minimal maintenance; capable of 2,500 HHP for pressure pumping applications.
Picture source: Halliburton

Image N

BJ Services Gorilla 2,700 HHP/20,000 psi max pressure stimulation pump unit, trailer mounted for high pressure and high rate pressure pumping applications.
Picture source: BJ Services

Image O

BJ Services Rhino 2,000 HHP/15,000 psi max pressure stimulation pump unit, trailer mounted for high pressure and high rate pressure pumping applications.
Picture source: BJ Services

Image O

Schlumberger Caterpillar unit SPF/SPS 343, trailer mounted pressure pumping unit; most common horsepower available by Schlumberger-Liberty Oilfield Services for high and low-rate applications.
Source: Schlumberger Oilfield Services

Image P

HT-1000 intensifier skid units, powered by HT-400s.

Image Q

Six HT-3000 intensifiers powered by eighteen twin HT-400 pumping units.

Image R

Dowell-Schlumberger duplex intensifier pumps, hydraulically powered.

Image S

US Well Services electric pressure pumping unit, powered by natural gas.
Source: U.S. Well Services

Image T

Dual Frac Stack Packs on a Trailer. 5,000 HHP driven by two 3,850 HP Turbines.
Source: Marine Turbine

Image U

Natural gas-powered, all electric pressure pumping fracturing fleet performing operations on a well pad; natural, gas-burning turbine pumping units developed by General Electric. Upper horsepower limit per pumping unit at 7,000 HHP.
Source: Evolution Well Services

Image V

Electric fracturing fleets are one of the latest trends in the shale market, being driven by lower fuel costs and lower CO2 emissions.
Source: National Oilwell Varco

CHAPTER 3

THE SAFETY MEETING: THE MOST IMPORTANT PART OF THE TREATMENT (1994)

Regardless of the size of the treatment, whether it is a single-truck cement job, a small acid cleanup, or a 20-truck, 4-million-pound sand frac, one thing ties all of them together. One thing must occur on all of these treatments-the safety meeting. All personnel, from the oil company and the service companies, should meet before the treatment to make sure that they understand the job and their responsibilities.

Come let us reason together

Like company safety meetings, this gathering does not just relate to safety. Safety considerations should be discussed in detail. Everyone should know where to go in the event of a disaster. Everyone needs to know the maximum operating pressure. The supervisor should count the people at the location, so he will know who might be missing in case a catastrophe occurs.

The major purpose of this meeting-and the one I consider most important is that each individual needs to know his part, his responsibility, what to do during the job. This may sound trite; but from observation over many years, I can tell just by looking at the organization of the safety meeting whether a treatment is going to run professionally and smoothly.

Who's in charge?

One thing that is required, especially on large treatments, is a leader. This person designates responsibility and sees that other personnel follow through with their jobs. On treatments that have failed because of poor quality control, personnel often were not given a specific responsibility, so no one took charge of a situation. Therefore, I think the oil company representative should ask who is responsible for monitoring various items.

The company representative should have a checklist (see Table 3-1) that he quickly runs through, not to harass but to question individual job responsibility. He should ask what individuals are in charge of monitoring the fluid· as it comes from the frac tanks. He also needs to know the blender operators, who is going to assist. them with running additives on the fly, and who is going to ensure that additional materials coming out of drums or trailers are monitored and supplied to the blender.

The oil company representative, of course, needs to know who is in charge of the treatment. In the oil service industry, the title of the person ill charge of the treatment may vary from one area to another. Usually the treater-the person running the computer or frac van-is in charge. Sometimes an engineer or other supervisor may assist and actually make the decisions for the treater. Dual leadership is asking for. trouble. One person needs to be in total command during the treatment. Engineers or technical representatives can assist and cooperate with the treater or the person in charge, but they should not make decisions for him. In a high-pressure treatment (i.e., cementing, acidizing, etc.), decisions sometimes must be made quickly but, for the most part, decisions must be made with all the facts. At times when a catastrophe occurs, decisions must be made instantly; at other times, a little calm and deliberation can make the difference in a successful treatment and a failure.

Most of the disastrous treatments are classic examples of the 40- or 50-car jobs. All of the management personnel arrive, and the oil company people from many areas arrive to observe. Typically, the treater or person in charge is intimidated, and unless he is a strong individual, he will not exert himself. Instead, he will rely on his managers to make decisions. Then many people become involved in the treatment; too much leadership results in no leadership.

During the safety meeting, one representative who has the final say for the service company and one operating company representative to whom the service company leader will respond should be appointed. It is imperative that in the lead-off frac van, computer van, or whatever the control point-it may be the tailgate of a pickup on ‘a cement job-that these two people will make the decisions. They are free to discuss operations with any others, but they should make the final decisions on the treatment. During the safety meeting, these people need to be pointed out. Then all representatives, service company and oil company alike, who want to make changes or who see errors or problems, should notify these individuals before any alterations are made. If one operating company representative and one service company representative are in charge, a much smoother, more controlled, and safer operation results.

Table 3-1 Fracture Stimulation Supervision Checklist

Well Name: ____ **Date:** ______________________

I. Equipment Needed on Job Site

____ 1. Oil company workover procedure with data sheet containing the reservoir properties, such as bottom hole pressure, porosity, permeability, temperature
____ 2. Service company stimulation recommendation
____ 3. Logs with perforations and collars premarked
____ 4. Tank strap
____ 5. Sand sieves
____ 6. Service company reference tables
____ 7. Containers for samples
____ 8. Calculator, pencils, and Quality Control Forms
____ 9. Hardhead and steel-toed boots
____ 10. Fann 35 or equivalent viscometer or availability of same from service company
____ 11. Water test equipment (or obtain from the service company)
____ a. pH meter or paper
____ b. Thermometer
____ c. Iron test kit
____ d. Phosphate test kit
____ e. Total dissolved solids tester
____ f. Reducing agent tester
____ g. Chloride test kit

II. The Day Before the Job

Tanks

____ 1. Are there enough tanks on the location to store all fluids? Figure that 50 bbl./tank will be unpumpable. Recommend at least 10% extra base gel on location.
____ 2. Have the tanks been cleaned prior to the job?
____ 3. Does the water have the proper amounts of potassium chloride, sodium chloride, and other compounds?
____ 4. Are all of the tanks full? Get on the tanks yourself-do not take anyone's word!
____ 5. Where did the water come from? Does it appear to be clean? Check each tank yourself. Do not pump dirty fluid down a well.
____ 6. Check the valves on the tanks to ensure that they are not leaking. If a valve has a trickle leak, replace it before the next job. If the tank has a large leak, consider having the tanks switched out prior to any pumping.
____ 7. Conduct pre-gel quality control on frac fluid.
____ 8. Fill out Quality Control Tables 1, 2, 3, 4, and 5. Parts of Tables 1 and 3 will be done either again or only on the day of the job.

Table 3-1 Fracture Stimulation Supervision Checklist (continued)

Sand Storage

____ 1. Get on top of the sand storage unit yourself and see if they contain enough proppant to do the job.
____ 2. Is the proppant in each compartment the correct size?
____ 3. Check the proppant for fines. Sieve samples from each compartment.
____ 4. Are you using the correct proppant?
____ a. Take samples and establish if the proppant is the proper quality (i.e.; Ottawa, Brady, Interprop, etc.).

Find the Service Company Treater and/or Treatment Supervisor

____ 1. Go over sand and fluid schedules with him.
____ 2. Are the proper additives going to be on the location?
____ 3. Is a standby blender going to be on location and in position to be usable? A standby is needed on treatments with pump time exceeding 1 hour.
____ 4. Make sure a sand densimeter is going to be available on the job.
____ 5. If the job is going to take more than 4 or 5 hours pumping time, request that a service company mechanic be on the location to repair any equipment that malfunctions. Also request an electronic technician to repair electrical problems on jobs with long pump times.
____ 6. Go over rig-up checklist with service company representative.
____ 7. Arrange for testing of all gelled fluids and test crosslink time if applicable. (Quality Control Table 1)
____ 8. Establish rapport with frac treater and give the person guidelines on what you expect before, during and after the treatment. ·
____ 9. Arrange to receive completed Tables 1; 3, 6, 7, 8, 9 and 10 at least two hours before pump time.
____ 10. Establish maximum treating pressure and calculate surface treating pressure with service company function charts.

III. Just Before the Job

Find the Service Company Treater and/or Treatment Supervisor

____ 1. Go over sand and fluid schedules with him. Receive and discuss Quality Control Tables 1, 3, 6, 7, 8, 9 and 10.
____ 2. Specify whether he will be recording volumes in clean or dirty volumes. Clean volumes will be displayed in barrels or gallons.

Dirty volume, bbl. – Clean volume, bbl. + (lb. sand x 0.00109)
-bbl. * (____________ lb. X 0.00109)
-bbl.

Or

Dirty volume, gal – Clean volume, gal + (lb. sand x 0.0456)
- ____________ gal + (______ x 0.0456)
- ____________ gal

____ 3. Finalize the pumping schedule on Table 2.

____ 4. Get on top of the tanks yourself and gauge all frac tanks using a tank strap. Have the treater present. Having the treater gauge the tanks with you will prevent any disagreements about fluid volumes after the job is finished. This stop should be completed only after all tanks have been rolled and viscosified. Finalize volumes on Table 1.

___ 5. Set up a system with the treater for numbering the tanks in the order that they will be drained. This helps keep track of the fluid volumes during the job.

____ 6. Fill in the Frac Tank Tracking Chart (Table 6). This will help you keep track of how much fluid is left at any point during the job.

____ 7. Arrange with the treater to have someone knowledgeable and dependable on top of the frac tanks and the sand storage units. He or she should be there all the time that the job is being pumped to insure a smooth uninterrupted flow of the proposed pumping schedule (Table 2).

____ 8. Impress upon the treater the adverse consequences if the pumps lose prime during the job because the tanks were sucked too low. When the fluid level in the frac tank drops below the suction valve, air is sucked into the pumps, causing the blender pumps to lose prime. The sand density goes extremely high and the rate has to be reduced. The sand-density then tends to drop very low while the pumps regain prime. This chaos normally takes 5-10 mins. to correct-10 mins. is a lot of fluid at 50 bbl./min.

____ 9. Get on top of the sand storage unit with the treater and gauge the volume of proppant in each compartment. Remember to check any 100 mesh sand that is being pumped in the pad as a fluid loss agent.

____ 10. Set up a system with the treater on numbering the sand storage unit compartments in the order they will be pumped.

____ 11. Complete the Proppant Tracking Chart (Table 7). This will help you keep track of how much proppant is left at any point during the job.

____ 12. Complete the Crosslinker Tracking Chart (Table 8). This will help you keep track of how much crosslinker is left at any point during the job.

____ 13. Complete the Breaker Tracking Chart (Table 9). This will help you keep track of how much breaker is left at any point during the job.

____ 14. Complete the Fluid Loss Additive Tracking Chart (Table 10). This will help you keep track of how much fluid loss additive is left at any point during the job.

Find the Service Company Field Chemist or District Engineer

___ 1. Have the chemist complete Table 1 for each tank of gel. This is in addition to your own quality control work. For legal and other reasons, always. have the service company confirm your tests- you might have erred!

___ 2. Check with the chemist to find which additives (such as cross-linkers, fluid loss additives, and breakers) will he added on the fly during the job.

___ 3. Check with the chemist to see that all tanks have been premixed with the necessary additives.

___ 4. If running a crosslinked gel, catch a sample of gel from each tank and add the appropriate amount of crosslinker to see if the gel crosslinks.

Equipment

____ 1. Is all equipment fueled up, and is there enough fuel on the location to complete the job?
____ 2. Were all pumps and lines flushed with clean water before the job started? ·
____ 3. Are all injection lines staked down?
____ 4. Is a standby blender rigged up or in an immediately usable position?
____ 5. Is the blender located close enough to each tank so that sucking the fluid at a high rate will not be a problem?
____ 6. To be assured of sufficient suction between the blender and the tanks, you should have 1 suction hose per 10 bpm for thin fluids and for thick fluids use 1 suction hose per 5 bpm. For example, a 40 bpm rate would require 8 suction hoses for 60# viscous gel.

Safety Equipment Checklist

_____.1. Locate pumping trucks and tanks crosswind and a reasonable foot from the well. Head all vehicles away from the well and keep access roads clear.
_____.2. Each discharge line should have a full swing at the well and at the truck manifold and be staked at each end. Additional staking should be based on judgement.
_____.3. Install check valves in each discharge linear as near wellhead as possible.
_____.4. No one should stand on or near discharge lines under pressure and never pass lines under trucks or other equipment.
_____.5a. Pressure test discharge lines from pump to well at 1000± psi greater than maximum treating pressure.
_____.5b. Inspect wellhead for any low pressure connections that may have inadvertently been added during well servicing.
_____.6. Bleed off lines should be staked and in a safe direction (downwind, downhill, and/ or to a pit).
_____.7. Insure that adequate firefighting equipment is in good working condition strategically located.
_____.8. Conduct pumping operation in daylight. Do not pump during electrical or severe dust storms.
_____.9. All personnel and equipment not necessary to the operation should be moved to a point at least 150 feet from well.
_____.10. If flammable materials (crude oil, diesel, xylene, methanol, etc.) are pumped, all persons within at least 150 feet from the well should remove matches, lighters, and cigarettes from their pockets.
_____.11. Prior to pumping, all company and contract supervisors and crew should meet to discuss job procedures, work signals, hazards, and safety precautions. At this time, an emergency assembly area should be designated in an upwind direction from well. Also, a head count and a buddy system should be established so that all personnel can be accounted for.
_____. 12. If pumping flammable material, have the service company wrap all discharge hoses from the blender to the pump trucks with canvas or other material. This will negate spraying of flammable material should the hoses leak or burst.

Note: All tables a referred to in Table 3-1 can be found in Appendix II

Communication is a must

One very important aspect to discuss during the safety meeting is the mode of communication. Service company representatives should report their means of communication to the oil company during the treatment and what they will do if that line of communication fails. I have seen electronic communication equipment go down during a rainstorm and tubular goods blow up, injuring or maiming personnel because of lack of communication.

For many years in our industry, personnel have used hand signals for communication. A treater on a stimulation job could talk to six, eight, or ten pump operators and control a frac job with hand signal art, much like an orchestra leader uses his baton. Without radio communication, personnel had to pay attention and always keep their eyes on the operator. As we got into longer treatments, this put a lot of strain on individuals.

Of course, we now have remote-control equipment, which moves people away from their units. Sometimes this can be detrimental if people are out of the treater's sight. One time in south Iran when I conducted several large acid frac treatments, our only means of communication were hand signals. The situation was difficult because the personnel did not speak English. A safety meeting might last an hour and a half. But with finger and hand signals, we told the people when to start and stop. And there was a real threat that projectiles might be thrown at the bodies of individuals on each truck if they did not pay attention during the treatment. I was very pleased when I received sound power communication sets in Iran.

Another story relates to the first treatment with good communication sets. I had been told that ball sealers did not work in Iran. This surprised me because even in that early period of my life I knew that ball sealers worked very well. While investigating the matter. with some maintenance personnel, I discovered why they had trouble on treatments involving ball sealers. The ball injector was in a state of disrepair. The maintenance supervisor and I repaired the ball injector and went out on our first treatment using it to stage a high-rate acidizing job.

On this treatment I was pleased; I had never felt safer on a treatment because I was doing a job using ball sealers and I had good communication. We were pumping approximately 40 bbl./min .. when the ball sealers hit bottom, and the pressure rose from some 500 psi up to 5,000 psi. When I spoke through the sound power communication system, I found that the Iranians-most of whom spoke relatively good English-were terribly excited and frightened and did not respond to my communication. Luckily, I remembered a few Iranian words that mean "stop" and was able to communicate that to the operators. From then on, I used the communication sets to get their attention. Once again, these men were instructed to watch me and to use hand or visual communication to decide what to do.

There is a real danger when we depend solely on headset communication during a treatment. The operators should be in a position where they can observe a pressure gauge either on a console or on the ground in front of them. They also need to be placed so that if communication fails for any reason, then either the treater or someone else can communicate visually with these personnel. The failure of headset communication should not be a cause for disrupting a treatment.

Communication, job responsibility, and safety are all factors that need to be discussed, confirmed, and clarified during the safety. meeting. A safety meeting should not be over and the job should not start until the oil company representative has verified that everyone on location has a job and a responsibility, or at least will stay out of the way of those people who do.

Since the writing of the first edition of this book, I have been on hundreds of fracture treatments and have attended a similar number of safety meetings. With the advent of Total Quality Management and the pressures that have existed from OSHA and other organizations, there has been in some areas more emphasis placed upon safety, and some of the service companies have done a better job conducting safety meetings. In other cases, I have found representatives to be quite lackadaisical and not take the safety meeting very seriously. I even had one service company tell me that they did fracture treatments every day and that a safety meeting wasn't worth the time. I instructed them that the next time they flew on American or Delta or other airlines to tell the pilot not to bother with a checklist – that the plane would be fine. The safety meeting is essential for safe operation during the treatment. Recently I repeated a safety meeting when I wasn't satisfied with the way the meeting was held. As stated earlier in the chapter, it is extremely important to go over everyone's responsibility. It is extremely important to find out who is responsible for what task. Equally so, you must have a head count of all individuals on location to be assured that if a catastrophe does occur that an individual isn't left behind.

I was on a large C02 treatment where a line ruptured early in the job and the location was flooded with carbon dioxide gas. Everyone moved up the hill and waited for the C02 to dissipate. We thought it wasn't a very dangerous situation until we returned to the location and found one of our fellow workers dead. He had become disoriented, knocked himself out, and had suffocated on location. You should have a sign-up sheet with a head count of all individuals on-site. All safety procedures should be set forth. These include having all fire extinguishers on location on the ground with ready accessibility. Two persons should be designated to take anyone injured to a hospital, and a vehicle should be set aside for their use. In the safety meeting, it should be stated where the hospital is located and the driver should be pointed out. Such safety considerations as no smoking, staying away from high-pressure lines, and so forth, should be discussed. In these safety meetings individuals other than service company people are present. Examples of the people are rig hands who happen to be on site, or just simply visitors. Obvious things need to be mentioned such as maximum working pressure and gathering areas in the event of a catastrophe, as well as pointing out who is in charge during the treatment.

After being· in so many safety meetings and seeing the response of various individuals, I think it is fairly important to get people's attention in relationship to their jobs and responsibilities. On every treatment, I have tried to remember to make the following points:

1. I tell the individuals that there are very few things that I get terribly upset about, but the first (and most important) is that if they run out of fuel on the treatment, I will have to kill the treater. Everyone laughs a little bit but generally everyone goes and checks their fuel tank.
2. I then let the individuals know that if they suck air during the treatment that is castration for the treater and possibly the blender operator. The point is that they should watch the tankage very carefully to make sure they don't suck the tanks down too low, which could cause a disaster during the treatment.
3. After consulting with the company for which I am working, I try to make a very strong point that if the individuals in the company will relate to us that they have mechanical, chemical, or other problems during the treatment, the operating company will work with that service company in relationship to cost. On the other hand, if I or oil company representatives find problems during the treatments or discover them without being told by the service company, then dark clouds are going to be seen on the horizon and their company is in jeopardy of losing work.

You should not be in a confrontational mode at any time with the service companies. You need to make it understood how important you consider the safety meeting and how important you feel this particular treatment is to your company.

Points for a safe treatment

1. Consider all safety rules.
2. Designate a gathering area in the event of a disaster.
3. Establish a maximum treating pressure.
4. Each person should be assigned a specific responsibility.
5. Assign leadership (one person in charge from the oil company and one from the service company).
6. The oil company representative should point out the persons responsible for each task.
7. Make sure a good communication network is set up.
8. Have alternate visual commands. Have a fire extinguisher placed on the ground for easy access.
9. Have two individuals given the responsibility of transporting injured persons to the nearest clinic/hospital.
10. Have a designated vehicle set aside for transporting injured personnel.
11. Do a complete head count of all personnel on location.
12. Set up a gathering area in case an accident occurs.

CHAPTER 3

SAFETY MEETINGS (2022)

If I were ever to brag on myself, it would be for this original chapter from the second edition. Most of the comments still hold. The checklist, however, is centered around conventional viscous fluid treatments. I will do my best to dispel the comment that slick water treatments are simple and do not require supervision. There seems to be a concerted effort to make shale stimulation and waterfracs a manufacturing process with little or no supervision required. We are moving more and more toward remote monitoring of fracture treatments. With the internet one can, with proper equipment, remotely oversee treatments around the world in real time. There is a constant move toward minimizing the number of people on location, i.e., one person running 25 pumps with a mouse as compared to 25 people running each individual pumping unit.

With all the above stated, I firmly believe that until we have multiple cameras and robotic equipment on site, some experienced human intervention is necessary. With the multiple downturns in our industry, lack of on-location experience is rampant. My company has been on location where there simply was no real industry experience present. Simple things like proper and safe rig-up was not applied, not to mention a basic lack of knowledge of execution, and sadly safety. 2020 is without a doubt the worst downturn I have ever experienced. We have lost what formerly was two of the big three pumping companies which have existed for more than 50 years. Others are coming forward, but it is almost universally impossible to properly execute a standard crosslink gel job. Without getting into QC, covered in other chapters, and needed real time simple diagnostics, one cannot be certain today, without human oversight, what the actual quantities of some simplistic items such as size and volume of sand, volume and rate of water pumped, and true addition of chemicals. For the next few years, I believe that experienced competent human oversight is required.

There are new companies that are building modern electric fleets with multiple safety factors embedded, but for the foreseeable future, our fracs will be conducted with more conventional diesel or natural gas-powered pumping and blending equipment. If you doubt my word, you should rent a plane as I did in 1982 and fly over massive fields of stacked pumping equipment. At the end of the chapter are multiple images of stacked equipment (past and present). I believe when a turnaround occurs, it would be foolhardy, but not impossible in the strange world we live in, for the service companies to not utilize this equipment for the coming years.

This chapter is titled Safety Meetings and I have digressed away from safety. Most of us realize that safety goes out the door when people are lacking training. The field operation of fracturing equipment is extremely dangerous. We have come a long way in emphasizing safety but without experienced trained people our present remote-control operations do leave operators with a lot of liability. I believe that artificial intelligence and robotics will for the most part replace on-site supervision, but that time is years in the future. Off-site diagnostics and control of operations is something that we are doing presently, and it is the future. This work is with very large companies with the resources to do so. Listed below is a checklist specifically built around large volume and or multistage slick water fracs that truly differs from items listed for smaller conventional fracs.

I mentioned in the short summary of the chapter that I would discuss PPE or basically what is worn on location. The service companies have for the most part stayed with coveralls which are purportedly fire retardant. In the early days of the new more visible uniforms I joked with the employees that they all looked like power rangers. For the most part the more visible uniforms are accepted and do make the employees stand out more in many of the dull backgrounds that we operate. Our employees have started wearing normal clothing which is fireproof, but we also supply more conservative coveralls and individual pants and shirts.

The reason I bring this up is that unless the fire protected clothing is properly cleaned it rapidly loses its fire protectant qualities. I see this as a major problem throughout the industry. Those who do not believe that fires occur while pumping water should review what has happened over the last few years with entire fleets destroyed. For the most part the fire protectant clothing is heavy, and in many cases hot, but it is really a shame when the uniforms are not properly cared for and are not protective in the event of a fire.

On one of my dumber jobs conducted near Marlow Oklahoma we had 9,000 barrels of diesel and crude on location to frac with gelled hydrocarbons. During the safety meeting the safety rep asked who was going to volunteer to handle fire extinguishers. I interrupted the meeting and told everyone to forget firefighting and have a path to run if a fire broke out. I also told them to take a good look around while they were running. As is usually the case someone asked why they should look around. I then told the crew that that was the last thing they were going to see on the earth-they were going to die in the explosion. Nine thousand barrels of condensate, crude and diesel would create a massive explosion that would kill everyone. Fire protectant clothing will help you in a small fueling fire, but please do not have some bright people decide to pump large quantities of hydrocarbons for fracturing again. I had a company man on location that was easily rattled. After the safety meeting I told him I had dropped my cell phone in one of the frac tanks and asked the treater to dial my number so that I could find my phone. The company man went nuts until I showed him my phone.

I do have another statement that I always made in the safety meeting before fracs. I stated that if we sucked air, i.e., lost prime, I would castrate the treater. In the old days we did not have lady treaters. I also said that if we ran out of fuel, I would kill the treater. In our world today I would probably be removed from the location for being politically incorrect, l would say that people did check fuel and fluid volumes after the meeting. One of my worst moments was a frac job just outside McAllen Texas which was run in the late 60's. At that time, we were expected to start pumping at daylight and would be severely chastised if we were not pumping as the sun came up. The crew was excellent and had fueled everything up the night before.

One of the worst sounds I ever heard was virtually all the equipment running out of fuel not long after we started the job. The company man was extremely outraged and could not believe that no one had checked fuel before starting the job. Some very professional thieves had done an excellent job of emptying all of the fuel tanks on location. I never forgot the evil looks of the company man, the weird sound of equipment running out of fuel, and for many jobs I checked fuel tanks myself.

Multistage, Large Volume Slick water Frac Checklist

1. Use whatever means necessary to verify the pumpable volume in the frac pond that you are drawing from. Countless jobs have been terminated due to miscalculations on pumpable volume.
2. Be assured that you have homogenous salinity in the frac pond. There is a plethora of polyacrylamide friction reducers and higher viscosity acrylamides that have varying ionic character are extremely susceptible to changes in salinity. The service company needs to be aware of a complete water analysis which is a representative sample of the pit.
3. Have a minimum of 100 % standby on transfer pumps and hoses to the holding tanks feeding the blender.
4. Have the service company be comfortable about the number of holding tanks if a leak occurs or other event occurs in the transfer line from the pond. This will allow for decision time to either displace the well or continue operations.
5. If you are using multiple suppliers of chemical, discuss with and have the service company assure of compatibility of mixed supplier chemicals.
6. Have on-site a means of doing sieve analysis on delivered proppant. It is unrealistic to do constant sieves but a sieve analysis every stage at a minimum.
7. Prior to pumping the first stage it is advisable to calibrate the flow meters, densitometers, and conduct bucket tests verifying pump rate on chemicals on site. A minimum of a daily test would be advised. This should be done on zipper fracs as well as single well operations.
8. Regardless of upright sand storage or sand box configuration, always monitor the loads already received and those arriving and do everything possible to ascertain the actual volumes of proppant per stage.
9. Not unlike conventional fracs, it is advisable to venture outside of the treatment van observing in safe manner the operations. If only one supervisor is present and wi-fi is available, it would be wise to set up cameras to monitor the operations with a cell phone.
10. During each stage have chemical rates monitored by individuals from the service company. This should be explained pre-job to assure safe working conditions for the individual monitoring the chemical.
11. Read over the checklist from edition 2. Some of the list is outdated but the comments at the ending of this chapter are very pertinent to our present and will be so for some time.

Image A

Haliburton manufacturing center, with stacked equipment behind the facility.

Image B

Stacked Western Company of North America equipment, 1982.

Image B

Halliburton equipment stacked at reclamation yard in Duncan, OK.
May 2020

Image C

Halliburton equipment stacked at manufacturing center in Duncan, Oklahoma
Google Maps 2021

CHAPTER 4

COMPATIBILITY (1994)

One problem that plagues our industry is that many jobs are conducted using 8-10 different additives simultaneously. I have never known specifically whether that relates to the adage, "A little is good; a whole lot is better," or whether it relates to good salesmanship. In the early days of cementing, we ran water and cement. Acid jobs required a little acidizing fluid, a little inhibitor, and a small amount of nonemulsifier. Fracturing jobs utilized, at most, some gelling agent for the oil. Now, we need a myriad of additives to cover the possible conditions that arise during the treatments.

Excess ingredients ruin the cake

Instead of running an acid job with hydrochloric acid, acid inhibitor, and a nonemulsifying surfactant, often we incorporate a fine suspender, an antisludge additive, an inhibitor intensifier, a fluid loss additive, a fluorosurfactant, perhaps methanol, and sundry other additives. I have reviewed customer proposals with as many as nine additives put together in one mixture. Some of the additives are anionically charged, some are amphoteric (both basic and acid properties) or zwitterion (carrying both a positive and a negative charge) materials, some are cationically charged, and some are nonionic.

There is tremendous confusion and exasperation in deciding which, when, and how much of the additives to use on a treatment. Indeed, some acid treatments require several additives. In other situations, you may incorporate ammonium bifluoride to create hydrofluoric acid, rather than conducting a hydrochloric acid treatment. A mutual solvent may prevent your surfactants from plating out. A nonemulsifier could reduce the fluid's surface tension as it goes into the formation. A water-wetting surfactant may be needed in the fluid. And, of course, an inhibitor is necessary to protect tubular goods and equipment.

Incorporating fine suspenders and clay stabilizers into a hydrofluoric acid treatment is controversial. A clay stabilizer is supposed to plate out on the surfaces of clays, tie them down, and inhibit their swelling or migration. Simultaneously, due to the nature of hydrofluoric acid, these same clays dissolve, which leads me to question downhole conditions when both these additives are recommended.

A classic example of mixing problems was a foamed acid job. Out of ignorance, the service company batch-mixed its foamer. Normally this is no problem, unless you go through a blender tub in which the agitation and the stirring paddles create severe foaming. Any fluid loss additives or other materials added through an educator add air to and agitate the system, creating even more severe foaming.

Service companies commonly have defoamer available on site. A tank containing surfactants or something causing severe foaming problems may require a silicone or phosphate ester-type defoamer to keep the pumps pressurized. On this foamed acid treatment, the service company started going downhole and was adding fluid-loss additives. A tremendous foaming problem occurred, but these well-prepared service. company hands added defoamer. That sounds all right until you consider the fluid's purpose. The foamed acid was, of course, to stimulate the well. Foam increases viscosity downhole, which controls fluid loss, thereby giving a wide fracture. Obviously, adding defoamer defeated the purpose of the treatment. In addition, too little defoamer was available, and the job was terminated earlier than expected.

This sort of thing happens too often. Frequently, companies other than the service company supply surfactants, biocides, and other additives directly to the oil companies. This is reasonable. There may be local expertise with a surfactant or biocide that indicates it is the best choice for the treatment. Commonly, service companies find out about the new additive just about the time they are to blend their fluids on the location.

In most cases, no time has been allotted for quality control of the compatibility of these additives. Several excellent biocides used in the oil industry are totally incompatible with cross-linkers or other surfactants. In fact, some biocides completely destroy the effectiveness of cross-linkers. Many surfactants, because of their charge structures, prevent acid inhibitors from functioning.

Additive compatibility tests

No fluids should be pumped into an oil or gas well that have not been checked for compatibility. Incompatibility of additives is often apparent: flocculation, precipitations, or materials floating out are visible. Other times, however, the incompatibilities are less obvious and thus extremely disconcerting. A competitor's treatment was completed, but the well did not respond. The customer could not understand why the treatment had failed. Similar treatments by other service companies had produced reasonable results. Evaluating the additives revealed that a local supplier had convinced the customer to utilize his biocide. It was cheaper and more efficient than the service company's choice. That biocide destroyed the effectiveness of the breaker mechanism of the fracturing fluid.

This situation was neither the fault of the service. company nor the biocide supplier. The biocide was indeed an effective material, but the service company did not know the biocide had been placed in. the tanks before the personnel arrived at the location. The service company people were told only that no biocide would be required from them. Well-meaning people caused serious problems for the oil producer.

Chapter 10, "Intense Quality Control," discusses at great length a pilot-testing operation that would have alleviated this problem and as a procedure that should be followed on all fracturing treatments. Incompatibility problems can easily be spotted by on site testing of fluids at in-situ conditions.

Too many cooks spoil the broth

On several occasions during the boom times, independent suppliers sold a variety of specialty items directly to the oil companies. Potassium chloride, not a big profit item for the service companies, is an example. Most service companies handle it in sacks, and because the hands do not like to add the material unless they get it in bulk, this was an ideal situation for independent suppliers to come in at cut-rate prices and supply KCI to the companies. A few independent suppliers got greedy and began to cut down on concentrations. In some cases, sodium chloride or calcium chloride was substituted for the most common potassium chloride. Serious failures on stimulation treatments took some time to trace to the fact that potassium chloride was not utilized to control clay swelling.

When using multiple suppliers on a location, the quality control is greatly complicated. Coordinating competitive equipment on a location is extremely difficult, but even more difficult is coordinating two or more chemical suppliers on a location. A lot of short-term and long-term problems can result from chemical incompatibilities.

I am not saying that specific chemicals cannot be purchased from suppliers other than the service companies. But if you do so, coordinate with both companies to make certain the products are compatible and free of detrimental effects. Such effects can be dramatic (i.e., precipitation or flocculation as shown in Figure 4-1). This type of incompatibility is easily recognized.

If you are going to utilize additives other than those supplied by the service company, conduct corrosion tests with the inhibitor. Conduct emulsion tests with nonemulsifiers, and carry out scale inhibitor or paraffin inhibitor tests. These tests, of course, are only required when these additives are included.

Figure 4-1 An example of chemical incompatibility. The flocculant precipitant indicates a reaction between incompatible components.

The oil company should never hesitate to ask the service company about the additives to be included in a treatment. It is not uncommon for well-meaning individuals to include incompatible chemicals. The most common error is using anionic acid retarders that are designed to be compatible with cationic inhibitors but will work with cationic surfactants. Only trained people can detect these problems.

Another error is combining mutual solvents with surfactants such as acid retarders that are to be plated out. Mutual solvents also tend to greatly undermine the effectiveness of acid inhibitors, so the acid inhibitor loading must be raised. Alcohol, widely used in acidizing, detracts from the effectiveness of the acid inhibitors, so loadings have to be increased.

Any combination of ingredients that has not been successfully run should be thoroughly tested, both for obvious incompatibilities and for those that affect the performance of the additives. A general rule is to utilize the minimum amount of additives necessary to accomplish the treatment.

Chapter 10 examines Intense Quality Control, which is basically pilot testing. One thing that has caused a great deal of consternation in our industry in regards to compatibility is reliability. Additives in systems that were once thought to be no problem with other materials are now being found as the major culprits in causing fracturing system failures. One of the most common ones is frac sand, particularly sand which may contain contaminants, or resin coated sand, particularly those that are curable resin-coated products.

I was on a treatment near Raton, New Mexico, some years ago, and was discussing the relative merits of batch-mixing a borate cross-linker in a low pH system or adding the borate cross-linker into fracturing fluids that had already been buffered. The service company personnel argued with me that they preferred to batch-mix boric acid into neutral pH gel and add an activator to cause the crosslinking to occur on the fly. I finally relented and allowed them to batch-mix the borate system utilizing a caustic material which activated the cross-linker during the treatment. During the fracturing treatment, everything was going well. The pad and the fluids were testing quite well, until we started proppant. As soon as the propping agent started, we lost prime in the blender tub and the job had to be terminated.

When we went to the blender and observed what was going on, we found that there was a 10 or 12 bbl. cross-link plug in the blender tub. There was just enough cement dust in the sand that had been transported to cause cross-linking of the gel in the blender tub. If we had tested the sand prior to the treatment and found the presence of the cement, then we would not have attempted the treatment in the manner of batch-mixing the cross-linker. On subsequent treatments, the cross-linker was added on the fly. This also taught me to always test for contaminants in the frac sand and in all additives to be used on the treatment.

When we run break tests doing our Intense Quality Control, which will be discussed at length in Chapter 10, I include not only sand, but also any diesel that might be added for fluid loss control, or any additives that might be added during the treatment. I have found significant effect from clean burning additives in the diesel and also effects caused by contaminants in the sand.

One of the most significant problem areas that a significant number of SPE papers have been written about is the effects of curable and procured resins on fracturing fluids. I have discovered that most of the oxidizer breaker systems used by the service companies are drastically effected by the curable resin-coated. sands. In many cases, we have had to run 5-10 times as much breaker to be able to break the fracturing fluids due to the presence of the curable resin-coated material. The suppliers of curable resin-coated sands have been working very hard to come up with products that are much more compatible with breaker systems and great progress has followed. The same thing has occurred from the standpoint of diesel and other additives. It is very important that pilot testing of the fracturing fluid with the actual water on location and actual additives to be used during the treatment are conducted. As will be shown in Chapter 10, on-site pilot testing, particularly for wells with bottomhole temperatures of 200°F or less, is relatively inexpensive and should be done on every job.

Guidelines on fluid compatibility

1. Use a minimum number of additives.
2. Never pump a fluid without testing it for additive compatibility.
3. Be very careful when mixing chemicals from different suppliers.
4. Always query your service company about the function and the compatibility of all additives in your treating fluid.
5. Test *all* additives to be used at in-situ conditions to determine compatibility.

CHAPTER 4

COMPATIBILITY (2022)

Like the previous chapter this one on compatibility is still relevant with a little discussion on specific problems with slick water treatments.

In the early days of large volume slick water treatments, there were significant problems with compatibility of cationic biocides and the friction reducers the service company was using. Primarily due to utilizing compatible, lesser anionic character acrylamides, copolymers or cationic acrylamides, this problem was solved. There also were multiple biocides which had little ionic character.

Another problem that developed in pumping very large volumes of water containing FR materials, was the production on flowback of massive amounts of something approximating high loading crosslinked gel. This slimy product fouled up downhole equipment and created a large research project to solve the problem. This work was coordinated by our company, Halliburton and the operator. The solution was the development of a controlled degrading agent for the acrylamide. We had designed and pumped single stage water fracs well in excess of 100,000 barrels in vertical completions. What was occurring was the anionic polymer was plating out in the near wellbore area and on flowback, the slimy polymer was pushed into the wellbore. Later work by other service provides indicated that ammonium persulfates also gave degradation. Some service companies used the perborate and others even today are using persulfates. The perborate, in my opinion, is superior to persulfates. These breakers in most service company product lines are similar to enzymatic and oxidizing degradation of the viscosity of guar and guar derivative gels. The reason that this problem has resurfaced has been the increase of stage volumes in laterals as well as the use of the higher viscosity acrylamide in the later part of large volume stages.

It was interesting that a recent front-page headline in the September 2020 issue of the Journal of Petroleum Technology (JPT) highlighted a similar problem that we had seen 18 years previous. The real dilemma was that our original work and finding were never published. Our industry is not in the habit of publishing our failures and mistakes but are enamored with only our successes. This was certainly a case where prior publication would have proved beneficial to the current industry.

An interesting, and to some, controversial finding in early waterfracs was the lack of need for clay control such as KCL or KCL substitutes. With the name "Shale" there was great trepidation on pumping fresh water into a "Shale". One of my early designs was for a 40,000-barrel single stage frac in a vertical completion. My customer who had rightfully accepted the need for KCL and not KCL substitutes in South Texas Wilcox formations wanted to run 3% KCL on the frac. When I explained that we were talking literally a trainload of KCL, he opted for addition of brine water. Many of our fracs that utilize recycling are simply using water at hand. If there were any possibility of swelling clays the use of brine water will negate any damage. As discussed earlier, virtually any nanodarcy matrix can be treated with fresh water.

As stated in the earlier edition obvious compatibility problems showing precipitation when an incompatible additive is used are easily identified. (see figure 4.1 2nd edition). For slick water fracturing, where high pressures are noted and you have conducted diagnostics to eliminate perforation friction and tortuosity, one should look hard at the compatibility and addition rate of the FR being used and the treating water. Other possibilities are interaction with clay stabilizer, biocides and scale inhibitors which caused problems in early slick water fracs, but this type of problem is rare today

As will be discussed in the proppant chapter, many time increasing pressures relate to the wrong size sand being pumped, A huge concern is the lack of a sieve of the actual sand run even per stage. If one is pumping 20/40 mesh instead of 100 mesh, there can be width problems and potential screenouts.

2nd Edition - Figure 4.1

Obvious incompatibility between chemical systems, captured during laboratory testing.

CHAPTER 5

WHAT DID THEY PUMP, HOW MUCH, AND AT WHAT LOADINGS? (1994)

For the 20 years of my career that I was a service company engineer and supervisor, I was continually disconcerted (and continue to be so today) by the lackadaisical attitude of the average company representative when he arrives at the location on the day of the job. These are the same people who will diligently shop all over town to get the best price on automobile tires. They will look far and near for the best bargain on clothes, cars, etc., and obviously, using intelligence, they only buy the very best they can afford. For some reason, which we will try to grasp later on, hydraulic fracturing treatments are another story.

Beware the frac van

Many times company representatives arrive on the location, walk directly into that black pit of isolation otherwise known as the frac van, and never leave it until the treatment is over. Their actions indicate a great deal of faith in their service company, the flashing lights, and the digital read-outs. However, the last place an oil company representative needs to be is inside the frac van!

Most of the real action, explosions, and flying bullets occur on the blending units, the pump trucks, and the frac tanks. If it were my dollar being spent, I would insist that a representative stay outside the van to monitor and observe the service company's operations. In the meetings after cementing and stimulation treatments, people are asked how much of a particular item was pumped. Other than those who delivered the material to the location, no one had any idea of how much product had been added.

Take inventory and samples of everything

The oil company representative must go to each tank and determine the amounts of fluid and the additives used in each tank. He should quality control the fluid viscosity and pH, and he should question the local service company to discover any problem with the fluid. He should further discuss any cross-linking problems or potential bacterial problems. At that point, the oil company representative should know the positions of all the fluids on the location. The representative should take samples of all these fluids and put them in reserve in case problems occur during the treatment. The oil company representative should also check the volume of the sand or cement storage equipment. Samples of sand, cement, or other additives should be taken for evaluation if a problem occurs during or after the treatment.

Exact volumes of fluid and proppant can be determined by simple calculations-length x width x height-or simply by asking the service company what volume is stored in assorted containers. By ascertaining exactly what is available on location before and after a treatment, you will know what was pumped into the well, even if all the electrical equipment fails. Many oil company engineers are careful to check what is on the location before the treatment, but they do not bother to check what's left after a treatment.

The best monitoring equipment makes errors, and an engineer needs to see that he is getting what he is paying for. I am not implying the service company is incompetent. It is simply good business sense to make certain you get what you pay for.

Since the first edition of this book, there have been some very serious lawsuits. The most significant one was the Parker and Parsley lawsuit in west Texas, where the service company was sued for well over $150 million. Parker and Parsley claim that the service company had shorted them of 300,000 lbs. of sand for over 600 jobs. As stated in many areas of this book, there have been times that there were indeed some shenanigans and shortages by the service company. In no way would I ever say that a major service company (or for that matter, a smaller service company) has ever openly tried to cheat a customer. As for myself, I did see various individuals in my career that for their own personal gain did cheat operating companies.

Today, I do not believe that there are any service companies that are actively shorting or cheating on any service job. However, reality is that there are human beings on the job, there is equipment, and there are electronic devices. All of these err and all of these fail. Quality control is about observing these people and equipment in a very stringent manner to overcome the times, as best we can, when they do fail. It may be of primary importance to know when indeed equipment, chemicals, or personnel fail in their task. What is terribly sad is that attempts have been made to stimulate untold numbers of wells with no success whatsoever. Because there was no serious quality control, because there was no serious monitoring, because there was no pre and post job inventory, we have no way of knowing what caused the failure. By using the technique "Intense. Quality Control," we will at least know when we have had a problem and perhaps be able to either solve the problem, if not on that job, then on future treatments. Typically, the geologist is blamed for placing the well in a bad spot, or sometimes (as is the case of screen out or failure) we say we fracced into a fault. In reality, these things rarely occur. In reality, there are real answers for all the problems that we have on stimulation treatments.

I have been on numerous treatments where there were indeed many problems. Just after I started consulting in 1985, I went out to a location just as they were about to initiate pumping. I had been called late to supervise the frac. I immediately did what I could in regards to conventional quality control, checking fluids, looking at the rig-up, checking viscosities, pH, cross-link times, fluids, and so forth. I then went back into the frac van and immediately realized that I had forgotten something. This was quite a large treatment with 600,000 lbs. of Interprop to be pumped into the well. Obviously, this was an expensive treatment, with more than half a million dollars proppant to be used.

What I had forgotten was to check the volume of proppant on site. I asked for the service company to delay the job, and I went over and looked in the proppant storage containers. I found that half of the proppant was not on location, and there was indeed no one headed that way with more propping agent. This particular finding caused quite a stir with the oil company for which I was working and, indeed, there were a lot of accusations about the service company. I believe that it was simply a case in which the people delivering proppant simply didn't get the job done. At that point in time, the service company was so afraid of delaying the job-so afraid of the problems of getting run off-that they were going to continue to do the job and modify the density readings to show the proper amount of proppant.

The point is that in the safety meeting, I explain to the service company that I want to know if there is a problem. I want to know if something is wrong and we will work together to solve it. Many of these service companies have long been under the threat that if, indeed, anything does go wrong, they will be run off. So if a problem does arise, they don't pass it on. This type of attitude must and can be changed by developing a good working relationship. This relationship may, in fact, entail the oil company paying for unforeseen breakdowns of equipment and perhaps paying the 'absolute cost of chemicals if unforeseen breakdowns do happen. Obviously, if breakdowns continue to occur, then it is time to change service companies, but developing that kind of rapport will, I believe, work to the benefit of the major and independent oil companies.

Regularly obtain status on fluids and additives

The oil company representative must also continually query the service company about the status of the fluids or proppant being placed in regards to the treatment's design. Maintain radio communication with an outside observer The oil company person must have radio communication with another representative outside on the treatment, apart from communication within the service company. One area that has always been debated is whether or not an open radio should be available to the oil company representative so he can listen to the service company conversations. One service company, as a matter of day-to-day work, has an open radio available in their frac van. I personally have) Mixed emotions about this. If I am inside the van, running the job, and by myself, I really don't have time to listen to both sides of a conversation between the treater and the people. However, if two personnel are available on location, it is not a bad idea to have your representative listen to conversations from the service company in order to detect any problems very quickly. I do not feel that the oil company representative should be able to discuss or to talk on the radio. To accomplish this, all that is required is to unplug the headsets from the radio and this individual can carry it around with him. I have not seen any problems lately with service companies having oil companies listening to conversations. In the past, the major ploy was that no radios were available. However, an oil company representative does need to be outside the van working alongside the service company personnel helping to solve problems that occur.

Panic decisions

Excitable oil company representatives who tend to be abrasive with service company supervisors are usually the same ones who make decisions such as prematurely flushing a frac treatment. They do not understand what is going on and act out of fear. This explains, in part, why oil company representatives maintain ongoing working relationships with personnel they can trust.

It should be noted that in today's litigious society, the oil company representative should always make the decisions for the service company. Based upon the ability of the equipment, it is the service company's responsibility to pump the stimulation treatment in a safe manner, exactly as the oil company representative tells him to do. No service company can take on the responsibility of making decisions such as when to flush, when to increase sand concentrations, or when to do things radically different from the design. It is my experience that it is highly unusual that a treatment is pumped precisely as planned. In fact, one of the things that I work on while communicating with treaters is to explain to them that I don't particularly care to have precisely 40 bbls./min. if the computer design states it. I want the pump trucks to be operating efficiently. I want all of the transmissions to be in lock up. If that means we are pumping at 38 or 43 bbls./min., it won't adversely affect the job. I believe the competent stimulation supervisor will indeed see the pressure responses that are available from the fracture treatment and increase the rate or decrease the rate based upon what he notes and based upon what he understands is going on in the formation at that particular time. It is indeed a big mistake either for the company man to allow or the service company to take charge of the treatment, as it can put him in serious liability should things go wrong.

Many times in my career I have seen the service company sent to the location without a representative. This is a very dangerous situation for the service company from the standpoint of liability, and it is also a very bad situation for the oil company in not being present for all phases of the stimulation operation.

In our business, many decisions must be made that do not directly relate to science. These decisions are based on experience. In most trouble situations, such as when a blender goes down or a pump truck loses prime, the oil company representative depends upon the service company to take specific action.

Temporarily flushing a well during the pad, until a trouble situation can be alleviated, will not necessarily hurt the final results of the treatment. Shutting down for short periods with sand in the pipe can be accomplished in many cases without negative consequences.

On the other hand, a shut-down can be extremely detrimental, as in a case of pumping such complex fluids as thixotropic cement. With some types of cement, a shut-down is not feasible. You may not be able to restart due to cement left in the pipe. With today's technology, cross-linked or delayed cross-linked fluids, gelled oils, or acidizing fluids, and brief shutdowns can occur without harmful results. Long periods of shutdown can cause acid corrosion problems or allow proppant to settle out in the sand-fracturing treatments.

To monitor the progress of fluids and solids during a moderate-to-large fracturing treatment, a check sheet like the ones shown in Appendix II can be utilized. (Use Tables 4, 8, 9, and 10 from the Quality Control and Job Supervision Guide.) From time to time, the oil company representative should check with the service company to find out the status and operational capability of the pumping units, blenders, and other equipment on the location. For example, it is imprudent to run high-sand concentrations on a treatment when a blender screw is malfunctioning or a particular blender is losing suction pressure. Similarly, you might not increase the pumping rate on a treatment if the standby horsepower is lost. It might be more profitable to maintain or even reduce the pumping rage to, assure the treatment's completion. Potential problems like these can be avoided by an alert, experienced service company representative.

Guidelines

1. Have a minimum of two oil company representatives on the location. If only one is available, he should spend 50% of his time outside the frac van observing operations.
2. Inventory everything on the location before and after the job.
3. Quality control all fluids. Take viscosity, pH, cross-link time, acid strength, and other relevant measurements.
4. Get status reports on all fluids and additives at regular intervals during the treatment.
5. Maintain radio communication with outside observers.
6. Don't make panic decisions.

CHAPTER 5

HOW MUCH, WHAT, AND RATE (2022)

The chapter, in the second edition, on what was pumped, how much, and at what loadings was again designed around more conventional treatments. The sad situation in our age of harvesting digital data and using that data for large geographical areas to optimize our ongoing stimulation, using water frac stimulation, has a major drawback. In the early days of fracturing, it was not unusual to batch mix the gel and most of the additives. Crosslinkers and breaker were typically stored on the blender for the job. As treatments got larger, we moved to chem-add trucks and then trailers holding multiple 3-400-gallon totes. The large volume of Horizontal lateral stimulation has virtually done away with small totes and typically tanks of chemicals with on-site volume of more than 6,000 gallons are common. At the end of the chapter are multiple size containers for chemical, large pits, & transfer hose. Shows pictures of smaller capacity chemical totes as well as chemical floats with additive pumps mounted. Acrylamides, biocide, scale inhibitor, biocide etc., are typically stored in individual containers on location. Although monitoring of this type of tank is feasible, it is much more difficult that monitoring a small volume chemical tote.

The push toward slick water fracturing has become to some operators simply a manufacturing process. This oversimplification has led in some cases to little effort in calibration of instruments such as flow meters, densitometer, and additive pumps. When multiple upright storage silos exceeding 400 thousand pounds each and water is being pumped from a 100,000 + barrel pit, absolute control of what is pumped is challenging at best. Another problem is that in many cases no quality assurance on relative size and proportion of proppant is conducted. The enormous problem of sand sieving due to massive volume is simply not accomplished due to the huge amount being pumped. As discussed earlier, I do believe lower quality sand with crush strengths around 4,000 pounds has worked well in deep ultralow permeability shales and ultralow permeability conventional rock. What has me more concerned is that we, many times, have no idea the actual sieve distribution of the sand we are pumping. Our company has seen many operators pump what they call 100 mesh when, instead of 70/140 mesh, it sieves out at 40/140 or in some cases 40/100 with a large variation in proportions of the different sieves. Also, we recently had sampled the 40/70 sand on location and found it to be 90 % between the 40 and 50 mesh screens. I recently published on a process called counter prop where larger sand is pumped first. We saw and continue to see promising results from the process but must be very diligent in ensuring what is being pumped. I would suggest that many of the screen outs that have occurred with 100 mesh at relatively low recorded concentrations may not have been the formation, amount of pad etc., but may have been that the sand was not 100 mesh but indeed much larger.

I do know that there is concerted effort by some of the service companies to have a better way of measuring volumes, sand distribution and absolute chemical volumes. Having observed thousands of pressure charts, a significant number of pressure anomalies are not a result of the formation but simply having the improper additive or size and volume of proppant.

I just discovered that a major service company has done away with densitometers using strap on sonic devices that have no capability of measuring sand concentrations above 2 pounds per gallon. I am very aware of the problems with radioactive densitometers but moving back to clean vs dirty rates or sand screw revolutions is a step backward. We do need something better than radioactive densitometers. I am very aware of the attempts to utilize correalis devises but suspect that we in our industry need to communicate with the coal industry other than just getting rid of the only device presently available to measure proppant concentration.

To summarize, the very same techniques used on conventional fracs need to be followed on large volume slick water fracs as was accomplished in smaller conventional treatments. Although challenging and requiring new technology, it can be accomplished.

An interesting example of the innovation within our industry is the Minion water tank, which greatly satisfy the need for much larger, above ground storage than the common 500-barrel tank. To my knowledge, they first resembled giant round swimming pools. Others also common were inflatable tanks. Pictures of various Minion are shown. These innovations are still in use but in most cases are now dwarfed by giant lined pits, with pumps transferring to holding tanks on location.

One huge mechanism for innovation is that some of the early Woodford jobs were executed with as many as 100 frac tanks on location. Transferring, filling, and worst of all the freezing weather caused nightmares for operators.

An innovative Landman went to one of the farmers where they were drilling, who did not own mineral rights and was not too happy and asked the farmer if it would be okay to dig a very large pond on his land free of cost. The skeptical farmer said what else do you want. The Landman asked if it would be ok to drill a nearby high-capacity water well to fill up the pond allowing the water to be used for the frac. While the farmer was considering the offer, the landman went on to say that his company would supervise filling the pond and lastly stocking the pond with nice bass. The deal was done and was duplicated many times thereafter. The interesting part is that the pond and well, given back to the farmer, were much cheaper than the rental frac tanks.

Image A

Bulk ISO containers on location; 6,000-gallon storage capacity when full, often kept at lower volumes due to overweight restrictions.
Source: Frac Chem

Image B

Bulk ISO delivery unit / 20-foot standard tank (rear view)
Source: Hoover Ferguson

Image C

Bulk ISO Delivery Unit/20-foot standard tank with drop-frame chassis
Nominal capacity of 6,340 gallons; utilized for large treatments.

Image D

Manual chemical tote strapping….

Image E

Manual chemical tote strapping….

Image F

Liquid chemical additive float, equipped with eight on-board liquid additive chemical pumps and storage totes; control cabin located in the rear for manual operation.

Image G

Enclosed, insulated liquid chemical additive trailer, with chemical totes and additive pumps enclosed within the body of the trailer; typically utilized for cold-weather operations to prevent additive pumps and totes from thickening or becoming un-pumpable due to low outside temperature.

Image H

Holding pit for large volume hydraulic fracturing operations.

Image I

Poly-line water hose used to transfer fluids from water pits to fracturing locations.

Image J

Haliburton mountain mover.

Image K

Nowsco Canada sand dumps.

Image L

Nowsco US Sand belt for multiple sand hogs.

Image M

Conveyor belt (aerial view) for upright sand storage silos.

Image O

'Sand-storm' modular sand storage
Source: Cig Logistics

Image O

Sand Box gravity-fed proppant storage and conveyor set up on wellsite; 40-45k lbs of sand per box
Source: Sand Box

Image P

Sandcan proppant storage and conveyor belt set up; storage containers are loaded onto conveyor system for gravity-fed sand delivery to blender
Source: Sandcan

Image Q

Sand silo, upright storage; proppant is gravity-fed from the bottom of the silo onto a conveyor belt system to supply fracturing blender with sand; typical total sand capacity, per silo, up to 410k lbs.
Source: Solaris

Image R

StimCommander Proppant system for automated operations; gravity-fed proppant system to blender; silo capacity up to 265k lbs.
Source: Schlumberger

Image S

'Sand Castle' upright storage, gravity-fed sand silos; capacity up to 260k lbs.
Source: Halliburton

Image T

'Minion' water storage tanks, for on-site water storage; capacity up to 20k bbls
Source: Campell Oilfield Rentals, LTD

CHAPTER 6

MONITORING FRACTURING FLUIDS (1994)

For 30 years, the service company industry has tried to develop a new language. Our speech is filled with terms, trade names, and codes; and it is terribly confusing to new engineers. For example, polysaccharide derivative, low-residue polymer, hydroxypropylguar, and derivatized guar all describe the same thing; namely, hydroxypropylguar. Up until a few years ago, hydroxypropylguar was the product that was the most common fracturing additive. It now is being replaced by our reversion back to guar gum and also by the more versatile carboxymethylhydroxypropylguar.

In the middle to late 1960s, guar gum was developed as the gelling agent of choice for hydraulic fracturing fluids. In the early years until about 1967-68, we had only used noncross-linked, water based linear gels and thickening agents for oil or just thick Newtonian oils. At that time, there were two choices for water based fracturing fluids. The first was a very clean nonresidues fracturing fluid, hydroxyethylcellulose, or other cellulose derivatives such as carboxymethylcellulose. The second, guar gum, which was not as refined as it is today, contained as much as 13% nondegradable product after the fluid was broken down with standard conventional breakers.

Late in 1967 and early 1968, the first cross-linked guar gels were developed. Early systems consisted of antimony cross-linked guar or borate cross-linked guar at fairly high concentration (i.e., 80 lbs./1,000 gals). These fluids dominated the industry between 1968 and 1973. In 1973, the much cleaner deravitized version of guar (hydroxypropylguar) was introduced to the industry. Because of its much lower residue (1-4% depending upon how the tests were run or what claims were made), this product was the dominant gelling agent in the industry for many years. The primary reason related to its "cleanliness." As will be discussed later, this proved to be an erroneous assumption in reference to damage, particularly for moderate-to-high permeability formations. ·

I have been fortunate enough to be around a long time in our industry and the old saying, "What goes around comes around," is extremely applicable to our oil industry. After some 70-80% of our fracturing fluids were reverting to derivatized guars in 1973, we have at the present day reverted back to the use of guar gum as the fluid of choice with the double-derivatized guar, CMHPG as the second polymer of choice in the industry.

One of the things that has continually caused a great deal of consternation and trouble in our industry is the use of "intuitive reasoning." Intuitive reasoning taught early scientists that the residue in guar gum was a terrible thing, and those colloidal solids and dirty material would be damaging to the proppant pack. In the mind's eye of the scientists, they could see that terrible things would occur if these dirty fluids were used in fracturing. In fact, there were major oil companies who would not allow anything other than ultra-clean fracturing fluids to be pumped in their wells. This kind of thinking carried on well into the 1980s until some new thinking and a more open-minded approach taught us that the most damaging fluids are the ultra-clean materials, such as cellulose derivatives and even the derivatized guars such as hydroxypropylguar or carboxymethylhydroxypropylguar. The reason for this lies in the primary damage mechanisms that occur in the fracture (i.e., the creation of filter cake). The main reason for loss of propped fracture conductivity is due to the creation of filter cake. In very clean fluids in moderate-to-high permeability formations, the gelling agent is literally filtered out, forming a very thick filter cake. This filter cake can indeed be thick enough to cover up half the sand grain on each side of the fracture. If you consider that in any 20/40 sand pack that 1 lb./sq. ft. of proppant is only 3 grains wide, you can easily see how filter cake may be the major factor in reducing proppant pack conductivity.

Table 6-1 Guar Gum

- A long-chain polymer composed of a mannose sugar backbone and galactose sugar side chains
- A product of the endosperm of guar beans (splits) ground into powder
- 7-10% insoluble residue
- Poor alcohol solubility
- Salt tolerant
- Limited thermal stability
- Sensitive to cross-linker concentration
- Easily dispersed
- Optimum hydration at pH of 6.5-7.5

Indeed, the way to minimize proppant pack conductivity reduction is to utilize a fluid that has excellent fluid loss control, such as the colloidal solids pre sent in guar gum. The colloidal solids make guar a much more efficient fluid; therefore, less filter cake is created with guar gum than with even derivatized guar, and this certainly would be the case for cellulose derivatives. We are not stating that you should never use derivatized guars or very clean fracturing fluids. It just should be noted that you need to control leak off and negate filter cake development. This is particularly important for moderate-to-high permeability formations. The guar gums to date, which typically contain 4-8% residue, are most probably the fluids of choice in which leak off is a problem. Incidentally, we have also found that with present date technology that the temperature stability of guar gum fluids can be extended to 300°F and perhaps higher. What we see in our industry is a complete reversal of opinion concerning what we considered the ideal fracturing fluids to use. We will discuss later the early adverse feelings toward high pH borate fluids and how they have taken over as perhaps the fluid of choice in the industry.

History of frac fluid development

The first fracturing fluid utilized in the late 1940s was war surplus napalm. Napalm is an aluminum gel used to thicken gasoline, diesel fuel, kerosene, and other such fluids. The first treatment was thickened gasoline carried in open top tanks along with a few sacks of sand. Obviously, luck was with us because we didn't burn down the location. The first fracturing treatment did not resemble the massive treatments conducted today. In fact, it was many years before relatively large treatments were even attempted. The great majority of treatments conducted in the 1950s and 1960s were damage removal treatments. The small sizes of the treatments did not allow deep penetration into the reservoir.

The procedure on these early napalm treatments was to gel up gasoline in the measuring tanks of cement trucks, and typically there would be anywhere from one to five trucks on location. Excellent quality control procedures were used (i.e., when the shovel would float in the measuring tub, the gel was ready for pumping). These gelled gasolines were pumped through a mixing tub on the ground, through a jet mixer and river sand or other proppant was added into that mixing tub, and the fracturing treatment was conducted until they ran out of gel or ran out of propping agent. I have had long discussions with individuals who were on such treatments. They described treatments being conducted at night where the sparks from the exhaust would fly through the air and land in the napalm and sputter and fizzle. We were lucky that many more people were not hurt and locations were not burned down while these treatments were being conducted.

For many years, gelled oils or refined oils were the only fracturing fluids in use. But a revolution occurred in the 1950s. The introduction of water as a fracturing fluid caused a great deal of consternation among those who felt water would seriously damage the formation. Such advancements in our industry took people who had nerve (and in many cases, a lot of luck) to push for fracturing some wells with water. I believe that the personnel who had utilized water in fracturing did it out of ignorance, not out of intelligence. This "intuitive reasoning" had led all of us to believe (and sadly some people today still believe) that there are terribly water sensitive formations and that water is the worst possible thing to put on formations. The people who did these early treatments were people who were either ignorant or actually frightened of oil or simply didn't have the money to find oil to fracture. These people were indeed ahead of their time. These people have led us to the point now where I don't believe that there are any formations that are water sensitive when one utilizes potassium chloride for control of clay swelling and migration. The wells we treated were not particularly water sensitive, and stimulation resulted.

If we had started fracturing wells with gelled water in the very early days of pumping aqueous fluids, we may never have gotten back to that point. Today, the formations that have been termed "water sensitive" were not water sensitive at all, but they were plugged with gelling agents without the proper use of breakers. Many scientific studies have been conducted that show you can have a great deal of fracture face damage, even as deep as 8-9 inches away from the fracture, and still have excellent production increase due to highly conductive propped fractures. The key to successful hydraulic fracturing is to achieve a highly conductive. propped fracture utilizing high-proppant concentrations and, coincidentally, degrade the fracturing fluids back to water viscosity. The lack of good breakers and the lack of high proppant concentrations were the reasons for thinking that formations were water sensitive. They were not water sensitive. They were sensitive to unbroken gel, and they were sensitive to not having proper placement of proppant across the productive formations and/or have sufficient concentrations to have enough relatively conductivity to give production increase. In the evolution toward the use of straight water, we have made great strides in technology. The first advancement was using starch as a viscosifier. I like to believe that a scientist while watching his wife prepare a starch solution for his shirts said, "That might be a good fracturing fluid!"

There is a story about the use of starch in fracturing. A service company, wanting to keep the fluid a secret, bought all the available starch from the surrounding towns. The employees opened all the boxes, re-sacked it, put an experimental number on the sacks, and went out to fracture the well. The customer noticed a familiar aroma and observed that he didn't know if he wanted his well starched! Starch does suffer serious problems with shear stability, temperature stability, and salt sensitivity. It didn't last long as a fracturing fluid. Early in the development of viscosifiers, people began to look at other thickeners that proved to be good fracturing fluids.

The guar bean and the meal produced from it are a polymeric substance. When combined with water, it yields a high-viscosity fluid (see Table 6-1 and Figure 6-1). The guar gums used today are similar to the earliest systems. A little more processing is done and typically the guin is a little cleaner, but these same guar molecules are simply processed from the guar bean and used as thickeners for water, acid, and other solutions. Water thickened with guar gum is an excellent fracturing fluid.

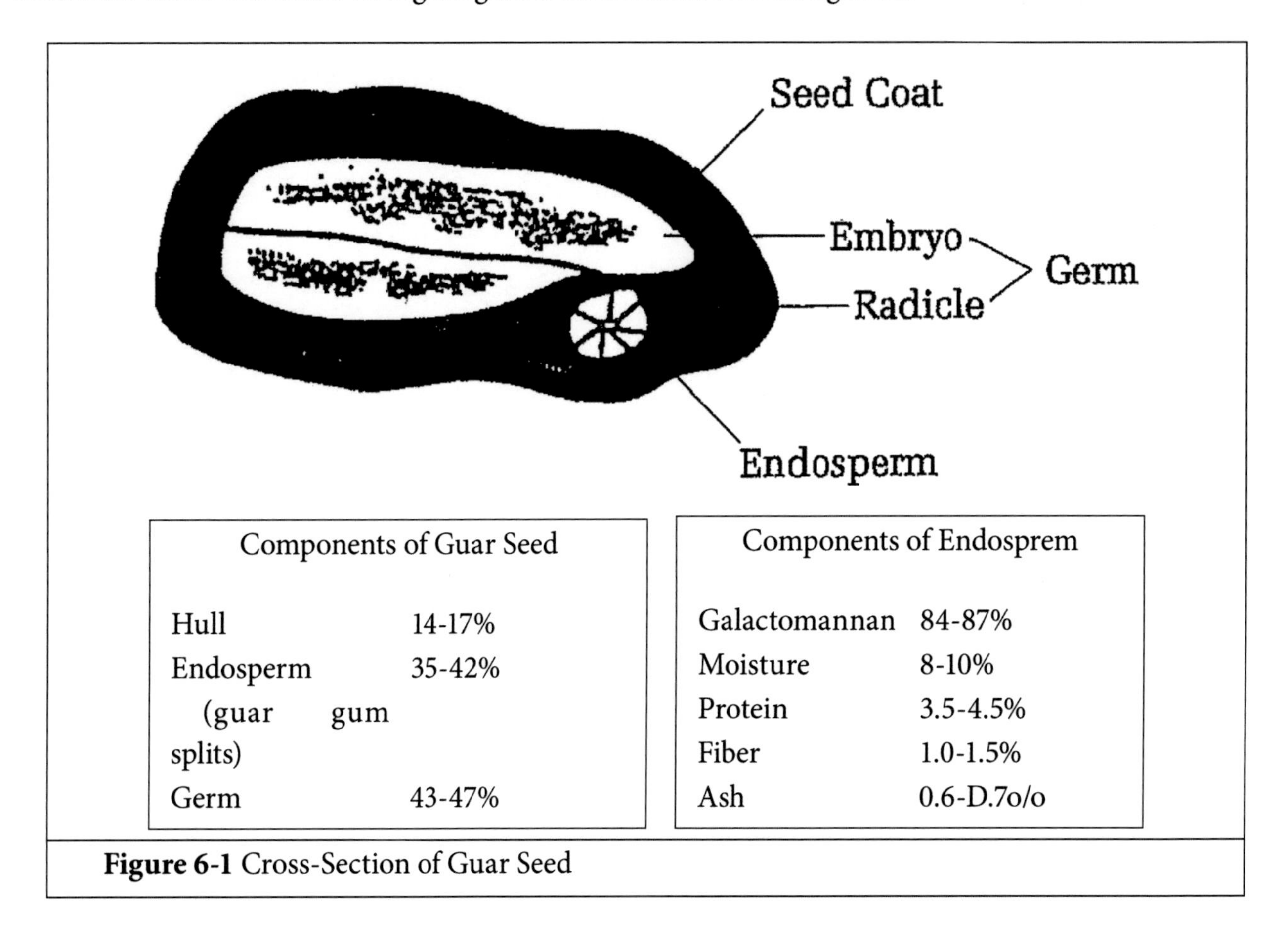

Components of Guar Seed	
Hull	14-17%
Endosperm (guar gum splits)	35-42%
Germ	43-47%

Components of Endosprem	
Galactomannan	84-87%
Moisture	8-10%
Protein	3.5-4.5%
Fiber	1.0-1.5%
Ash	0.6-D.7o/o

Figure 6-1 Cross-Section of Guar Seed

Earlier, it was discovered that toll oil fatty acid combined with caustic created a soap material that could be used as a fracturing fluid. This fracturing fluid, which was utilized until the mid-1980s, was a cheap fluid with controllable viscosity. Although it did not have particularly good friction properties, it was an efficient medium. It was called *Halliburton's Vis-0-Frac or Dowell's PetroGel.*

Some new gelling systems for oil were developed in the late 1960s and early 1970s. These systems were aluminum gels including aluminum octoate in combination with a toll oil fatty acid. These systems have limited application today, but the most successful and still the most widely used is the aluminum phosphate ester reacted with sodium aluminate. This product is the frontline-gelling material used by most service companies. Table 6-2 shows typical oil-gelling agents utilized by the service companies.

Some slight differences in aluminum esters, which relate to reactivity, give enhanced temperature stability. These gels do not approach water-based systems in temperature stability and handling ease. Where they are used, however, aluminum esters have been successful. But they require a great deal of quality control, which we. will discuss in a later chapter.

Table 6-2 Oil Gel Systems	
Aluminum Phosphate Esters	
Four Types	
• Batch-mix, low-temp stability	150-180°F
• Batch-mix, moderate-temp stability	225°F
• Batch-mix, high-temp stability	300°F
• Continuous-mix, low-temp stability	200°F

The cross-link revolution

In the late 1960s, a new development enhanced the viscosity of guar gum. One service company began using potassium pyroantimonate to create a cross-linked gel (See Figure 6-2). This low pH system is still in use but to a lesser degree than some of the new products. Another service company introduced a high pH borate system for cross-linking guar. Other service companies selected similar materials or systems utilizing chromium, aluminum, or other metal additives for crosslinking.

Viscosity is enhanced by tying together the low molecular weight strands, effectively yielding higher molecular weight and a rigid structure. These water-based, cross-linked fluids were developed in response to an oil based fracturing fluid developed by a major oil company during that period. This oil-based fluid utilized very high-viscosity emulsified oil, resembling heated road tar. This viscous fluid was pumped down tubing inside a physically developed water ring. This process was very successful in placing large proppants at high concentrations at fairly low-pumping pressures.

The obvious problems with this system were the need for diluting hydrocarbons, quality control, and preparation of the fluid on the surface. Several hundreds of these treatments were pumped. However, the driving force for the development of cross-linked fluids resulted from cleanup operations following such a treatment. Whole fracturing crews quit after breaking down their equipment and finding the solidified tar left after this type of treatment.

A salesman's dream

The first cross-linked fracturing fluids were a salesman's dream. One of the inherent problems in selling stimulation fluid is that you have to describe and sell the product without something to see or touch. The first sales approach involved carrying samples of the liver-like material, placing the sample on a customer's desk, and letting him touch the product.

Many times in technical presentations or in sales calls, customers' eyes lit up as they played with the material and envisioned its ability. Interestingly enough, this material is also marketed as the toy called "Slime." Anyone could see that the cross-linked gel could carry large concentrations of sand. It seemed to be the panacea for all of our problems. It was a pumpable fluid, it could carry large amounts of sand; and, with proper addition of degrading materials, it could be removed from the formation. However, in the early days, there were problems with pumpability.

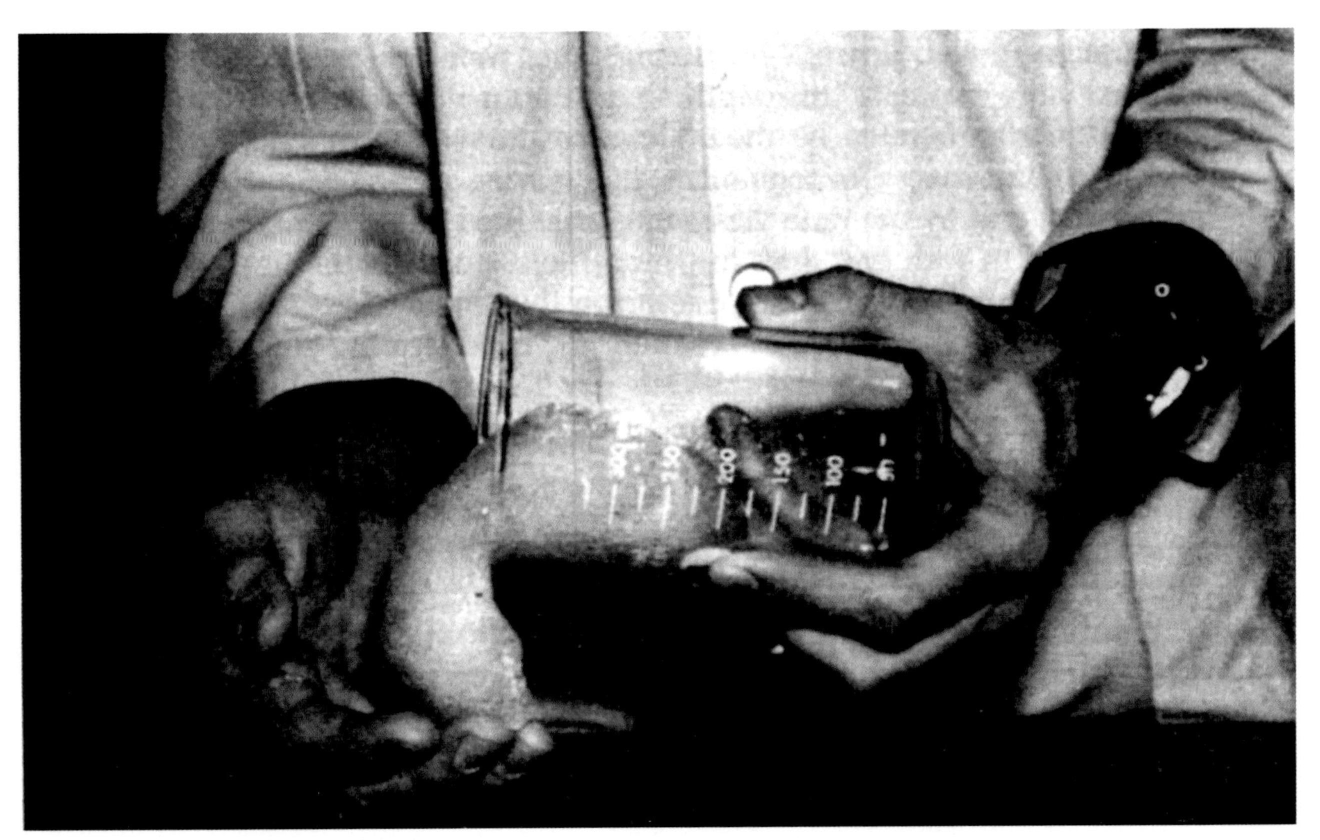

Figure 6-2 Rigid structure of a cross-linked gel. Cross-linked gel shown is a cross-linked hydroxypropylguar using zirconium.

"Teething" problems with early cross-linked gels

The first cross-linked fluids utilized concentrations of guar gum approaching or exceeding 1% (i.e., 80 lbs./1,000 gal). An antimony or a borate cross-linked gel with this high concentration would become so rigid that if pumping ceased you could not restart pumping. Often such problems resulted from inappropriate concentrations of buffers and/or cross-linker loadings. The nature of these products is, however, that under relatively high shear, the cross-linking is negated, resulting in something akin to the base gel. Under very low-shear conditions, these gels approach rigidity. These properties can cause problems, particularly when pumping at low rates or in large casings.

One particularly interesting job took place in Louisiana. A service company was fracing a well down 7 in. casing. This well had been much deeper and had utilized large casing to this depth. Large casing normally allows low-treating pressures at moderate rates. As the 80 lb. cross-linked gel entered the casing, pressure started to increase. Consequently, the service company personnel lowered the pump rate; and by the time the casing was filled, they were unable to pump. They were forced to wait two days for the gel to degrade so the fluid could be pumped out. This gel out of early day fracturing fluids has been duplicated today, while using high-temperature borate, which just reiterates my comment that what goes around comes around. Since the industry has moved back to borate as a fracturing fluid of choice, some very deep wells in Oklahoma have been fracture-treated with borate. On several occasions, the service companies and the operating companies have attempted to pump delayed borate systems with concentrations of borate equaling or exceeding 40 lbs./1,000 gals. Due to the inability to delay these borate fluids properly, and with large casing size, all the service companies were able to accomplish was filling the casing with cross-linked borate and shutting down. We have found that by limiting borate concentrations to 35 lbs., the viscosity achieved (although certainly acceptable to fracture treat wells to temperatures to 260° and below), is not a high enough viscosity to achieve the gel-out problems that are seen with 40 lb. gels and higher. There is almost an exponential increase in viscosity of the borate at concentrations above 35 lbs. per 1,000 gallons.

With a B-2 bob at 37 reciprocal seconds, the apparent viscosity of a 30 lb. cross-linked gel is in the 1,000 cps range, for a 35 lb. gel you approach 4,000 cps and for a 40 lb. gel the viscometer is pegged with viscosities well over 10,000 cps. As a matter of course, unless we have a completely delayed borate, and some of the service companies do not have these presently available, we do not ever pump borate treatments with polymer concentrations exceeding 35 lbs. Conversely, along that line, we typically don't go much below 35 lb. gel systems because of problems with being very near the critical polymer concentration. We have found when we get much below 30 lbs./1,000 with conventional borates that we do not have a stable gel in the presence of any breaker. Therefore, as a matter of course for virtually all fracturing treatments, we run only 30 and 35 lb. gels. As a recommendation to operators who do not have personnel to do intense quality control, you would be well-advised not to attempt to run concentrations much below 35 lbs. per 1,000 gallons.

I 'have been on many treatments where pressure was increasing due to low shear rate, and we increased the pump rate, reducing the friction pressure from the high-concentration cross-linked gels. Today, this problem does not normally occur, as we seldom exceed 50 to 60 lbs. per 1,000 gals with delayed metallic cross-linkers; and as previously stated, we stay below 40 lbs. for borates. We are also using cross-linkers that do not activate as rapidly as earlier products. Stopping and starting with 40-, 50-, and 60-lb. gels is relatively easy and causes few problems.

One U. S. field in which 30 to 40 wells were treated had no available pipeline. The wells were simply treated, flowed back for a short time, and shut-in. When a pipeline to these wells became available, it was determined that productivity had not been enhanced but had, in fact, been decreased. The decrease in productivity of all these wells was not due to water sensitivity, water inhibitor, or swelling clays. Decrease in production was due to the pumping in of large concentrations of polymer without proper breaker additions. We have come a long way in being able to pilot tests and use intense quality control, which will be discussed in a later chapter.

Rampant "breaker" problems with early fluids

In the late 1960s and early 1970s, we learned much about properly degrading concentrations of polymers that were pumped into wells. Service companies using high pH gel systems suffered because no means were available to degrade the polymer at temperatures lower than 120°F.

Typical low-temperature breakers in those days were enzymes. Conventional enzyme breakers do not function at pH levels much higher than eight. Many jobs were conducted with high pH borate gels over flushed with acid. High concentrations of oxidizer breakers were run, but tremendous problems resulted from the use of such high pH systems. In addition, contacting the gel for degradation was extremely difficult because the gel itself works as a good diverting agent. These early problems have been solved by the development of catalyzed oxidizer-breaker systems. Catalyzed oxidizer breakers combined with the delayed and encapsulated breakers allows good control for breaking of both low and high pH fracturing fluids.

For many of the same reasons, there were also problems with the low pH systems. You had to be just as certain that the enzyme breakers utilized at low temperatures got into the fluid properly and that nothing occurred in the formation to change the pH of the fluid itself. Enzymes and oxidizers are not very discriminating in what they attack! They can attack the polymeric components of fluid loss additives. Many times concentrations of breakers had to be run much higher due to obvious changes in the pH when the fracturing fluids were commingled with formation fluids. In some cases, large-sized, slow-dissolving, flake buffers were run with the fracturing fluids to keep the pH in the proper range.

Very high-concentration gels are unforgiving when oxidizers and enzymes are not properly added. An ongoing quality control problem involves the contamination of these solutions as they go into a well. We will discuss the mode of addition of breakers in Chapter 7. The recent developments of the aforementioned encapsulated breakers, both oxidizers and enzymes, have allowed us a great deal of latitude in being able to put excess breaker in while allowing for stable fracturing fluids during the job.

Sweet wells turned sour by guar

High-polymer concentrations make excellent agar plates for bacteria. Many wells (and whole fields in areas of the United States), were turned sour because of the large polymer concentrations pumped into them without biocides present. It is always our recommendation that a proper biocide be pumped on every fracture treatment. This biocide is present not only for protection of the gel from degradation in batch mix operations, but also for protection of the formation. Biocide supplied by the service company should be put in the frac tank prior to filling with water, even if it is on a continuous-mix treatment. This allows for a high concentration of the biocide to come in contact and allow for a quick kill of bacteria. Specific biocides will be discussed more in a later chapter. It should be noted that bacteria are not really the problem on the surface in degradation of gel; it is the enzymes that are secreted by bacteria. That is why it is so important to put biocide in the tanks prior to or just as the water is being added to the fracturing tank. If bacteria grow and enzymes are secreted, killing of the bacteria at a later time will not negate breakdown of the gel, even on continuous-mix operations.

High-success ratio through perseverance

Even with all the previously mentioned problems, there was a lot of success with high-concentration guar cross-linked fluids. This success resulted from oil company personnel and service company representatives performing a tremendous amount of quality control. The quality control ensured that breakers were properly added or that fluids were applied to formations that had adequate bottom hole pressure or bottom hole temperature to completely degrade the polymers for removal from the well. Although pumpable, these systems exhibited high-friction pressures, and their temperature stability was not really applicable to high-temperature formations.

It has always been my basic feeling that somehow oilfield people and research people are relatively slow to pick up on things that appear to be obvious. I have always felt that borate fluids had little or no temperature stability, particularly when one approached 200°F. In approximately 1990, BJ Services went to work to develop a high-temperature borate fluid. Most of us would have stated that they didn't know any better. What they did was effectively add higher concentrations of borate and adjust the pH of the fluids higher than had been done before. What they found was a fracturing fluid that was stable to temperatures approaching 300°F. We had previously felt that due to the hydrogen bonding nature of the borate cross-link system that it would have no stability due to thermal agitation of the polymer at these temperatures. These assumptions are correct if the borate concentration remains low and pH is dropped due to insufficient buffer. Since that time, Halliburton, Dowell, Western and other service companies have done a tremendous amount of work and shown that borate fluids are very stable systems and can be used for fracturing at higher temperatures. In fact, one of the major considerations in borate fluids is that they need to have internal breakers, even when temperatures approach 300°F. With some of the low pH metallic cross-link fluids, at temperatures above 200°, there is a hydrolytic mode of degradation. This mode obviously does not function for the high pH borates.

The low pH antimony systems were very limited in their stability due to the quite low pH which allowed hydrolytic degradation and the instability of that metallic cross-link under these conditions. Today, we now have low pH fracturing fluids in the range of 4-7 that will stand up to temperatures approaching 275°F. These typically are either HPG or more dominantly CMHPG fluids cross-linked with either titanates or zirconium, with zirconium being the dominant cross-linker. At 200-250°F these fluids are relatively fail-safe from the standpoint of having sufficient stability to maintain viscosity for job time, yet they are self-destructive because of the presence of the hydrogen ion.

Early development in temperature-stable fluids

In the late 1960s, Shell Oil met with Halliburton Services Research and indicated they required a temperature-stable fracturing fluid, one which would be functional at temperatures approaching and exceeding 300°F. Prior to this, fracturing treatments in deep horizons had been done utilizing water with friction reducer and low-proppant concentrations. The initial fracturing fluid with high temperature stability was a combination of a primary gelling agent and a secondary gelling agent that would viscosify only after entering the formation. Early fluids contained a hydroxyethylcellulose (HEC) polymer base gel and a secondary HEC gel treated with glyoxal. Glyoxal is an aldehyde that prevents hydration of the polymer until temperature activation occurs in the formation. The concept was to have a base gel system that would carry the sand through the perforations; when the temperature of the formation began to thin the initial base gel, a secondary gelling agent would start to viscosifier.

This system allowed us to have any required viscosity on the surface and any viscosity downhole. The major disadvantage was its high cost. Though still available, the use of this product has almost ceased since the service companies have developed high-temperature, stable cross-linked fracturing fluids good to 350°F.

Improved fluid systems

In Table 6-3 through Table 6-7, the standard fracturing fluids supplied by the service companies are shown. In my opinion, the majority of fracturing fluids in the future will be guar with borate and/or CMHPG, cross-linked with various ligands of zirconium, either for low pH systems, self-degrading somewhat, and/or high pH, very temperature stable fracturing fluids. If we do have a requirement to have a fracturing fluid with temperature stability beyond 350°F, we will probably have to utilize synthetic polymers. Although there are still large numbers of treatments conducted with hydroxypropylguar and titanium, as well as some treatments with borate utilizing hydroxypropylguar, these are rapidly decreasing. Borate and guar have excellent leak-off control and excellent temperature stability. CMHPG is a low pH, delayed cross-linked fracturing fluid that is very applicable in tight formations, as well as the same polymer being used for high-temperature stable fluids for long pump times above 300°F.

Table 6-3 Service Company Frac Fluids

BJ Services		
Medallion Frac	CMHPG	Zr and Ti
Terra Frac RXLII	HPG	Zr and Ti
Krystal Frac XL	CMHEC	AL
Spartan Frac	Guar	Borate
Spartan Frac HT	Guar	Zr
Spectra Frac G	Guar	Borate
Spectra Frac H	HPG	Borate
Super Allo Frac	Phosphate Ester	Oil
Polyemulsion	Guar/HPG	

Table 6-4 Service Company Frac Fluids

Halliburton		
MY-T-Gel	Guar	Ti
Versagel	HPG	Ti
Thermogel	CMHPG	Zr
KleenFrac	CMHPG	Zr
Purgel	CMHPG	Zr
MY-T-Oil	Aluminum Phosphate Ester	Oil
Boragel (G) (H)	Guar/HPG	Borate
Hybor (G) (H)	Guar/HPG	Borate
Superemulsifrac	Guar/HPG	Polyemulsion

Table 6-5 Service Company Frac Fluids

The Western Company		
Saturn I	Guar	Ti
Saturn II	HPG	Ti
Saturn III	CMHPG	Zr
Viking I	Guar	Zr
Viking II	HPG	Zr
Apollo I	Guar	Oil
Apollo II	HPG	Borate
Super K	Guar/HPG	Borate
Maxi 0-74	Aluminum Phosphate Ester	Polyemulsion

Table 6-6 Service Company Frac Fluids

Acid Engineering		
CXB	Guar/HPG	Borate
CXD	Derivatized Guar	Zi
K-Mul	Guar/HPG	Polyemulsion
Sure Gel	Derivatized Guar	AL

Table 6-7 Service Company Frac Fluids

Dowen· Schlumberger		
YF-100	Guar	Borate
YF-200	HPG	Borate
YF-300	Guar	Ti
YF-400	HPG	Ti
YF-500	Guar	Zi
YF-600	HPG	Zi
YF-GO	Aluminum Phosphate Ester	Oil
Super Sand Frac K-1	Guar/HPG	Polyemulsion

Complexity breeds confusion

Perhaps the major problem in the implementation and design of hydraulic fracturing treatments is the persistent incidence of pumping fracturing fluids without internal breakers. Excellent research work indicates that unless you have a bottom hole temperature in excess of 280°F, fracturing fluids will not degrade. Oryx actually cored through a vertical fracture with a horizontal well and retrieved from an Austin Chalk formation a part of the vertical fracture where the fracture was stuck together with polymer. The fracture treatment had been conducted 9 years prior to the core operation, and it was still plugged with polymer. The bottom hole temperature of that well exceeded 220°F. It is, therefore, not difficult to believe that most of the formations in west Texas and other areas that have lower bottom hole temperatures contain a great deal of very stable fracturing gel. Perhaps most of this problem occurs in formations with somewhat low-temperature gradients, where at 10,000 ft bottom hole temperatures may be less than 170°F. Because of the depth, well-meaning service companies pump temperature-stable fracturing fluids without breaker. Also, there is always the ongoing fear of screen out. To negate screen-out problems, the breaker is left out. This leads to an untold amount of unbroken gel plugging up formations and causing many people to believe that the formations were indeed water insensitive. They were not water sensitive because many formations such as the Frontier and Wyoming are now very successfully being treated with simple delayed cross-linked, water-based fracturing fluids. I believe that most of the foam fracturing treatments succeeded where the gels failed because of lower polymer concentrations and the ability to clean up, where the more stable gels failed because the improper breaker was placed.

It should be noted that formations with bottom hole temperatures of 170°F and above cannot typically be treated with conventional breakers. If you run enough conventional breaker, then the stability of the polymer of the cross-link gel is quite short. A good example for formations such as this is the Canyon sand in southwest Texas. I believe that the majority of the wells fractured are still plugged today, since many. people "intuitively believed" that the gel would certainly degrade with time. That is not the case. We are having tremendous success in hydraulically fracturing the Canyon sand today by utilizing encapsulated breaker. Encapsulated breakers allow us to place the sand and do the job with stability and still have excess breaker to be absolutely sure that complete breakdown occurs. This, combined with forced closure to be discussed later, has greatly enhanced productive. capacity of these reservoirs, and it indicates a great opportunity for independents and majors to go back into areas where previous stimulation attempts have been fruitless.

Quality control on a stimulation treatment frequently begins upon acceptance of the proposal. The oil company representative must familiarize himself with the service companies' products and how they are utilized. A service company that will not tell you what is being used on your well is a service company not to use. There are few (if any) secrets in our industry. Most service companies would not specifically name their front-line cross-linker; but due to changes that have occurred over the past few years, they will tell you generically the type of crosslinker, the base gel, describe the buffering agents, and help you to understand and how to degrade and control the stability of the fracturing fluid.

Advances in fracturing technology

Table 6-8 lists 12 advances in fracturing technology since the first fracturing treatment in 1949. Though all improvements are not included, these are the most significant ones.

"Foam" -a new twist

One of the most significant developments in the 1970s was the advent of *foam fracturing.* Though early research basically proved that foam fracturing would never succeed, some farsighted individuals attempted the initial foam-fracturing treatment. For those of us accustomed to fracturing with either viscous gels or the liver-textured cross-linked fluids, it was difficult to believe that the stuff we used on our faces could be a successful fracturing fluid.

Negative vibes toward nitrogen foam

Foam fracturing suffered a great deal of negativism in its early days. The most apparent reason for this was its detrimental effect on service companies' profits. The single source nitrogen supplier on a typical foam frac treatment received 60% of the gross revenue from the entire job. Needless to say, many major service companies were not pushy about putting foam fracturing in front of their customers.

Though there were many early failures, the usefulness and advantages of foam won in the end. Rather than fighting it, the service companies decided to join in. Today, virtually every service company in the industry has nitrogen pumping equipment.

Table 6-8 Advances in Fracturing Technology

1. First treatment gelled napalm
2. Viscous lease oil
3. Gelled oils, soaps (fatty acid and caustic)
4. Water as a fracturing fluid
5. Viscosified water starch and then guar
6. Improved gelled oil (aluminum phosphate esters)
7. Cross-linked guars
8. Secondary gel system
9. Hydroxypropylguar and other derivatives
10. Foam fracturing with nitrogen and carbon dioxide
11. Improved cross-linkers (temperature stable and delayed systems)
12. Delayed release and low-temperature catalyzed breakers

Advantages of foam

Foam fracturing allows the customer to place high concentrations of proppant and utilize only about 25% of the fluid placed with a conventional linear cross-linked gel. Most importantly, the fluid itself contains a large amount of energy through the entrained nitrogen or $C0_2$. Advances in foam fracturing in the last few years allow highproppant concentrations to be placed downhole.

The difficulty with placing large concentrations of proppant is that all the proppant has to be added to the liquid phase. Five pounds downhole on a 75-quality foam requires 20 lb./gal at the blender tub. Service companies have developed various techniques to place high-sand concentrations downhole. The primary ones involved special valves and blending equipment. It is not uncommon to have foam-fracturing treatments where proppant concentrations equal or exceed 12 ppg. This is accomplished with the use of a technique called "constant internal phase." Under this procedure, the assumption is that the sand becomes part of the internal phase (i.e. the gas phase). You just decrease the amount of gas proportionally to the volume of sand. By doing so, the foam quality is reduced, but the viscosifying properties of the sand allows you to conduct an adequate stimulation treatment properly even with a foam fracturing fluid. An example of the constant internal phase treatment is shown in Table 6-9.

Carbon dioxide foam

In the early days of foam fracturing, the only gaseous medium used was nitrogen. Today a large portion of foam-fracturing treatments. use carbon dioxide as the gas medium. Carbon dioxide is pumped as a liquid and converts to gas at critical temperature downhole. This critical temperature is 88°F; therefore, on most treatments, the medium where the proppant is carried and pad (etc.) is an emulsion of $C0_2$ liquid and water. Table 6-10 shows the pros and cons for C02 and nitrogen as mediums for creating energy to help produce wells.

Over the past few years, I have developed some very strong opinions about the use of foam-fracturing fluids. They are as follows:

1. Foam should only be used as a medium for moderately under pressured wells. If 70 quality foams (based upon calculation) are insufficient to unload the well, then you are probably wasting your money pumping a foam into such a well. The use of foam in normal and geopressured wells is based upon more "intuitive reasoning" whereby water was seen as a very serious problem in fracturing. Today, we understand that except for the case of extremely under pressured wells where wells can water out that the presence or lack of water has very little to do with the capacity of the reservoir to produce as long as the viscosifying agents have been removed.

Table 6-9 Constant Interval Phase-70 Quality Cross-linked Foam

Foam Volume Gals	Liquid Volume Gals	Proppant			Slurry Volume		Rate			
		Foam PPG	Liquid PPG	Total PPG	Foam Gals	Blended Gals	CO_2 BPM	Liquid BPM	Sand BPM	Total BPM
40,000	12,000	0.0	0.0	0.0	40,000	12,000	21.0	9.0	0.0	30
2,500	450	2.0	5.3	5,000	2,728	1,178	18.6	9.0	2.4	30
4,000	1,867	4.0	8.6	16,000	4,721	2,597	16.0	9.0	5.0	30
6,000	2,760	6.0	13.04	36,000	7,642	4,402	13.8	9.0	7.2	30
8,000	4,853	8.0	13.20	64,000	10,918	7,771	11.8	9.0	9.2	30
10,000	6,733	10.0	14.85	100,000	14,560	11,293	9.8	9.0	11.2	30
15,00	10,901	12.0	16.5	180,000	23,208	19,109	8.2	9.0	12.8	30
1,800	540	0.0	0.0	0.0	1,800	540	21.0[2]	9.0	0.0	30

Table 6-10 Nitrogen vs. Carbon Dioxide

Feature	N_2	CO_2
Relative Density	Low	High
Solubility	Low	High
Surface Tension	No Effect	Lower
Compatibility	Good	Limited
Reactivity	Inert	Acidic

2. We recommend strongly that cross-link foam be utilized to allow for better proppant transport and fluid loss control. Whereas it was previously thought that foam was an excellent fluid-loss control medium, we now know that foam in many cases has relatively high leak-off, and you must include bridging or plastering fluid loss agents with the foam in order to be able to place high concentrations due to relatively high leak-off.
3. On selecting C02 or nitrogen as the carrying medium, there are many people in the industry who feel that due to the solubility of C02 and its low pH effects that it has many advantages. In fact, the inertness (lf nitrogen makes it a choice in. many situations, particularly when dealing with higher temperature where the carbonic acid acts as a breaker for the fracturing fluids. I make the decision to

[2] at BHTP conditions
If practical, prefer to flush with no C

use the foam medium based on cost of the product in the general area and the capability of the service company in "the area to pump that particular system. $C0_2$ can be a very dangerous fluid to pump if the service company is not comfortable with pumping this material under high-pressure conditions.

It should be noted that there have been various attempts to come up with viscosifiers for carbon dioxide by itself, and some operators have even used straight nitrogen as a fracturing fluid. All of these attempts have ended in failure. The only successful stimulation ever achieved with this type of fluid is removing near wellbore skin damage. Huge volumes of nitrogen or C02 are indeed expensive materials to use for skin damage removal. To achieve production increase with undamaged reservoirs, you must either etch the fracture for carbonate reservoirs with an acid or prop it. To achieve this, you need to use an excellent proppant transport fluid that is highly efficient. We will now discuss quality control of each particular fluid individually, then go into more detail on the technical specifics of various fracturing fluids, gels, etc.

CHAPTER 6

MONITORING FRACTURING FLUIDS (2022)

Not unlike the previous chapter the second edition concentrated on not only monitoring conventional frac fluids but also reviewed the history of fluid development. Most of you know that slick water fracturing is not new. One of my early tasks with Halliburton was to evaluate friction reducers from various suppliers. The FR materials available in the later 60's were fairly comparable in performance to what we have today other than available only in dry form. Improvements have been made with copolymers of acrylamide and better forms of cationic acrylamides have been developed for high salinity water. Dry acrylamides were widely used in the 60's and 70's, but were problematic in mixing, and typically created lots of lumped polymer causing service companies to run 5 pounds per 1000 when only 2 was required.

As an interesting aside, crosslinked acrylamide was pumped in the early days of crosslink gel by some service companies and crosslink systems were sent to the field but failed miserably due to lack of breakers even for moderate temperatures. During the early 70's Dowell developed a high temperature system to compete with Halliburton's Hy-Frac which was termed YF-P which was stable at temperatures approaching 400F. The system consisted of 100 pounds of dry acrylamide and was indeed very temperature stable. The system only lasted for a very short time as it was impossible to mix in the field. When the high concentration powder was added to blenders, gigantic lumps of acrylamide were created. This was a very good case of products that can be mixed in laboratories but not in the field. Dowell later came out with a product virtually identical to Halliburton's Hy-Frac termed YF-HC.

Not unlike guar suspensions with hydrocarbons, latex suspensions of acrylamide allowed for better mixing. Some of the early day suspensions required non-emulsifiers to allow for the quick hydration required in continuous operations. As time progressed through the late 90's quick reacting acrylamide suspensions became readily available.

As has occurred so much in our industry, we now, in many areas, have gone back to dry powder with enhanced mixing capabilities. Acrylamide suspensions are still available and are being used by several smaller pumping companies.

The other classifications of chemicals used in slick water are virtually the same as for conventional systems. Biocides, scale inhibitors, surfactants, and acrylamide breakers pretty much account for all the additives presently pumped. There are still some hangers on doing what I term imitation hybrid treatments. There has been a lot of developments in scale and corrosion inhibitors as well as biocides.

Monitoring technology has progressed a great deal since 1994 with service companies moving away from stroke counters and in some cases just visual monitoring. Multiple images are shown at the end of the chapter, offers vintage monitoring equipment and up to date digital readouts for your viewing. As was the case in 1994, intermittent monitoring of storage tanks for chemicals should be overseen by unbiased on-site supervisors. There will come a day where the use of monitoring cameras and improved volume control will minimize the need for human oversight. We are a significant time away from that time and for the present, robots will not replace us all. The major problem our industry faces in the massive amount of conventional equipment that will have to be used during the restart (hopefully) of full out stimulation of shales and ultralow permeability conventional rock. Presently there is simply not enough of the next generation turbo powered electric equipment to reestablish our lead in worldwide oil production.

The 1994 version of this chapter was basically a history of frac fluids. Very little has changed in the use of type and use of fracturing fluids until the explosion that occurred relating to shale stimulation. Slick water treatments albeit smaller and poorly designed have been around for as long as I have been involved in stimulation. The San Juan basin truly never got far away from slick water treatments in the Mesa Verde formation. I tried very hard to convince a major operator that limited entry stable crosslink systems would outperform the dominate slick water treatments in the area. I failed for the same reason that slick water is taking over most of the stimulation in the domestic US as well as worldwide. There was an independent company operating in the Texas Panhandle, Serfco, that was famous for their 3 % gelled acid fracs with low sand concentrations. In hindsight, since there was no real viscosity in the acid, these were just smaller volume slick water treatments.

In creating this book, I have had a great deal of time to remember many clues about what was occurring with poorly executed fracs, linear gel fluids, foam fracs etc. What was ultimately wrong with our inability to understand why these worked, although marginally better than conventional crosslinked gels due to small size and large sand, was due to a basic misunderstanding of what sort of fracture geometry was generated.

I am convinced that low viscosity fracs by their very nature create complex fracture systems translating into high surface area and the larger the treatment the greater potential for creation of long-lasting production in ultralow permeability rock. There is a very real point in which the economics of huge treatments fail. Based on NPV analysis, there is a real economic limit and jobs such as "Fracageddon", pumped by a large slick water operator, produced well but probably will never pay out in my grandchildren's lifetime. The other counter intuitive part of successful slick water fracturing is the inclusion of very small, 100 mesh and sometimes smaller, sand which was portrayed for years in conventional fracturing as a plugging agent. The success of the small sand further highlights the facts relating to our misunderstanding of frac geometry. It is my belief that indeed the small sand is a proppant holding open fractures rather than the conventional packed proppant pack perceived for conventional single planar fractures. Additionally, the use of low strength local sand is in complete disagreement with conventional closure packed proppant fractures for the perceived single planar fracture based on 60 + years of modeling. As we will discuss in the short proppant chapter, we do not believe there is any significant closure on sand from nanodarcy reservoirs allowing for use of much lower strength proppant. Sand has been used in ultra-tight reservoirs at conventionally calculated closure pressures of more than 12,000 psi without any indication of crushing for more than 20 years. There was a giant battle more than 25 years ago relating to ceramics versus white sand in Rocky Mountain stimulation. As is the case in almost every situation, engineers had absolute proof that their proppant, be it sand or ceramic, was better. In reality, the curse of reservoir heterogeneity throughout Wyoming made a definitive answer on what was working impossible. This very large argument went on while we were still pumping crosslink systems in the Western Wyoming reservoirs. Today there would be no argument as small sand with slick water would be used.

In writing this chapter it saddened me to realize that there has not been any real developments in conventional frac fluids since the 1994 book was written. Conventional jobs are still conducted in moderate permeability reservoirs, but it is very difficult to find service companies who have the experience or even want to conduct the smaller conventional fracs. Although fracturing has a very bad name in Europe, there are vast numbers of reservoirs which would produce much more economically with conventional frac fluids. The entire world where hydrocarbons exist is open to hydraulic fracturing.

Image A

South Texas; bad picture, good guy.

Image B

Sound-power in use, during frac operations. No frac vans available.

Image C

Advanced, outside control of frac.

Image D

Dowell command center.

Image E

Small, early van for monitoring frac jobs.

Image F

Inside Nowsco frac van, monitoring with strip-chart recorder. Captures treating pressure, pump rate, and density.

Image G

Nowsco early frac van.

Image H

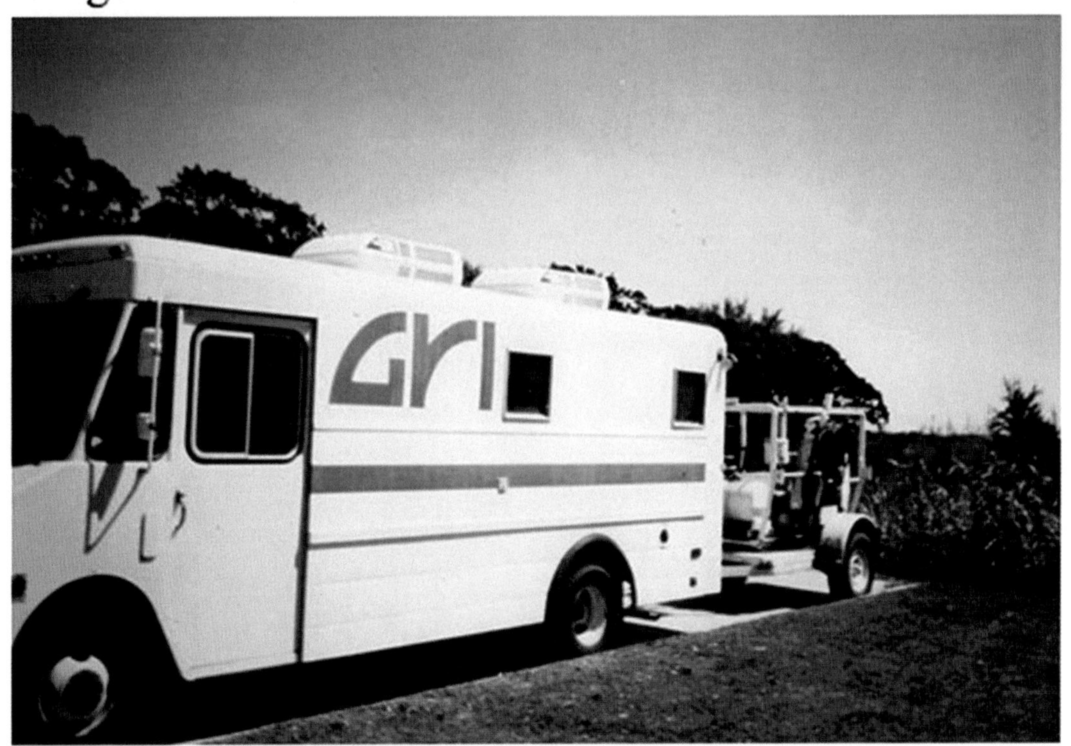

Gas Research Institute (GRI) frac quality control van, with generator trailer.

Image I

Early Dowell frac van.

Image J

Haliburton frac van, transducer cabinet.

Image K

Haliburton frac van in the Piecance Basin.

Image L

BJ Services pump operators at their panels.

Image M

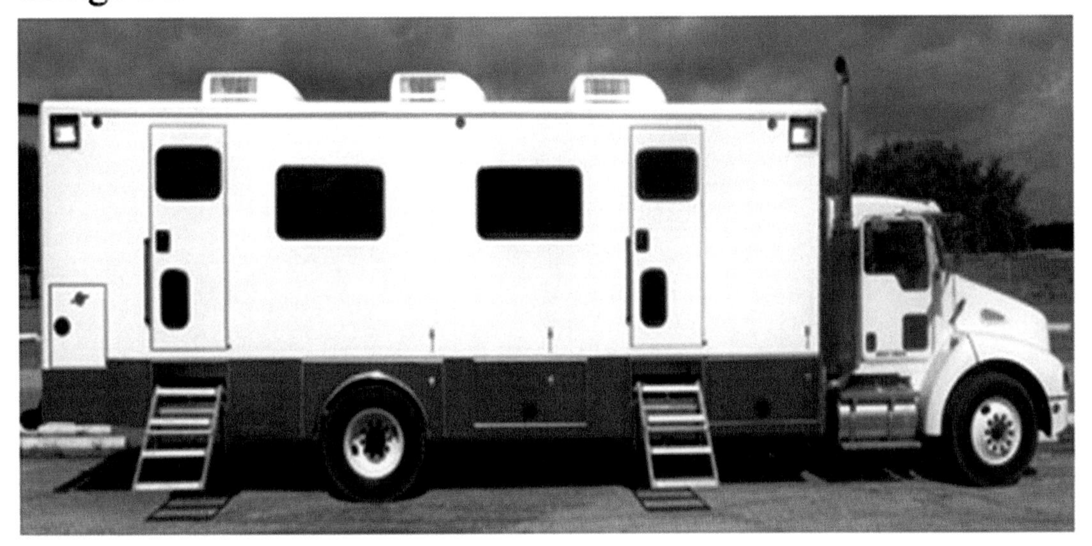

Body-load style fracturing command center; units typically divided into two halves, with one half serving as the pump operator control room and the other half dedicated for treater/engineer/customer viewing area.
Source: Stewart Stevenson

Image N

Trailer mounted command fracturing command centers provide more space and seating than earlier styles and boast on-board technology such as internet access and satellite communications.
Source: Liberty Oilfield Services

Image O

Interior view of modern style fracturing command center; larger space capacity of trailer mounted command centers provide multiple viewing screens and onboard technology that enhance customer and service company supervision during operations; Service Supervisor controlling all pumps from pump screen.
Source: AP Photo/Keith Srakocic, File

Image P

Interior view of modern style fracturing command center; service company treater/engineer station during wellsite operations.
Source: GOES

Image Q

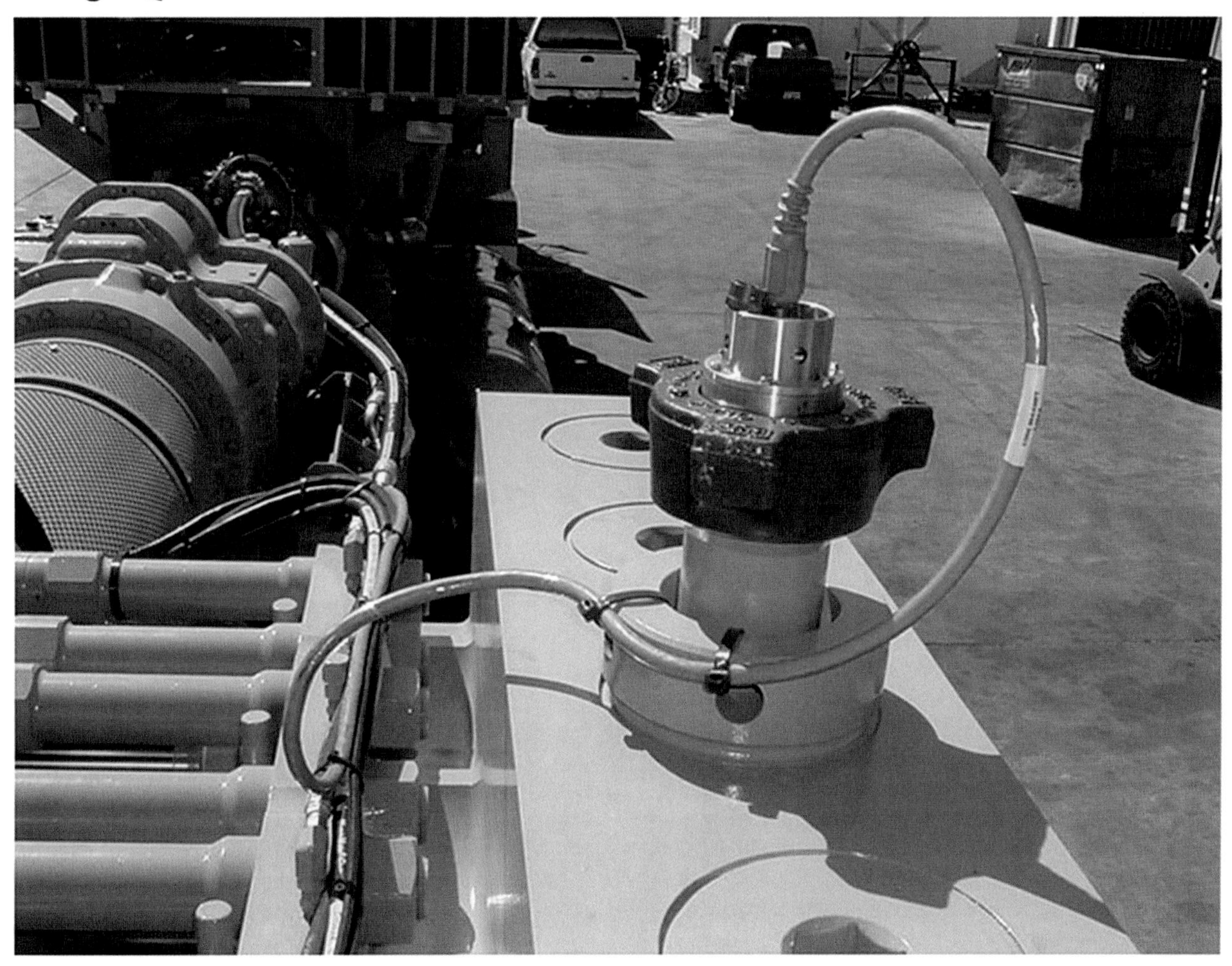

Pressure transducer rigged into fluid-end of hydraulic fracturing pump truck; pressure transducers monitor and allow for recording of pressure within the fluid-end and wellhead. Source: Industrial Diesel Manufacturing and Services, Ltd. Catalog

Image R

Illustration of strain-gauge pressure transducer and Martin-Decker gauge, as well as two different flapper-style check valves.

Image S

Modern, advanced hydraulic fracturing software utilizes on-site pressure transducers and sensor arrays to graphically represent pumping parameters during operations. This data can be transmitted offsite to customer offices, or homes, with very little delay to real-time. Advanced appliacations also provide a means to broadcast this information to phones, tablets, etc.
Sources: FracPro

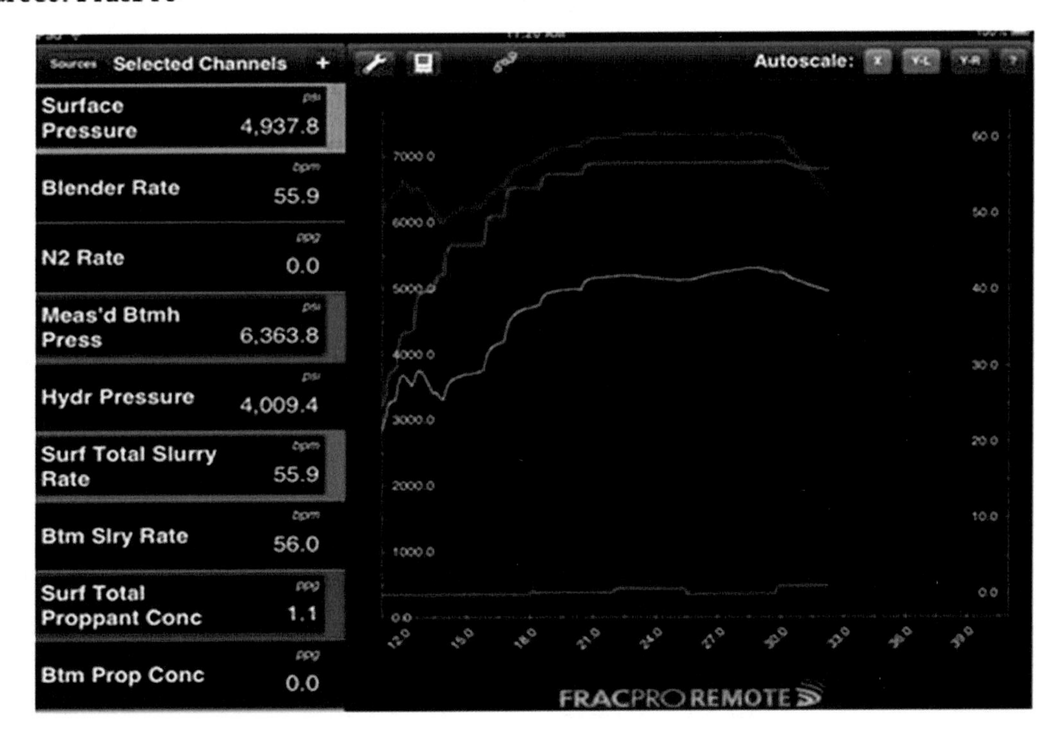

Image T

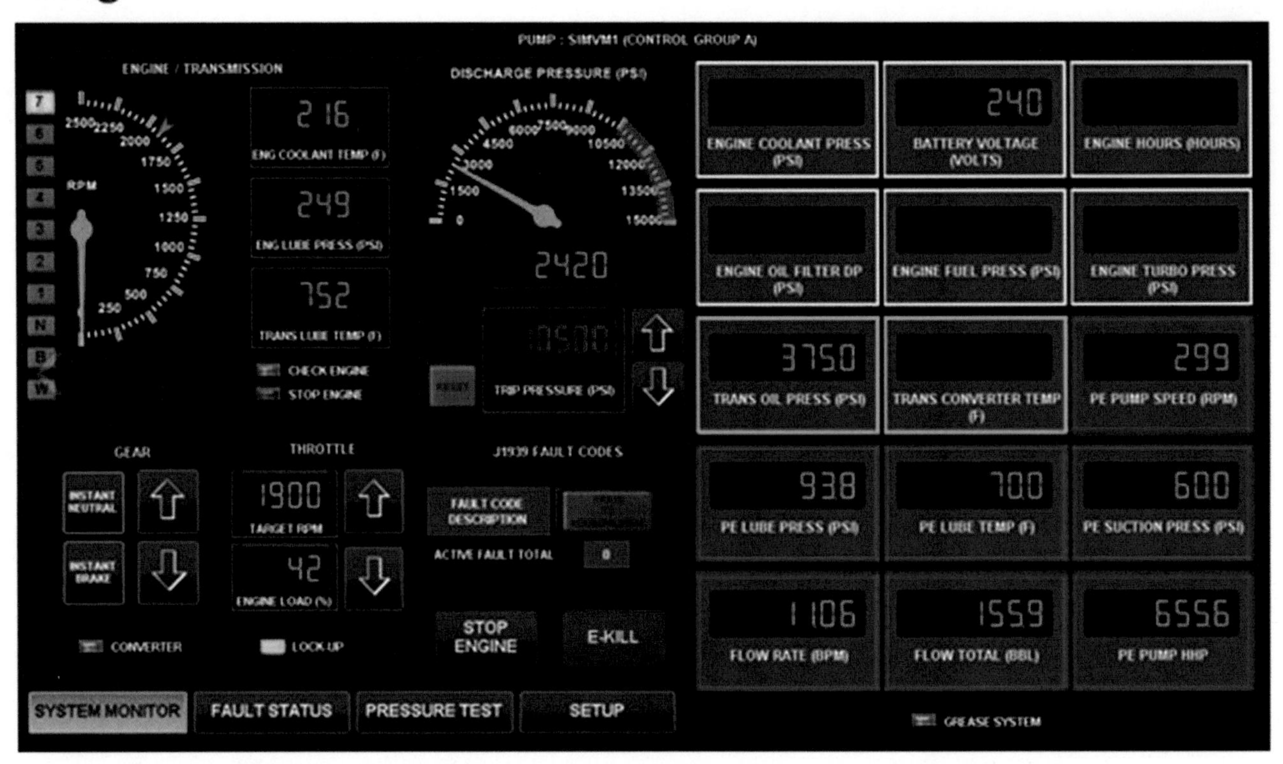

Typical user interface for digital pump control system; can toggle between pumps during operation with use of touch screen or computer mouse. Software allows for single operator to command 20+ pumps at one time.
Source: Mobile Tech Data-Lion

CHAPTER 7

QUALITY CONTROL ON WATER-BASED FRACTURING FLUIDS (1994)

An overwhelming quantity 'of fracturing treatments are successfully completed with just thickened water gels. The gelling agents used for linear gel-frac treatments are guar gum, hydroxypropylguar (HPG) gum, carboxymethylhydroxypropylguar (CMHPG) and, in some cases, cellulose or cellulose derivatives. Listed in ·Figures 7-1 through 7-5 are apparent viscosities of various base gels taken at 300 rpm on a no. 1 bob and sleeve of a Fann viscometer. These viscosity measurements can (and should) be made in the field prior to pumping downhole. These same measurements are taken for the base gels when utilizing cross-linked fluids.

Linear gel systems

One area of discussion that has always been troubling, and for many company personnel discouraging, is that various service companies report viscosities of the fluids at different shear rates. Typically, in the field the standard portable viscometers produced by either Fann, Chandler, or other supplier, with a no. 1 bob and sleeve have two settings: one for 300 rpm and one for 600 rpm. On rare occasions when electrical hookups and multispeed viscometers are available, you can obtain readings at various shear rates and calculate n' and K' values.

For comparison and quality control of fluids, I strongly suggest that the 300 rpm dial reading for Newtonian fluids be utilized to compare high-viscosity guar, hydroxyethylcellulose HPG (HEC), CMHPG or other fluids in the field. The curves of Figures 7-1 through 7-5 are good references. Another source is 300 rpm data from your service company fluids for the purpose of quality control in the field.

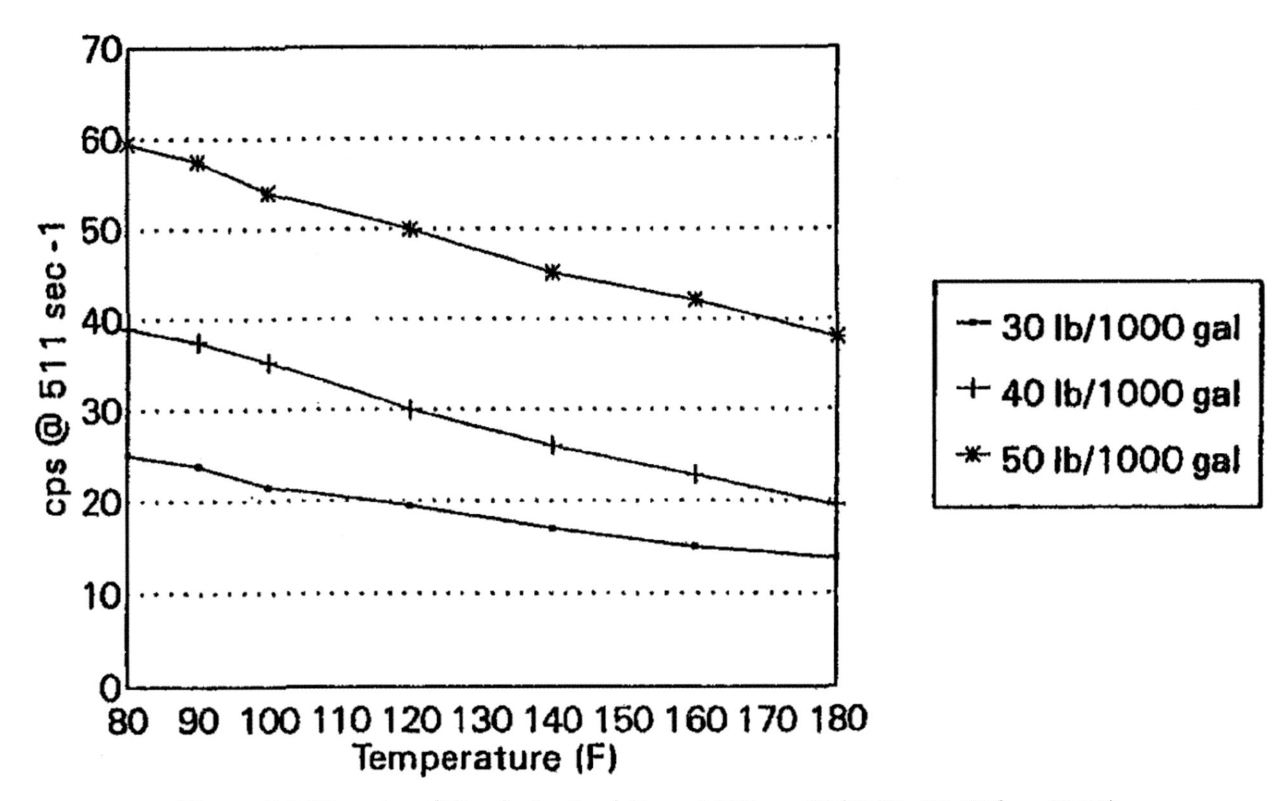

Figure 7-1 Viscosity of Nonderivatized Guar. 1/5 Fann 35 (R1B1, 2% KCl = pH 5.8)

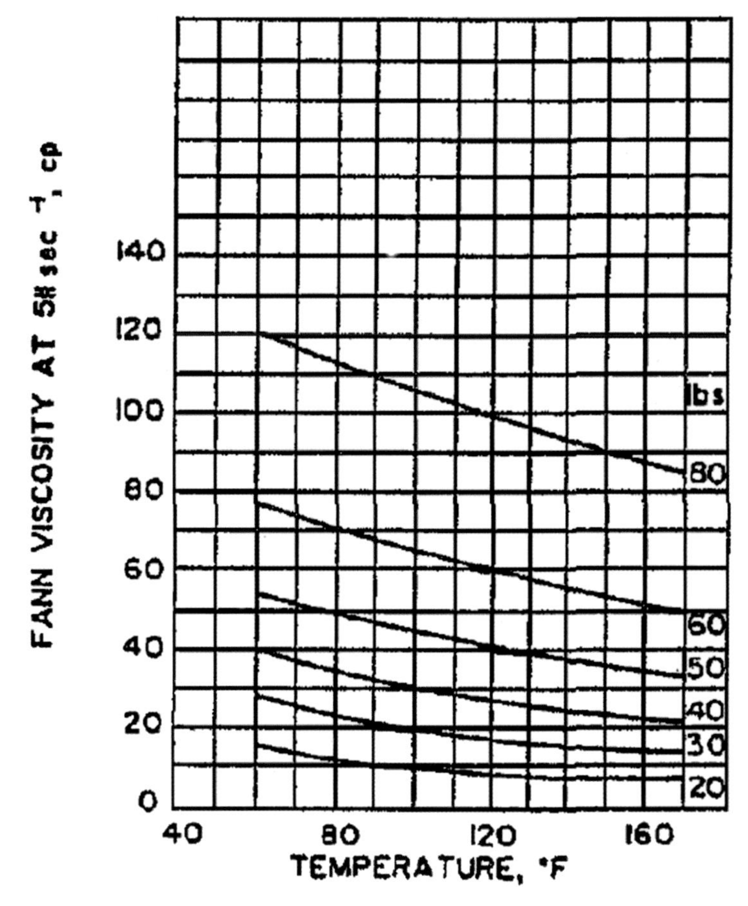

Figure 7-2 High-Viscosity Hydroxypropyguar. Fann viscosity vs. temperature for various concentrations of polymer (2% KCL).

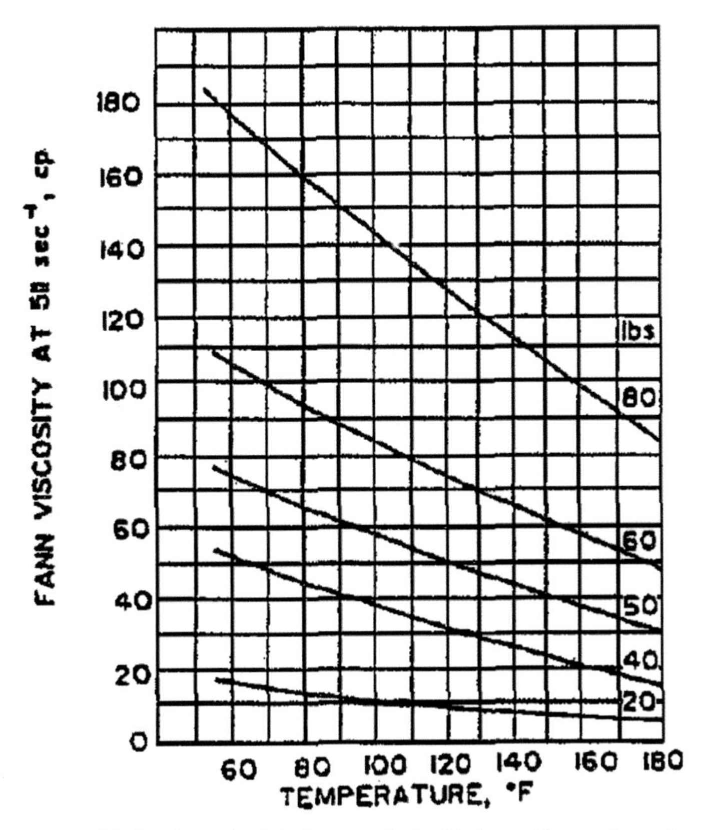

Figure 7-3 High-viscosity Hydroxyethylcellulose. Fann viscosity vs. temperature or various concentrations of polymer (2% KCL).

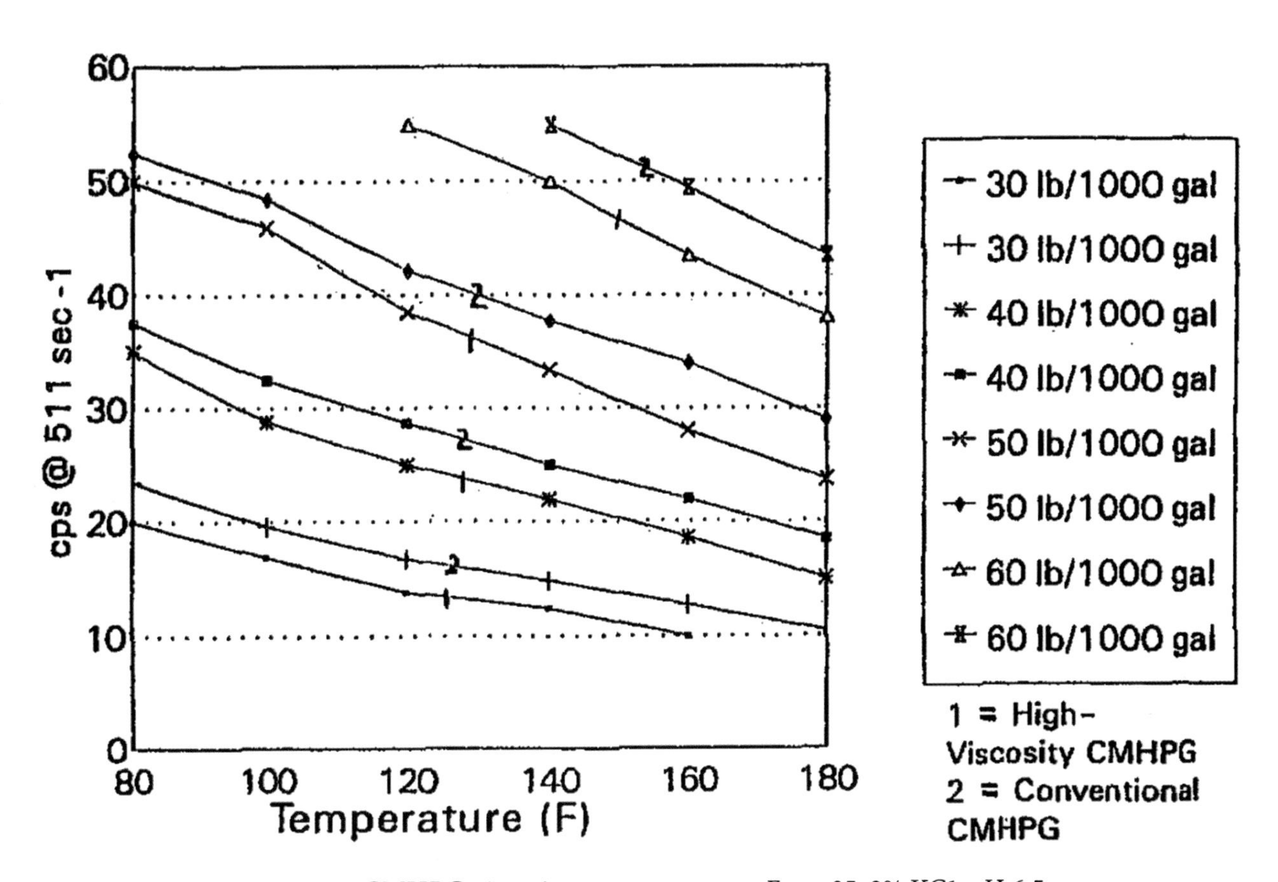

Figure 7-4 CMHPG viscosity vs. temperature. Fann 35, 2% KC1, pH 6.5.

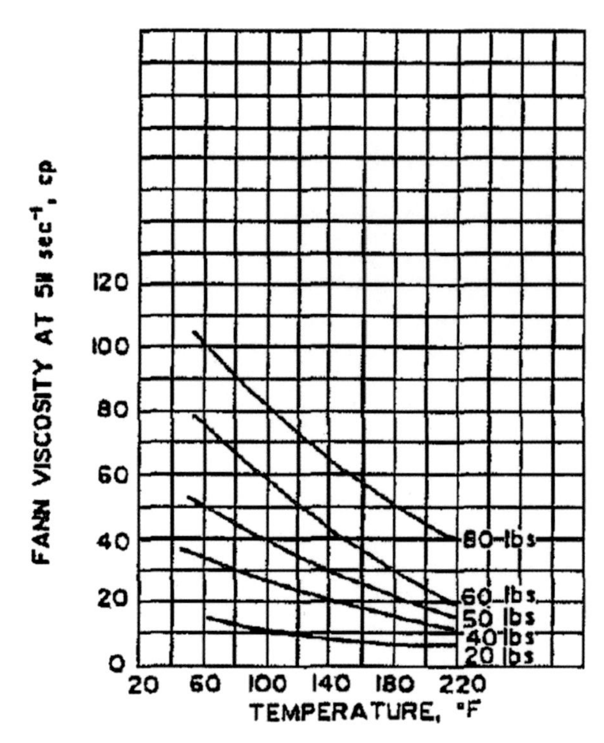

Figure 7-5 High-viscosity Carboxymethylhydroxyethylcellulose. Fann viscosity vs. temperature for various concentrations of polymer (2% KCL).

What pH is the base fluid?

Viscosity is an important measurement made on fracturing fluids in the tanks on fracturing treatments, but it is certainly not the only one. Check the pH of the fluid. This is particularly important in relatively low-temperature wells, depending on the type of breaker mechanism for the treatment. If a conventional enzyme breaker is used, then the pH of the fluid going downhole needs to be in the range of 3-8 but preferably from 4-6. This allows optimum use of the breakers so the gels can be completely degraded.

The pH and viscosity of the gels in the tanks need to be measured at two points at least: on the top of the tank and at the bottom. This ensures there is good mixing of the gel systems. If these viscosities differ, have the service company roll the tank to be sure there is uniform viscosity in the tank. Also check for the presence of lumps, which indicates poor mixing of the fluid. If there are lumps on top of the tank, your service company may or may not be at fault. There may have been quality control problems from the supplier, or the lumps may be from a previous job. If lumps are present and viscosities are within acceptable ranges, then take care that the tank fluids are not pulled down to the extent that the lumps of highly concentrated gel are pumped into the well. Lumps or clods of polymer are extremely difficult (if not impossible) to break, and they can plug pore channels in the formation and can negate flow in the proppant pack.

With the vast majority of fracturing treatments today being conducted using polymer, which is slurried and suspended in diesel, lumping or fish eyes is not nearly the problem that it was in the past. There are, however, many occasions where service companies are still mixing with powdered materials. Even with powdered materials, there is really no reason for lumping or fish-eye problems to occur in gel, since dispersible, powdered gels are readily available. The technology to mix lump-free gel with powder has been around for more than 20 years.

What temperature· is the base fluid?

Another item related to the viscosity that needs to be evaluated is the temperature of the fluid. This temperature can be cross-referenced to correlate with the viscosity charts. The fluid temperature is an important factor in the actual design of fracturing jobs. Always remember that if you have very cold fluids pumped down long tubular goods, you must consider possible tubing shrinkage. If you use a polished bore receptable, be sure you have enough seal length. If you use a retrievable packer, set down enough weight to be certain that you don't pull the packer loose or pull the tubing in two during the treatment.

If a permanent packer assembly is in place without a polished bore receptacle, then, again, it is possible to pull the tubing in two. This, of course, can result in an extremely dangerous situation. Most service companies have computer programs to evaluate potential tubing shrinkage or buckling. These problems occur because of temperature differential between the fluid and the bottom hole temperature. Computer programs should be run if there is any possibility of either shrinkage or buckling.

Are surfactants and biocides present?

A further consideration for quality control is to ensure surfactants and biocides are present in the gel system. Biocides should have been placed into the tank before it was filled. You won't have much trouble with the service company from this standpoint. The biocides are there to protect the operating company and to negate bacterial degradation of the gel on the surface.

With the increased emphasis today on the use of continuous-mix fracturing treatments, many service companies do not insist upon the presence of biocides in the fracturing fluids. We have just recently noted, however, bacteria problems for continuous-mix operations particularly in warmer weather. This has occurred because gel was not properly removed from the intermediate holding tanks supplied by the service companies. These units at many points will contain fairly large concentrations of polymer, which act as an agar plate for bacterial growth. We feel very strongly that a combination aerobic and anaerobic bacteriacide should always be placed in the frac tanks just prior to filling. This properly kills any bacteria that might be pumped into the well. It should be noted that you may be able to kill all of the bacteria and still have enzymes present in the tank from previous bacterial growth. Destruction of the gel systems is not from bacteria but from secreted enzymes.

Additionally, many times we have noted the field destruction of the fracturing fluids because of the interrelationship between bacterial growth and the secretion of enzymes and/or iron reducing bacteria that cause catastrophic degradation of linear and cross-linked gels. It is always a good practice to place bactericides in the frac tanks just prior to filling so that a high concentration of biocide is present to achieve a rapid kill of any bacteria that may be present from any previous jobs or just present in the water itself.

Over the past five years, the use of anything other than simple nonionic surfactants in hydraulic fracturing has been minimal. Although the part played by oil-wetting or nonwetting surfactants has been somewhat overplayed, it is extremely important to be sure that if serious emulsification problems exist or even potential emulsification problems that prejob emulsion testing is necessary.

To check for the presence of surfactants, mix produced oil from the well with the fracturing fluid to ascertain if nonemulsifiers will break any emulsions that are created. This may be an unfair test, since viscosity tends to stabilize emulsions; usually emulsion break tests are conducted with broken gel. When fracturing a dry-gas well or one that produces only a small amount of condensate, testing produced fluids may not be practical or possible. Ask the service company about the amounts and types of surfactants used in a well. Good questions are whether the surfactants used are water-wetting and are compatible with the formation to be treated. Reservoirs have been seriously damaged by using oil-wetting surfactants simply because the service company or operating company did not know the particular product being pumped.

Review schedule of additives

As stated earlier, quality control on the treatment starts with the design. Always have the design proposal that lists the products present on the location. Check and work with your service company to be sure that all additives are there and are included in the fluid.

Check for proper viscosifier

It is important that quality control measures be utilized by the operating companies to determine if the gelling agent on the proposal is the one used on the job. A quick test can be used to tell if hydroxypropylguar is present or if it is straight guar or a blend of guar and hydroxypropylguar. One simple test is to take some of the gelling agent on the location and add it to a 50:50 mixtures of alcohol and water. Hydroxypropylguar quickly viscosifies in low pH conditions and makes a thick gel. Guar will not viscosify the fluid at all. A mixture of the two results in part of the material precipitating out.

The previously described paragraph was an important procedure utilized when the majority of fracturing systems were hydroxypropylguar. As stated earlier, there was a time when our industry switched from the infamously dirty and residue-containing guar to hydroxypropylguar. This occurred approximately in 1973. As more and more treatments were conducted and hard times began in the 1980s, researchers and other personnel began to take a better look at the reasons for damage in hydraulic fracturing. It was discovered that the guar containing considerably more residue than the hydroxypropylguar or the marginally cleaner carboxymethylhydroxypropylguar could be a less damaging fluid because of its enhanced efficiency and lack of filtercake development. Presently, more than 60% of the fracturing fluids are guar gum, and carboxymethylhydroxypropylguar is replacing HPG as the high-temperature fracturing gel of choice. In fact, and stated earlier, the industry is moving to the time when it will have only guar and carboxymethylhydroxypropylguar or perhaps some other new product developed in the future. A carboxymethylhydroxypropylguar system allows for a great deal more diversity for having a low pH system to cross-link on the carboxymethyl side and for higher pH, higher temperature conditions, utilizing the hydroxypropyl group for cross-linking. Guar gum cross-linked with borate has been used up to bottom hole temperatures of 300°F and higher. As the service companies develop better delayed breaker systems, the use of guar throughout the entire spectrum of temperatures may become more common.

Continuous-mix systems

Since the mid-1980s, the use of continuous-mix fracturing has (sadly) become the norm rather than the exception. Service companies term these systems liquid gel concentrate, slurry gel, SPM, and so forth. Great expenditures have been made by service companies to develop equipment and technologies to slurry the polymers and to add all the other additives as liquids wherever possible. Halliburton Services has made it a company policy that virtually all of their additives will be added as liquid.

Obviously, doing jobs continuously is the mode of choice for service companies. Their personnel simply arrive on location with the gel concentrates and other additives, hook up their equipment, and start mixing on the fly and pumping downhole.

Quite frankly, I am not in favor of such systems. For small-to-moderate sized treatments, there is enough difficulty with batch-mixed operations. The added complexity of adding several systems on the fly greatly decreases the chances for a successful treatment. There are many reasons, including environmental ones, why you would want to run treatments continuously. Since there is a failure rate on fracture treatments (see Chapter 10) of more than 70% for batch-mix treatments, I see no reason-other than for very large treatments-for doing the fracturing treatments on the fly. In the following paragraph, reasons are listed why the vast majority of treatments should be batch mixed until all service companies have microprocessor-controlled additive systems backed up with 100% redundancy.

1. In many situations, there is insufficient time for complete gel hydration. If I wanted to pay for partially hydrated gel, then I would batch mix up lower concentration and save myself money.
2. If you add breakers, particularly low temperature breakers, to gel that is partially hydrated, you see a very rapid breakdown of fluid, and you never develop full viscosity, even on linear gel treatments.
3. Consequences of human error or equipment malfunction in any part of the treatment (particularly where you're running high-sand concentrations of a wide variation in gel concentration, cross-linker loading, etc.) are indeed disastrous. By batch mixing all the materials, you can create a system that eliminates many of the potential problems on the treatment.

Even with the previous objections, I have seen a few very well-executed, continuous-mix fracturing treatments. These treatments were run by very dedicated, qualified personnel with up-to-date equipment utilizing microprocessor control for metering of slurry gel, cross-linkers, breakers, and so forth. The problem with doing this overall is that the type of equipment used on these successful treatments is not available universally, nor is it even available for a small percentage of the treatments. We have a long way to go before we are ready to run all of our treatments continuously on a daily basis. If you evaluate Chapter 10 and see the very high-failure rate noted on batch-mixed, more controlled treatments, you can surmise that continuous-mix treatments have a much higher chance of failure.

Double-check breaker addition

I mentioned earlier the importance of pH in gel degradation. A further consideration is utilizing the proper breaker during the treatment. Candidly, most service company personnel and operating companies did not like to run breaker in the early stages of the treatment because they feared it might cause a screen-out. In low-temperature wells or in wells with bottom hole temperatures less than 280°F, linear gels and certainly cross-linked gels have an infinite life span. It is imperative, particularly in low-temperature wells where there isn't any thermal thinning, that breakers be added uniformly and properly.

Oil company personnel need to be involved intimately with the service company concerning breaker loadings and addition techniques. Typically, the best technique for adding enzymes or oxidizer breaker systems is in a liquid form. The enzymes are usually powders diluted with sugar. This combination can be dissolved easily in water just prior to the job, and it can be uniformly added during the treatment. Two other common breakers are used in linear water gels. One is a granular oxidizer breaker and a high-temperature oxidizer breaker.

One of the best things that has occurred in our industry has been the development of delayed or encapsulated breaker systems. As mentioned previously; many service companies did not like to run breakers in early stages of the treatment because of the detrimental effects upon the viscosity of the fracturing fluid, which sometimes could cause screen-outs. With the use of delayed or encapsulated or controlled-release breaker systems, you now have the ability to add the proper amount of breakers and still maintain a stable fracturing fluid.

It should be noted that there are many formations that cannot be properly treated with conventional breaker systems. Examples of these are the Frontier formation in the Rocky Mountains and the Canyon Sands in west Texas. Typically, these are large intervals that require large treatments. Many of these treatments, particularly when treated down tubing, required pump times of 2 to 4 hours. If a conventional breaker was used, and sufficient product was used to break the gel, there was very little stability beyond a pumptime of more than 1-1.5 hours. Therefore, many treatments were conducted where little or no breaker was run in the early stages of the treatment, and a small amount of breaker was used in the final stages of the treatment. For many personnel, this was a successful treatment (i.e., it did not screen-out). In fact, this was the reason that some personnel did not want to run water-based fluids in many formations. The problem was due to unbroken gel in the formations. At temperatures of 170°F (in fact, temperatures approaching 280°), fracturing fluids will not degrade without external breaker systems. Today, virtually all service companies have delayed release breaker systems that allow them to treat wells into the 200°F range, allowing stable fracturing fluids that do break totally. You need to take a great deal of care in evaluating the service companies' breaker systems with their fluids. As will be discussed in Chapter 10, actual on-site testing should be done to be sure not only that the fluids will break, but also that they are stable throughout the pumptime.

A particular service company may use different breaker concentrations for varying systems. Work with the service company to be sure the proper breaker loadings are always used so the gel degrades in the time required. Normally, the loadings are staged to start out with the minimum required to break early in the treatment and may be increased for later stages of the frac.

Note that proper functioning of the breaker is very dependent on the pH of the fluid. For linear gel systems, I recommend a pH of 4-7. The optimum pH for conventional enzyme breakers is approximately 5. If the pH should drop below 3, conventional enzyme and oxidizer breakers are dysfunctional. If the pH rises above 8, conventional enzyme breakers are ineffectual.

This again emphasizes checking the pH of the fluids before pumping into the hole and understanding possible pH changes that may occur in the formation. If the job is short, you may add extra breaker in order to take into account any rise or fall in pH level.

Tables 7-1 to 7-5 show breaker systems for the major service companies. Note the dependency upon pH and effective temperature ranges within these breaker systems. You should always query the service company to be assured the proper breaker is being used.

Recently, I was contacted by a major operator in the Texas Panhandle. They had ten wells that were not performing that were direct offsets to another operator, which we had worked for, which were performing quite well. It was discovered quickly that the wrong breaker had been used. In fact, an oxidizer breaker had been used for bottom hole temperatures of 90°F. This mistake was not made due to malice or anything else. It was just a simple mistake from a service company. Communication with the service company and conducting pilot tests or intense quality control, which will be described in Chapter 10, can negate these kinds of problems from occurring.

Table 7-1 Oilfield Service Company Breaker Systems-Dowell Schlumberger

Low Temperature		High Temperature	
J-297 J-134 J-218 J-218, J-318, J-466 J-475	Weak acid 70-100°F Enzyme pH 3-8 70-130°F Oxidizer pH > 3 120-200°F Catalyzed oxidizer pH > 3 70-130°F Encapsulated oxidizer pH > 3 70-2oo°F catalyst reg. @ low temp.	J-481	High –temperature oxidizer pH > 3 180-260°

Table 7-2 Oilfield Service Company Breaker Systems-Halliburton Services

Low Temperature		High Temperature	
GBW-3, GBW-30 SP, AP SP, AP, CATLTJ- OptiFlo AP OptiFlo E	Enzyme pH 3-8 70-130°F Oxidizer pH > 3 120-200°F Catalyzed oxidizer pH > 3 70-130°F Encapsulated oxidizer pH > 3 70-2oo°F with catalyst reg. @ low temp. Encapsulated enzyme pH > 3 < 8.5 70-130°F	Breaker HT	pH> 3 180 - 260°F

Table 7-3 Oilfield Service Company Breaker Systems-BJ Services

Low Temperature		High Temperature
GBW-40 GBW-12 GBW-5	Enzyme pH > 3 < 8 70-130°F Enzyme pH> 3 < 10 for pH> 9 70-125°F for pH< 9 70-175°F Oxidizer pH > 3 120-200°F	BJ uses a proprietary composition SpectraFrac G with an internal breaker.

Table 7-4 Oilfield Service Company Breaker Systems-The Western Company

Low Temperature		High Temperature	
B-11 B-33 B-5 B-5, B31 B-9 UltraPerm CRB	Enzyme pH > 3 < 8 70-130°F Enzyme pH > 5 < 10.5 70-120°F Oxidizer pH > 3 120-200°F Oxidizer & catalyst pH> 3 70-130°F Oxidizer pH > 3 150-240°F Encapsulated oxidizer pH > 3 120-350°F, can be used 70- 120°F with catalyst	UltraPerm CRB	Encapsulated oxidizerpH> 3 120-350°F

Table 7-5 Oilfield Service Company Breaker Systems-Acid Engineering	
Low Temperature	
B-1	Enzyme pH> 3 < 8 70-130°F
B-5	Oxidizer with catalyst pH> 3 70-130°F
	Oxidizer can be run separately-range then 120-250°F pH > 3

Pilot tests are always advisable

Another recommended point is that the service company should, as a matter of course, run break tests before the job. For treatments with linear or cross-linked fluids, this simply amounts to getting the treating fluid, mixing the gel and other additives, adding breaker, putting this combination into a heat cup or Fann 50, and observing it for the period that you want the gel to break. Insist that your service company do this for every job.

We will discuss on-site, pilot testing in Chapter 10. The prejob testing in Fann 50's and heat cups would be preferable, and it will eliminate a lot of on-site iterations, assuming that you have been testing the proper fluids. It is imperative that on-site testing be done to be assured that the quality of the fluids pumped are similar to those tested in the laboratory.

It is common practice to use methanol as a stabilizer in combination with breakers. Methanol is used in cross-linked fluids more commonly than in linear gels. In our discussion of cross-linked fracturing fluids, comparative loadings of breakers are shown with and without various percentages of alcohol. There is a dramatic difference between loadings with and without methanol. A high concentration of methanol will totally destroy the usefulness of enzyme breakers and/or oxidizers. If methanol is used on a linear gel frac, make sure break tests are conducted prior to the treatment.

Breaker dilutant must be compatible

Particular precautions need to be taken with breaker dilution fluid. The water that you mix your breaker needs to be neutral (pH of 7) and without contaminant. If you mix an enzyme breaker in a highly acidic or a highly alkaline medium, you can denature the enzyme and effectively destroy your breaker system. A similar reaction occurs when mixing oxidizer breakers, such as sodium persulfate or ammonium persulfate with water. If you mix these in a low pH water, they will dissociate and you will not have a breaker to degrade the frac fluid.

There was a time when I added oxidizer breaker in large quantities to a viscous gel system. Proud of myself, I very carefully metered the material and made sure the breakers were all being pumped into the well uniformly. To my chagrin I found out at the end of the treatment that the breaker solution that I had prepared was mixed into a tank with raw hydrochloric acid in the bottom. I didn't have any breakers, and it took a great deal of breaker and time to remove the gel from the formation we were treating.

Fluid loss additive affects breaker concentration

Many of the fluid loss additives utilized today contain polymers. These polymers may be Karaya* or starches or guar gum. Often in selecting breaker concentration, you should choose the, breaker based upon a 30, 40, or 50 lb. base gel. From past experience, you need to account for the fluid loss additive used. If you plan to use a fluid loss additive that contains some polymer, then a good rule is to start testing using one-half of the concentration of the fluid loss additive. For example, if you run 20 lbs. of fluid loss additive with a 40 lb. gel, select the breaker concentration based on a 50 lb. gel. This, of course, does not hold if you use inert products such as 100 mesh sand, silica flour, and so forth for fluid loss additives.

Obviously, when selecting breaker concentrations, you are selecting the concentrations with which you are going to initiate your pilot testing on-site or the prejob tests the night before in the service company laboratory. Never run a fracturing treatment just reading breaker concentrations from a table and utilizing that on the treatment. That is tantamount to assuring a disastrous result.

Prevention far exceeds the cure

I want to make a strong point about using breakers. If prevention is better than the cure, it's in the case of gel breaking and potential emulsions. Once a gel is present in the fracture system or in the matrix, it is extremely difficult to reach and degrade. The gel itself acts as a diverter; it's very hard to reach out and start moving the fluids out of the well. Typically, low-temperature wells translate into low pressure wells. The viscosity of the gel system tends to make the gels good blocking materials. The same goes for nonemulsifiers. Do everything possible to be absolutely certain that surfactants that act as nonemulsifiers are used if there is any danger of emulsions forming from your stimulation treatment.

Another pertinent point about surfactants is that many of the commonly used surfactants are not inherently nonemulsifiers. By their very nature, most of the fluorosurfactants are excellent surface tension reducers (i.e., they reduce the pressure required to flow fluids through the pore spaces). They are not, however, nonemulsifiers and cannot break emulsions if they form. Again, work carefully with your service company to make certain breakers and nonemulsifiers are present where needed for your well.

Items to check on linear gel systems

- Base gel viscosity
- pH of base gel
- Presence of lumping
- Temperature of fluid
- Presence of surfactants, including nonemulsifiers
- Presence of biocides
- Have design proposal with list of additives and confirm their presence
- Presence and proper addition of breakers
- Consider fluid loss additives in breaker addition
- Utilize Intense Quality Control procedures

Monitoring fluids from cross-linked gel systems

Basically, all of the criteria mentioned for linear gel systems must be followed to quality control the fluids for cross-linked systems. In addition to carefully monitoring the viscosity of the base gel, the presence of biocides and/or nonemulsifiers, the pH, and the temperature, you need to further test the cross-linkers used.

Check for contaminants affecting cross-linkers

Tests must be run to detect the presence of metals or reducing agents. Also, each tank of gel must be checked for its cross-linking with recommended concentrations of the cross-linker. The reducing agent test usually indicates the presence of foreign metals that may interfere or negate cross-linking. The presence of these materials in the water will most probably result in dumping the fluid from the tank. This often occurs when you have acidic water or water that has been stored in iron tanks. Heavy metals can be particularly detrimental if you use cross-linkers such as titanium, zirconium, or those that are directly affected by other metals.

The test for reducing agents on cross-linking is extremely important. Just being in the proper pH range and having the fluid appear to be in good shape does not guarantee that cross-link time and/or texture and final viscosity of the cross-link will be adequate. This test for reducing agents is rather simple by using potassium permanganate. Your service company representative can explain it to you. Basically, it involves adding permanganate solution to the water. If the water is pink, then all is well. If the water becomes discolored or if it turns white, then you may have a problem.

Quality control appearance and time of cross-link

Ever since the advent of cross-linked fracturing occurred in late 1967 or early 1968, hordes of individuals have claimed the ability to look at a cross-link gel sample and determine its stability or its ability to transport proppants into the formation. The only thing that can be determined by looking at a cross-link sample is whether it is indeed cross-linked. The appearance at ambient temperature, unless that coincides with downhole temperature, has little to do with the stability of that fluid after it has undergone shear or once it is at in-situ conditions of shear and temperature.

Statements made about chunkiness indicating over cross-linking and wetness indicating undercross-linking are true to some extent. But particularly with the advent of delayed cross-linkers and borate systems, nothing is ascertained by the sample caught at the blender or prejob cross-link samples. What you must determine with this test is the time that cross-linking occurs.

If you meet with any number of personnel in the service industry and ask them how they measure cross-link time, you will find that same number of different techniques. They talk about such things as vortex closure, they discuss lipping, and they discuss rigidity or dryness of the gel. What you are looking for is a dramatic increase in viscosity and the effects of that viscosity upon pumping friction and shearing problems with the fluid.

The technique I follow to determine cross-link time is to simply put the gel into a blender and measure the time that the vortex closes in the blender at a given shear rate. This can be arbitrary, depending upon whether there is a rheostat control on the blender. If I do not have a blender available, I will do a shake jar test and visually observe the time that the fluid becomes much more viscous. With the advent of delayed cross-link gels, I try to have cross-link times well in excess of pipetime. For borate systems, where there are not very deep wells, I typically allow the fluids to cross-link rapidly and run low enough base gel concentrations so I don't have high-friction pressures. The borate systems do not suffer from shear, and I would prefer to suffer a little bit of excess friction pressure in order to be sure that I have a good fluid during the treatment. We recommend that each tank be tested for cross-link time; and as will be discussed in Chapter 10, we test the viscosity of that cross-link and its stability and break properties at in-situ conditions prior to pumping the treatment.

Cross-link time: slow or fast?

What is the difference between a delayed cross-linking system and a conventional rapid cross-linking mechanism? Interestingly, some of the very early crosslinkers had delayed cross-linking. By varying the concentration of the cross-linker or the buffer in the borate systems, you could pretty much achieve any cross-link time you wanted.

Most service companies in the late 1960s and early 1970s tried to get crosslink time halfway down the pipe or certainly before the gel reached the bottom of the hole because everyone thought uncross-linked gel could not carry sand through a perforation. This was a bit silly, since uncross-linked gel had carried high-sand concentrations for many years.

The *antimony system,* by varying the pH, could also give variable controlled cross-link time. However, unless friction pressure problems caused the need for delay, a very rapid cross-link time was accepted. With the antimony at low pH, a good cross-link time in a relatively short period was thought to be the best approach. Some of the early titanium cross-linking systems were removed from the field because their cross-link time was too slow. One service company had a winter cross-linker and a summer cross-linker. The winter cross-linker was a titanium triethanolamine because it had a rapid cross-link time, even in cold water. In the summertime, a titanium acetylacetonate was used because it was considerably slower, and the service company personnel were afraid they would have screen-outs resulting from long cross-link time delays.

The most common early cross-linker, or the one used more than any other, was titanium triethanolamine, which typically has a rapid cross-link time. Its crosslink time is instantaneous in hot water with moderate-to-high pH. In the past, any variation in cross-link time was achieved by lowering the concentration of the cross-linker or diluting the cross-linker with water, which tended to poison the system somewhat and slowed the cross-link.

It is obvious to anyone knowledgeable in chemistry that there must be a precise band within which an optimum ·concentration of cross-linker is achieved. Coincidentally, you would think an optimum pH and variation of cross-linker loading would be detrimental to the treatment. Overcross-linking results in a loss of stability because syneresis (i.e., bleeding of the system), a reduction in fluid loss control, and loss of viscosity tend to occur. Undercross-linking gives lower viscosity and instability. Variations of pH could do the same things.

For 13 years, these systems were accepted. Then researchers determined utilizing delayed cross-linker systems would indeed be beneficial. You could use either the inherent nature of some cross-linkers or you could develop techniques to delay cross-link time without affecting pH or optimum loading. The advantages of delaying the cross-link time were that the cross-link fluid would not undergo undue shear through centrifugal pumps or high-pressure pumps and that there would be no additional shear down the tubing.

Delayed cross-link enhances stability

Subsequently, work conducted independently at Texas A&M University by a group headed by Dr. Steve A. Holditch indicated that final viscosity of cross-link fluids was greatly enhanced if the cross-link mechanism took place under low shear conditions (i.e., cross-linking was delayed until the gel was past the pumps and through most of the tubular goods). There is still an industry debate for the real or hypothetical reasons for this. Most people agree that slow cross-link time allows adequate dispersion of the metal ions to migrate uniformly within the polymer strands. This permits a more nearly uniform structure to form, which lends itself to higher viscosity and enhances temperature stability.

Very fast cross-linking, when the metal ions grab the polymer strands as they enter the solution, can result in undercross-linking in some areas and in over crosslinking in others, which yields nonuniform texture, low viscosity, and poor temperature stability.

Techniques that control cross-link time

Today, we are able to control cross-link time by a number of methods. Some companies still control cross-linking by utilizing pH, a dangerous technique that can be detrimental to the final stability of the gel. Other companies vary cross-link time by changing cross-linker concentration or by blending cross-linkers. This, too, is dangerous if one gets outside of the optimum range of cross-link stability. Other companies utilize accelerators or retarders that control cross-link time without changing the cross-linker or the pH. This last procedure is the technique that I prefer.

Final stability of gel systems is dependent upon accurate metering of the crosslinker and upon a precise pH at which the gel is most stable. This creates some quality control problems and requires a great deal of monitoring during a crosslink gel treatment.

Old habits die hard

There are still large numbers of conventional cross-linked gel jobs carried out across the country. These treatments are rapid cross-link systems using metal cross-linker and place sand in a fracture in a workaday fashion. Ultimately, however, controlled cross-link time systems will prevail; the advantages are apparent.

Troubleshooting problem tanks

If tanks of gel do not cross-link properly, then the service company needs to find out if this is due to buffers or contaminants. In many cases, you may have to utilize that particular tank as a prepad or a flush, and gel up another tank. Sometimes changing buffers or increasing cross-linker concentration allows you to use an off-spec tank of gel.

Check all cross-linker on the location

If your service company is using multiple drums of cross-linker and/or activator or retarder, I highly recommend that samples from each drum be taken and checked. If one tank is used, an aliquot sample from the tank can be evaluated and proved functional. Cross-link systems, retarders, and activators are very complex chemically, and the service companies need to maintain rigid quality control on these systems. Some systems shelf degrade or can degrade if they have been exposed to air or water contamination.

Check cross-link times before initiating treatment

On a very large job, quality control work takes only a few minutes, but it is certainly needed to make sure everything is functioning before the treatment. It is apparent that adequate final cross-link needs to be checked as well as checking cross-link time in each tank against the amount of additives required.

Until just the last few years, there was a tremendous amount of concern over cross-link time. Service companies developed all sorts of means to delay or accelerate cross-linkers. Presently, with the exception of borates, the majority of treatments are done utilizing fluid that cross-links past pipetime. In fact, the only way to check in a fairly rapid period of time, if the cross-linker is present, is to put the gel into a microwave and heat it up very quickly to be assured that indeed it is cross-linking. These are what is termed temperature-activated cross-linkers and do not really ever cross-link the fluid until temperatures approach 120-130°F, This technique has been very successful in placing proppant and has negated the need for precise control of cross-link time (as discussed so much in my previous book).

I might add that this is not the case for coalseam fracturing. In those particular reservoirs where near wellbore tortuosity and dendritic fracture are the rule rather than the exception, very rapid cross-link times are a necessity, and you must have very high viscosity near wellbore to get the proppant put away. This has been proven by hundreds of field jobs. We have been utilizing a cross-linked CMHEC with a dual cross-linker with which we are seeing a lot of shearing with the high early viscosity, but we are also using a delayed zirconium that allows us to have good viscosity even after a great deal of shearing has occurred at the near wellbore area. For the majority of treatments, though, we feel that delaying past pipetime is the best procedure for having high-viscosity, temperature-stable fracturing fluids.

Take samples constantly during treatment

Once the job begins, a technician or an engineer must constantly take samples from a sampling point downstream of the blender to evaluate cross-link time. This is a dirty and often thankless job, but it keeps you out of trouble time after time.

The blender operator and service company engineer should be evaluating the pump rate and the concentration of the cross-linker constantly. The oil company engineer should spend a great deal of time at the sampling point, conferring with the service company engineer. Nothing is more critical to the job's success.

Some service companies have large trailer tanks of cross-linker and/or other additives, and these are monitored from the tank units, which usually have positive displacement pumps. Constant monitoring of these pumps gives you the fluid levels in the tanks and is necessary on this type of treatment to be certain that cross-linker is added correctly. Most service companies have flow meters and digital readouts, but visual observation is the best way to be sure of what is taking place during the treatment.

I have seen people get tired during a long treatment and forget to fill reservoirs or put the wrong fluids in the reservoirs. As you might expect, this can cause tremendous problems. If you are not quality controlling by taking samples continuously, you're going to have a difficult time recognizing your problem. The only way to keep out of serious trouble is to recognize problems in the early stages and rectify them quickly.

Standby cross-link additive systems are necessary

I recommend on long duration treatments (i.e., treatments that last more than 1-1 1/2 hours) that both standby blenders and standby cross-linker pumps be available and manifolded with reservoirs with cross-linker and cross-linker additive ready to go. This setup involves a small amount of plumbing from the standby blender. The most disastrous occurrence is to lose your cross-linker pump and not be able to take over immediately with a standby unit. You certainly do not want to change blenders because a tiny hydraulic pump was lost. Having specific plumbing set up to activate a standby cross-linker pump can prevent this problem.

Communication with quality control engineer is essential

Varying concentrations of gel are commonly utilized, particularly on large treatments. On a very long treatment, you might have 60, 50, 40, and 30 lb./1,000 gal gel systems on a single job. This means different cross-linker loadings, and different activators or retarder loadings take place as you switch tanks throughout the job. The critical nature of many of these delayed cross-linkers could create real problems. In other words, you could switch from a 60 lb. to a 50 lb. gel and see a large increase in friction pressure because you did not decrease your cross-linker loading. The pressure rise from excessive cross-linker could be misinterpreted, and you might terminate the treatment because of the rising pressure.

There must be good communication among the person monitoring cross-link time and the treater and the oil company man. Experimenting with cross-link times and loadings is a common occurrence during a job. Loadings should be varied only with good communication and with some preplanning.

How long should cross-link time be?

One of the topics discussed later (see Chapter 12) is the use of computer vans to monitor bottom hole treating pressure. Those techniques can be used advantageously to evaluate what's going on downhole by utilizing some technology developments by Amoco. One thing that's essential to evaluating downhole bottom hole treating pressure is to know the exact friction pressure during the treatment. It is, therefore, a necessity that cross-link time be relatively constant throughout the treatment. Obviously, a much faster cross-link time (i.e., much higher in tubular or casing goods) results in higher friction pressures. Varying cross-link times during the treatment makes it difficult, if not impossible, to evaluate friction pressure values accurately.

Temperature-activated cross-linkers have simplified this process by having essentially the same friction pressure as linear gel. Presently, service companies have developed borate systems that can be controlled so that cross-linking occurs past the pipetime. Obviously, you need to have some means on-site, such as a microwave oven, to heat the fluids rapidly so you can be sure that cross-linking is going on. We used to control the cross-link so that it just started to give us friction pressure in the pipe and that we detected a small amount of friction pressure during the job. Although this was a viable technique, it still caused unneeded and unnecessary degradation of the gel system during the treatment.

Treatments have been conducted where the slurry rate increased as the sand concentration went up. This technique is viable on small foam-frac treatments and may be applicable on large cross-link gel treatments. Most service companies have automatic systems where the concentrations of additives are controlled by the clean fluid rate. The increasing rate technique has been widely used in east Texas, where the treatments are conducted down casing. Be sure the increased friction pressure from higher rate and higher viscosity, due to the slurry effect of the sand on the gel, will not exceed pressure limitations. This technique has no viability on tubular treatments where you typically operate at or near maximum pressure. It is a difficult technique to quality control.

Items to check on cross-linked gel systems

- All "Items to Check on Linear Gel Systems"
- Presence of metals and/ or reducing agents
- Appearance and stability of cross-link on each tank
- Cross-link time for each tank of fluid
- All containers of cross-linker, activator, and retarder
- Cross-link appearance and time throughout job
- Gel loading and cross-link loading during treatment
- Conduct Intense Quality Control

Humans required

My basic contention is that nothing can take the place of a person observing the flow of additives going downhole and cross-checking all of the complex electronic equipment in either a computer van or a standby frac van. The combination of new technology through monitoring systems and good quality control procedures results in excellent treatments. Tables 7-6, 7-7 and 7-8 illustrate the effect of sand, interprop, and bauxite on density and volume.

Table 7-6 Sand-fluid Slurries

Sand-fluid Ratio, lb. sand: gal fluid	Sand slurry Ratio, lb. sand: gal	Fluid-slurry Ratio, gal fluid: gal slurry	Slurry-fluid Ratio, gal slurry: gal fluid	Slurr Density, lb.gfal
0. 5	0.489	0.978	1.022	8.743
1.0	0.957	0.957	1.045	9.034
1.5	1.404	0.936	1.068	9.304
2.0	1.834	0.917	1.090	9.573
2.5	2.245	0.898	1.112	9.824
3.0	2.643	0.881	1.135	10.079
3.5	3.024	0.864	1.158	10.316
4.0	3.388	0.847	1.180	10.537
4.5	3.740	0.831	1.202	10.754
5.0	4.080	0.816	1.225	10.967
5.5	4.406	0.801	1.248	11.166
6.0	4.728	0.788	1.270	11.379
6.5	5.023	0.773	1.294	11.547
7.0	5.315	0.759	1.317	11.721
7.5	5.601	0.747	1.339	11.906
8.0	5.872	0.734	1.362	12.067
8.5	6.137	0.722	1.385	12.231
9.0	6.399	0.711	1.407	12.400
9.5	6.641	0.699	1.430	12.541
10.0	6.890	0.689	1.452	12.705
10.5	7.119	0.678	1.475	12.841
11.0	7.349	0.668	1.498	12.986
11.5	7.567	0.658	1.520	13.121
12.0	7.776	0.648	1.543	13.245
12.5	7.988	0.639	1.566	13.381
13.0	8.190	0.630	1.588	13.507
13.5	8.381	0.621	1.611	13.622
14.0	8.571	0.612	1.633	13.736
14.5	8.755	0.604	1.656	13.853
15.0	8.935	0.596	1.679	13.965
15.5	9.110	0.588	1.701	14.073
16.0	9.281	0.580	1.724	14.176
16.5	9.445	0.572	1.747	14.273
17.0	9.610	0.565	1.769	14.381
17.5	9.766	0.558	1.792	14.476
18.0	9.923	0.551	1.814	14.576

18.5	10.071	0.544	1.837	14.665
19.0	10.215	0.538	1.860	14.753
19.5	10.361	0.531	1.882	14.930
20.0	10.499	0.525	1.905	15.011
20.5	10.633	0.519	1.928	15.011
21.0	10.769	0.513	1.950	15.097
21.5	10.897	0.507	1.973	15.176
22.0	11.028	0.501	1.995	15.259

Note: Based on sand's having a true density of 22.1lb./gal (specific gravity= 2.65) and 2% KC1's having a density of 8.44 lb./gal @ 60°F (specific gravity= 1.011).

Table 7-7 Intermediate Proppant-fluid Slurries

Proppant-fluid ratio, lb. proppant: gal fluid	Proppant-slurry ratio, lb. proppant: gal fluid	Fluid-slurry Ratio, gal fluid: gal slurry	Slurry-fluid Ratio, gal slurry: gal fluid	Slurry Density, lb. gal
0.5	0.490	0.981	1.020	12.133
1.0	0.962	0.962	1.039	12.336
1.5	1.416	0.928	1.059	12.529
2.0	1.855	0.911	1.078	12.722
2.5	2.278	0.895	1.098	12.908
3.0	2.686	0.880	1.117	13.093
3.5	3.080	0.895	1.137	13.262
4.0	3.460	0.880	1.156	13.433
4.5	3.828	0.865	1.176	13.606
5.0	4.184	0.851	1.195	13.763
5.5	4.529	0.837	1.215	13.922
6.0	4.862	0.823	1.234	14.075
6.5	5.185	0.810	1.254	14.230
7.0	5.499	0.786	1.273	12.133
7.5	5.803	0.774	1.293	12.336
8.0	6.098	0.762	1.312	12.529
8.5	6.384	0.751	1.332	12.722
9.0	6.662	0.740	1.351	12.908
9.5	6.932	0.730	1.371	13.093
10.0	7.194	0.719	1.390	13.262
10.5	7.449	0.709	1.410	13.433
11.0	7.698	0.700	1.429	13.606
11.5	7.939	0.690	1.449	13.763
12.0	8.174	0.681	1.468	13.922
12.5	8.403	0.672	1.488	14.075

13.0	8.626	0.664	1.507	14.230
13.5	8.844	0.655	1.527	14.372
14.0	9.056	0.646	1.546	14.508
14.5	9.262	0.639	1.566	14.655
15.0	9.464	0.631	1.585	14.790
15.5	9.660	0.623	1.605	14.918
16.0	9.852	0.616	1.624	15.051
16.5	10.040	0.608	1.644	15.172
17.0	10.222	0.601	1.663	15.294
17.5	10.401	0.594	1.683	15.414
18.0	10.576	0.588	1.702	15.539
18.5	10.746	0.581	1.722	15.650
19.0	10.913	0.574	1.741	15.758
19.5	11.076	0.568	1.761	15.870
20.0	11.236	0.562	1.780	15.979
20.5	11.392	0.556	1.800	16.085
21.0	11.545	0.550	1.819	16.187
21.5	11.694	0.544	1.839	16.285
22.0	12.360	0.562	1.780	17.103

Note: Based on intermediate-type proppant's having a true density of 26.0948 lb./gal (specific gravity = 3.13) and 2% KCl's having a density of 8.44 lb./gal @ 60°F (specific gravity= 1.011].

Note: As intermediate proppant is a heterogeneous substance, verify the density before using this table.

Table 7-7 Intermediate Proppant-fluid Slurries

Bauxite-fluid ratio, lb. proppant: gal fluid	**Bauxite-slurry ratio, lb. proppant: gal fluid**	**Fluid-slurry Ratio, gal fluid: gal slurry**	**Slurry-fluid Ratio, gal slurry: gal fluid**	**Slurry Density, lb. gal**
0.5	0.492	0.984	1.016	8.797
1.0	0.968	0.968	1.033	9.138
1.5	1.430	0.953	1.049	9.473
2.0	1.877	0.938	1.066	9.794
2.5	2.311	0.924	1.082	10.110
3.0	2.731	0.910	1.098	10.411
3.5	3.140	0.897	1.115	10.711
4.0	3.536	0.884	1.131	10.997
4.5	3.921	0.871	1.148	11.272
5.0	4.296	0.859	1.164	11.546
5.5	4.659	0.847	1.180	11.808
6.0	5.013	0.836	1.197	12.069

6.5	5.358	0.824	1.213	12.313
7.0	5.693	0.813	1.230	12.555
7.5	6.019	0.803	1.246	12.796
8.0	6.337	0.792	1.262	13.021
8.5	6.647	0.782	1.279	13.247
9.0	6.949	0.772	1.295	13.465
9.5	7.243	0.762	1.312	13.674
10.0	7.530	0.753	1.328	13.885
10.5	7.810	0.744	1.344	14.089
11.0	8.083	0.735	1.361	14.286
11.5	8.350	0.726	1.377	14.477
12.0	8.611	0.718	1.394	14.671
12.5	8.865	0.709	1.410	14.848
13.0	9.114	0.701	1.426	15.030
13.5	9.357	0.693	1.443	15.206
14.0	9.594	0.685	1.459	15.375
14.5	9.827	0.678	1.476	15.594
15.0	10.054	0.670	1.492	15.709
15.5	10.276	0.663	1.508	15.872
16.0	10.493	0.656	1.525	16.030
16.5	10.706	0.649	1.541	16.182
17.0	10.911	0.642	1.558	16.408
17.5	11.118	0.635	1.574	16.480
18.0	11.321	0.629	1.590	16.629
18.5	11.512	0.622	1.607	16.764
19.0	11.707	0.616	1.623	16.906
19.5	11.890	0.610	1.640	17.038
20.0	12.077	0.604	1.656	17.175
20.5	12.261	0.598	1.672	17.308
21.0	12.433	0.592	1.689	17.429
21.5	12.610	0.587	1.705	17.564
22.0	12.776	0.581	1.722	17.680

Note: Based on bauxite's having a true density of 30.4878lb./gal (specific gravity= 3.66) and 2% KCl's having a density of 8.44 lb./gal @ 60°F (specific gravity = 1.011).

Quality control checklist

The fracture stimulation checklist in Appendix II [3]will aid the engineer supervising fracture stimulations. The key to this checklist is that the engineer should check everything. If he doesn't, he will not really know what has occurred before, during, or after the frac job. The gauges in the frac van do not ensure that the right. amounts of sand and gel are pumped, that the gel is properly cross-linking, and that the sand and frac fluid are uncontaminated. The engineer also needs to track the amount of fluid and proppant actually pumped at every point during the progress of the job.

The best way to use this list is to read it before going to the location. Be sure you understand everything on. it. Call the service company salesman into the office and ask if equipment that you want, such as backup blenders and viscometers, will be on the job. This gives him time to make sure the equipment is at the location. You might give the salesman a copy of the checklist so his field people will know what you are going to check. Get to the job with plenty of time to go through your checklist.

When you ask the treater questions, do it in an off-hand manner; mix a lot of small talk into the conversation. This puts the treater more at ease than if you drill him with questions, and you will still get the information you want. You need to know exactly what is going on at all times, so keep a running conversation going with the service company hands without getting in their way. Another tip is to check as many of these items as you can by yourself without bothering the service company personnel. Give them time to do their jobs. Try to be relaxed and informal while getting the facts you need. Be firm about what you want, but also be flexible.

The principal thing to keep in mind is that the average service company hand is interested in the following items, ranked in the order of their importance:

- Getting the ticket signed
- Getting off the location as quickly as possible
- Pumping some kind of frac fluid with some concentration of sand
- Doing the job right

The bottom line is that you, as the oil company production engineer, should *check everything.*

Methanol and methanol water-based fracturing fluids

Many people dream of using nonaqueous fluids for hydraulic fracturing. Even with the old guar gum, a little methanol added to the fluid was felt to be beneficial in some water-sensitive formations. With the advent of hydroxypropylguar and other polymers, you can employ 50% and higher percentages of methanol. Development of new polymers allow use of 100% methanol. In some cases, pure methanol fracturing gels have been utilized to fracture treat oil and gas wells.

It was felt the major application for methanol-based fluids was dry gas reservoirs or reservoirs containing high concentrations of clay. It is my opinion that it is a big mistake to run methanol in virtually any situation. The problems that one has with breakers (i.e., requiring very high concentrations to degrade the gel, and the inherent danger of methanol to personnel) negates its use for most fracturing applications.

---

[3] The original version of Appendix II was prepared by William J. Ditter of the Superior Oil Co. Since the time of publication of the book, this checklist has gone through a number of modifications. The one we are using today is continuing to be modified at least on a monthly basis. The original work put forward by Mr. Ditter is appreciated.

It is indeed very expensive, dangerous to handle, and the benefits therein have not really been proven in the industry.

Safety considerations with methanol

The oil company man and the service company always need to keep in mind some safety considerations. Of primary importance is the inherent danger of methanol or isopropyl alcohol systems. Methanol and isopropyl, of course, are flammable. With a large amount of water, that danger tends to disappear. Flames from methanol and these types of fluids are nearly invisible because they burn with neither smoke nor color. Therefore, great care must be taken (as with oil based fracturing) in covering hoses and having fire-fighting equipment available. Breathing apparatus need to be at hand so your blender operators and people on the blending unit do not breathe the vapors. Pure methanol fracs are more dangerous than oil-based treatments. The danger is not necessarily from fire but from irreversible brain damage to the people who breathe fumes on the location.

Breaker considerations with methanol

In addition to these precautions, be aware that adding methanol to polymer gel drastically changes the ease of degrading the base fluid. Much higher concentrations of all breakers are required for degradation if you are using methanol.

If utilizing an enzyme, you will see denaturing of the enzyme by methanol, and much higher breaker concentrations will be required. Methanol acts as a reducing agent by counteracting the effect of the oxidizer breakers. Sometimes from 50 to 100 times the amount of oxidizing breakers is required when alcohol is used. When methanol concentrations get into the 40-80% range, the enzyme and oxidizer breakers simply do not function. For comparison, refer to Table 7-9 and Table 7-10, which give breaker concentrations for cross-linked gel with and without alcohol. Most service companies list breaker recommendations in their service manuals.

Quality control for water methanol and methanol gels

The criteria given earlier for linear water-based gels basically hold for water methanol gel or methanol gel systems. Utilize the checklist and tables for water based fracturing fluids. Get the expected viscosities of the gels from your service company and have them checked in the tanks. Follow all procedures in the preparation of tanks and the gel mixing. Your service company should give you recommended pH ranges as well as viscosity for their benefit and yours.

Of course, the proper safety measures need to be taken. Be aware of the appearance of the cross-linked gel. Be sure that the cross-linkers are added during the treatment.

Biocides and surfactants are still critical when utilizing methanol. The reduction in surface tension for methanol at concentrations up to 20% is minimal. In fact, the surface tension achieved by simply adding the gelling polymers is almost equivalent to that with methanol. Guar gum or hydroxypropylguar polymers lower the surface tension as much as concentrations approaching 20% methanol.

Obviously, these polymers do not affect surface tension once they degrade. In any case, surfactants, biocides, and other additives must be added even in the presence of methanol. This is particularly true when you are talking about 20% or less methanol. Surfactants for surface tension reduction and emulsion prevention should always be added. Although a fairly good biocide, methanol is not strong enough to keep the gel from degrading on the surface in the presence of very persistent bacteria, nor is it strong enough to inhibit anaerobic bacterial growth downhole.

Percentage of methanol is critical

Another very important consideration in quality control is making certain that the recommended or requested methanol concentration in water is judiciously followed. Many polymers (excellent viscosifiers for methanol in water) work up to a specific concentration. Above that, they give no viscosity, and some of them precipitate at higher concentrations. Guar gum, for instance, gels up to about 25% methanol, but it gives no viscosity at concentrations higher than that. A hydroxypropylguar works up to about 60%, with little or no viscosity above 60%. Higher substituted hydroxypropylguars work into the 90% range but will not work in 100% methanol. Polymers such as hydroxypropylcellulose, which is a viscosifier for pure methanol, do not work in low concentrations of water and methanol and, in fact, precipitate, which can cause severe plugging.

Table 7-9 Four Hour Breaker Data for Batch-mixed, Cross-linked Gel

Base Fluid: 1% KC1 water
Additives: Indicated breaker and pounds HPG/1,000 gal.
HPG gelling agent/1,000 gal. base fluid

Temp °F	20 lb			30 lb			40 lb			50 lb			60 lb		
	E	O	HTO	E	O	HTO	E	O	HTO	E	O	HTO	E	O	HTO
80	0.50	—	—	0.75	—	—	1.00	—	—	4.0	—	—	>4.0	—	—
100	0.30	—	—	0.40	—	—	0.60	—	—	>2.0	—	—	>2.25	—	—
120	0.25	—	—	0.35	—	—	0.50	—	—	0.50	—	—	2.0	—	—
140	—	0.30	—	0.30	—	—	0.15	0.50	—	—	0.50	—	1.0	1.0	—
160	—	0.15	—	—	<0.20	—	—	0.30	—	—	0.50	—	—	0.60	—
180	—	0.05	—	—	1.10	—	—	0.10	—	—	0.20	—	—	>0.20	—
200	—	<0.05	—	—	0.05	—	—	0.05	—	—	0.10	—	—	0.15	—
220			—	—	—	<0.40	—	—	>0.5	—	—	>0.5	—	—	>0.75
240							—	—	0.25	—	—	>0.25	—	—	>0.2
260							—	—	0.20	—	—	0.15	—	—	>0.10

E = Enzyme breaker
0 = Oxidizer breaker
HTO = High-temperature oxidizer breaker

Table 7-10 Four Hour Breaker Data for Batch-mixed, Cross-linked Gel

Base Fluid: 20% methanol in 1% HC1 water
Additives: Indicated breaker and pounds HPG/1,000 gal. base fluid
HPG gelling agent/1,000 gal. base fluid

Temp °F	20 lb			30 lb			40 lb			50 lb			60 lb		
	E	O	HTO	E	O	HTO	E	O	HTO	E	O	HTO	E	O	HTO
80	>1.50	—	—	>1.50	—	—	2.00	—	—	>2.25	—	—	>4.0	—	—
100	1.15	—	—	>1.00	—	—	1.85	—	—	>2.00	—	—	>3.00	—	—
120	<1.25	<3.00	—	0.75	5.00	—	>0.75	>7.00	—	1.75	5.00	—	>3.00	>6.00	—
140	<1.25	<2.00	—	>0.75	>3.00	—	1.00	5.00	—	>1.75	>5.00	—	>3.00	>5.00	—
160	—	2.00	—	—	>2.00	—	—	3.00	—	—	3.50	—	—	3.00	—
180	—	0.85	—	—	1.00	—	—	1.25	—	—	2.50	—	—	1.50	—
200	—	—	>3.00	—	—	>4.00	—	—	>7.00	—	—	>7.00	—	—	>7.00
220			>1.00	—	—	>1.00	—	—	>1.50	—	—	>4.00	—	—	>2.00
240			0.75	—	—	0.75	—	—	>1.00	—	—	>1.25	—	—	>1.50
260				—	—	>0.25	—	—	>0.30	—	—	>0.50	—	—	>0.60

E = Enzyme breaker
0 = Oxidizer breaker
HTO = High-temperature oxidizer breaker

Pure methanol solutions require even more care in degradation and handling. The danger of pure methanol is much greater than alcohol water solutions because of flash point considerations and fumes. Degradation of pure methanol solutions is limited strictly to the use of acid solutions at elevated temperatures. There is no known way (save acid hydrolysis) to degrade pure methanol-based gels at temperatures below 180°F.

Nonviscous fluids containing friction reducers

Perhaps the simplest treatment, and one once widely used, was water or KCl water containing small quantities of gelling agent or polymers. The best friction reducers on the market are latex polymers or copolymers of acrylamide, which are added at concentrations of .25-2.0 lb./1,000 gal Many treatments are conducted, particularly in the eastern United States and in the Rocky Mountains, that use no additives other than surfactants, nonemulsifiers, and friction reducers.

In shallower formations, the simple treatments can place a fairly high concentration of proppant by utilizing only friction reducers. These treatments may be pumped out of pits instead of frac tanks. Usually, large quantities of water are pumped at high rates, and the friction reducers are added continuously. This puts a great deal of pressure on the service company and oil company people to make certain the additives are properly proportioned.

"Complex" additive systems

With today's advanced technology, it is surprising that the system for adding powdered friction reducer can be a man standing over a hopper shaking a sack. Or he may have a 5 gal pail and be pouring additive into a blender tub. This type of addition may be practical or functional on small jobs, particularly if the individual shaking the sack or pouring the material is conscientious. I like to think our industry has progressed a little further than this crude technique.

When I teach schools across the United States for service company or oil company people, one thing I state as a major cause of job failure relates to the tobacco chewer or the person who dips snuff. I have seen blown off fluid ends, destroyed pumps, screen outs, and blown up tubing because someone. decided to take time out from shaking his sack or adding liquid to chew tobacco. Regardless of the reason, if an individual stops adding material or if there is mechanical failure on a continuous-mix treatment (e.g., a feeder stops turning or a liquid additive pump stops pumping), catastrophe can result.

The service company can do some things to help overcome potential problems. Nearly all service companies have feeder systems. Most of these are backed up. A little preplanning, like having someone start to add the material manually, could keep the job going and prevent a disaster. The same thing goes for feeding liquid additives. A backup pump or simply material in a reservoir accessible to the blender operator, so he can add it manually, could assist in completing a job that might otherwise be terminated.

The cessation of friction reducers or gelling agents on the fly does not necessarily destroy equipment or terminate the treatment. However, the potential to do so is there. Most sloppy treatments, though, relate to poor metering when adding chemicals.

Batch-mixing is preferred

By now, it may be obvious that I do not particularly favor continuous-mix treatments. When you are adding a slurry of chemicals, dry powders, or liquid emulsions, there is always a chance of error. That possibility at the best can allow sloppy treatments; at the worst, it can cause a disaster. There are many instances, however, where continuous mix is the only approach. When that is the case, all those involved are under a great deal of pressure to ensure the job is completed satisfactorily.

Alternative to straight continuous mix

One alternative to a straight continuous-mix treatment is a semi-continuous technique. This technique utilizes a working tank: one blender mixes into a holding tank and another blender sucks out of that, pressurizing the pumps going downhole. This gives you time for the polymers or friction reducers to hydrate properly, and it also gives you some buffer time if additive systems malfunction. You will probably mix at least one 500-bbl. tank before initiating the treatment. I recommend this technique because it gives you the opportunity to quality control and observe your fluids before going downhole. A little delay time also allows much efficiency from the fracturing fluids.

If you will note Figure 7-6, there is a Dowell Schlumberger specially designed holding tank for continuous-mix fracturing treatments. This particular device holds more than 400 bbls. of fracturing fluid, which takes the place of a premix fracturing tank. Most of the service companies, including Halliburton, Western, BJ, Nowsco, and others, have volumes of premixed fluid between the downhole blender and the frac tanks. This allows for what I term "semi-continuous" in a modern fashion.

Remember, when dealing with the acrylamides or copolymers of acrylamides used as friction reducers in slick water treatments that most of these materials are ionically charged. One of the more common errors is that these additives are added in conjunction with incompatible materials. An anionically charged acrylamide may be incompatible with a cationically charged surfactant. Sometimes both malfunction with a precipitation of the friction reducer, thereby giving no friction reduction and concurrent loss of the surfactant. This reaction, therefore, allows a fluid to be pumped into the well that has no nonemulsifier. The consequences are apparent.

Figure 7-6 Side view of Dowell precision continuous mixer. This piece of equipment is primarily used on continuous-mix treatments as a gel preparation and holding area just prior to the frac blender.

Most of the water-based friction reducers are anionic or slightly anionic. Their compatibility with other additives needs to be checked or questioned by oil company representatives. The friction reducers used for acid-friction reduction are usually cationic. With cationically charged polymeric clay stabilizers, many cationic friction reducers are used in fresh water. The basic difference in the anionic and cationic friction reducers is that the cationic materials are not quite as efficient as the anionic ones. For compatibility, cationic friction reducers must be run with polymeric clay stabilizers. Once again, refer to the chemical cross-reference for your service company and the additives they are using.

Compatibility check of additives is essential

It is extremely important that oil company and service company representatives carefully evaluate the additives combination for a well. In fact, these additives with the base fluid should be put together and, if nothing else, visually observed prior to treatment. Unfortunately, in our industry we sometimes live by the philosophy that a little is good and a whole lot is better. We can get into trouble with fracturing fluids if we mix too many additives at once. This is discussed at length in Chapter 4.

Troubleshooting problems in water-based gel systems

Let me develop a common fracturing treatment scenario. A very competent service company or oil company engineer surveys the viscosities of the gels in the tanks. Of the five tanks on the location, two have substantially lower viscosities than the acceptable range shown on the service company chart or on Tables 7-1 through 7-5. The first thing to check is whether the gel has totally yielded (i.e., has complete viscosity development occurred?). After the tanks have been thoroughly rolled, if you need to, check with the service companies to make certain they added the proper amount of polymer. You should also check to see if viscosity values were higher earlier.

Dropping viscosities in the low-yielding tanks indicate possible bacterial problems from enzymatic degradation. This usually occurs in the hotter months, but it can happen in the winter if the tanks were not cleaned well. If there is fairly rapid bacterial degradation, make a fast decision: either beef up the viscosity of the gels in the tanks and proceed with the job quickly or dump the gel and start over. This decision takes experience and some work with your service company. If the fluid is going to be used on a high-temperature well, a critical well, or a well with a long pumping time, it may be prudent to start over, steam-cleaning the tanks and refilling them with fresh water.

This problem can be alleviated by quality control measures that take place prior to putting any fluids in the tanks. The tanks should be checked by both the tank supplier and service company when they are on the location. Once the tanks are on the location, the biocide should be put in the bottom of the tank before adding water. This tends to shock kill any bacteria from leftover gel or from anything left in the tank. The biocide should be mixed into the tanks of water at least 6-12 hours before adding the gelling agent. Most biocides need time to destroy bacteria.

If something happens to the wellhead or the job is delayed, you may want to add extra biocide to maintain low bacterial counts in the tanks. Tanks of gel have been held for weeks at elevated temperatures with use of biocides. When gel is to be kept for extended periods, add caustic soda to raise the pH to above 12. This negates enzyme degradation, but it complicates later work because the pH often needs readjusting before the treatment.

I have conducted treatments where I beefed-up gels on the fly and successfully completed treatments where bacterial degradation was taking place. In this situation, we need to look at economics and possible results. There are potentially many problems, but if you are going to treat at a high rate in moderate temperature formations, bacterial attack is greatly lessened once you get beyond 140°F, Enzymes denature at approximately that temperature, so you probably can complete the treatment. If there is fairly fast degradation of the polymer, communicate with your service company engineer and perhaps with his local chemist to get their recommendations for utilizing the fluid.

If some tanks have low viscosities that are not due to bacterial decay, then you need to beef-up the gel on the location or perhaps use those tanks for the latter stages of the treatment. This decision depends on design experience and capability. If you design a treatment for a particular gel concentration, add the gel and continue the treatment.

Today, enhancing the viscosity of the gelled fluid is a relatively easy task. By utilizing the slurry gels, you simply add the slurries to the gel blender tub as the tank is rolled. Early on when you tried to add powdered gels to gel systems, you would always have fish-eyes or lumping occurring, and it was a serious problem. If no slurry gels are available, you can simply add powdered gel to the diesel while agitating it, and add it to the system to beef up the gel.

Do not mix breakers and stabilizers

Breakers and stabilizers work fundamentally in opposition to one another. In the case of methanol, you must usually add very high concentrations of breaker to achieve complete degradation, compared with loadings without alcohol. For thiosulfate stabilizers, the addition of oxidizer breakers can cause serious problems.

Adding oxidizer breakers to gel containing sodium thiosulfate or other reducing agents may result in a controlled break system, but it can cause premature break and potentially disastrous results-a screen-out. I have seen job designs that utilized thiosulfate stabilizers and recommended adding a breaker. This can work in moderate temperatures, 200-230°F. The service company may want long life out of the gel system (it might add gel stabilizer for long pump-type treatments), yet may need a breaker for total, complete viscosity reduction. Query your service company representative to see that he does not add either a persulfate or a peroxide breaker system to a gel system containing thiosulfate stabilizer.

Quite recently; there have been cases of successful addition of small amounts of thiosulfate or other reducing agents where encapsulated breakers have been pumped with high temperatures. This type of situation should only be done utilizing intense quality control. In other words, you should test this prior to pumping. This type of thing is typically only done where there is leakage of the encapsulated breakers, and this is compensated for by small amounts of reducing agent. This can be a somewhat dangerous procedure that can result in a very rapid break if there is contamination in the water with such things as intermediate oxidative level iron or other catalysts to oxidizers.

Methanol decreases the activity of breakers

The other perplexing problem with methanol concerns its use as a stabilizer. Methanol greatly lessens the efficiency of breaker systems. Whereas it takes something less than 0.1 lb. persulfate/1,000 gal fluid to degrade either a cellulose or a guar gel at 160°F, you might need 100 times that much in the presence of 5-10% methanol. The problem is that the methanol acts as a reducing agent. If you are running methanol in your fluid for surface tension or for other reasons, greatly increase the breaker loadings. These are the kinds of questions you need to ask the service company representative. He can show you breaker data for specific gel systems at particular temperatures in the presence of a mixture of methanol and water.

Other additives that are utilized .in linear-based fracturing fluids-- in addition to surfactants, nonemulsifiers, biocides, and breakers-are fluid loss additives. These may be bridging materials such as 100 mesh sand, 100 mesh oil-soluble resin, and silica flour; or they may be plastering materials such as blends of starches, talc silica flour, and clay. The plastering materials go under trade names like Adomite Aqua® and WLC-5®. The oil-soluble resins, silica flour, and 100 mesh sand usually have little or no effect upon the viscosity and the break of the fluid.

Do not batch-mix fluid loss additives

Another vital consideration is whether fluid loss additives should be batch mixed with the gelling agent before the treatment or whether these should be added continuously. I recommend that fluid loss additives not be batch-mixed in gel of concentrations less than 40 lb./1,000. If a 40 lb. gel is going to be left on the surface for a long time, you may not want to batch-mix in that situation. The reason for not batch-mixing is that some portion of the fluid loss additive might settle out and lose its effectiveness. In weak gel systems, such as 30 lb. and 20 lb. gels, I seldom recommend batch-mixing the system unless the tank could be circulated continuously. Fluid loss additives are extremely important for increasing and enhancing the efficiency of the fracturing fluid and should be run as design.

Clean fluids + dirty fluid loss agents = waste

An interesting aside when discussing fluid loss agents in conjunction with gelling agents relates to the use of the products with clean viscosifiers. Many times in low- to moderate-temperature wells, customers select the cleaner gelling agent, such as hydroxypropylguar or cellulose derivatives. In doing so, particularly for wells with moderate permeability, they must include "dirty" fluid loss additives to enhance fluid loss control. Basically, if customers utilized lower priced guar, which has residue in the vicinity of 4-6%, they might run considerably less fluid loss additive. The nondegradable materials in guar act as fluid loss additives.

Let's consider this a little further: 6% residue on a 40 lb. gel is about 2.5 lbs. of residue. If you add, for instance, 25 lbs. of silica flour to compensate for what you lose by switching to hydroxypropylguar with 1-4% residue, you may lose the battle of nonresidual fracturing fluids.

Base fluid must be dean

Another consideration for the company representative inclined toward ultraclean fluids to reduce potential formation damage is that he must quality control the base fluid for cleanliness. The water should be carefully monitored to see that mud and other debris are not present in the fracturing tank. I have been on locations using clean, high-cost viscosifiers and had from 6 in. to 1 ft. of sludge in the bottom of the tank before gelling up. These sorts of things make no sense from the standpoints of cost effectiveness and well productivity.

Turnaround time or flowback time after the treatment

In Chapter 11, we will discuss a rapid turnaround of the fracturing fluids. The use of the forced closure technique has probably caused a great deal of confusion in the service company industry. This is due to the assumption by all concerned that all of the fracturing fluid needs to be completely broken by the time we are flowing the well back. In the first edition of this book, a great deal of time was spent discussing tapered breaker schedules and emphasizing the need for gels to be totally degraded before the well was opened. We have since learned that we can flow back wells at very aggressive rates without having the fluid degraded even minimally from very high cross-link viscosities. The reason for this is because these aggressive rates do not exceed 2 bpm; and in most cases, they exceed less than 1 bpm. At these rates, we achieve proppant bridging at perfs inside the pipe. We are able to flow unbroken gel, even cross-linked gel, out ofthe formation.

We feel that very aggressive breaker schedules, particularly with conventional breakers, has been the cause of a great deal of near wellbore proppant settling and loss of communication with the fracture to the wellbore. As we will discuss in the chapter on intense quality control, and also in forced closure, we believe in adding enough breaker to completely break the gel back to water. In addition, we believe in adding excess breaker to move filter cake buildup, but not so that you have too rapid a breaking phenomenon, causing a settling of proppant prior to the fracture closure. We will emphasize again and again the necessity to do pilot testing on site to be sure that the fluid is stable throughout the treatment and also breaks back to water. All of this testing will be done at bottom-hole static conditions. We feel quite strongly that a minimal amount of cool down actually occurs in the formation. The large number of treatments where we have stayed on location when doing forced closure and have noted rapid increases in temperature of the fluid when we get bottoms up from the formation indicates a minimal amount of cool-down going on. Some years back, I read a paper by a Russian who had done a study indicating exponential increases in heat transfer by increasing the surface area on heat transfer tubes. It is obvious that we do not have smooth rock faces on the surface of our hydraulic fractures. This very large amount of surface area, combined with the infinite heat sync, which is the formation itself, creates (I believe)

Table 7-11 Linear Gel Break Schedule for 24-hour, Shut-in Time (Break Data Applicable· for Guar, HPG, HEC, CMC, and Carboxymethylcellulose, CMHEC)

Lb. Breaker/1,000 gal Fluid Gelling Agent/1,000 gal								
Bottom-hole Temp °F	20 lb		30 lb		40 lb		50 lb	
60	0.40	____	0.7	____	1.200	____	3.00	____
80	0.10	____	0.2	____	0.400	____	1.00	____
100	0.06	____	0.1	____	0.250	____	0.50	____
120	0.06	____	0.1	____	0.175	____	0.30	____
140	0.06	0.25	____	0.500	0.175	0.800	0.25	1.000
160	0.06	0.175	0.1	0.375	0.150	0.375	0.20	0.550
180	0.06	0.08	0.1	0.200	0.150	0. 375	0.20	0.550
200	____	0.075	____	0.150	____	0.275	____	0.425
220	____	0.075	____	0.150	____	0.200	____	0.300

much more rapid heat up than is predicted by most models in our industry. We feel quite strongly that most fracs do not even approach parallel plates with smooth walls. Based on our experience with forced closure, we feel that the fluid is heated at a very rapid rate and that the majority of fluid is near bottom-hole temperature conditions. Our field experience has also shown that you are using the most intelligent and proper approach by planning the job built around bottom-hole static temperature conditions. Therefore, the breaker table shown on Tables 7-9, 7-10 and 7-11 are only for guidelines, and you should do testing on-site to be assured of proper quality, not only in being sure that gels are broken but also that they do not break back prematurely.

Pilot tests required

It is necessary for the proper break schedule to be planned for a treatment and that actual break data be gathered on the gel system, base fluid, and additives to be utilized on that well. Most of you are aware-we talk about quality control in cementing in Chapter 11-that great care is taken to ensure that the cement does not set up prematurely during a job, which is indeed a catastrophe. Many additives change break time for various gel systems. Most service companies have extensive research laboratories and field laboratories that can quickly perform a break test to evaluate what will happen in your treatment. It is prudent and important to gather specific break data on your particular well with its fluids. This does not take much work or planning, and it can prevent expensive consequences if (1) the gel does not break or (2) the gel breaks prematurely, causing screen-out.

Most of the break data, like those shown in Table 7-11, are representative, and have taken in some city water containing KCl and additives. Always run a break test on the specific fluid used. In the field, we often use produced water or fluids that may contain strange organic contaminants. It is always better to discover these things before pumping the job rather than after.

Catch samples during the job

A final suggestion in monitoring linear-gel fluids is to collect samples of the pad fluid and of each stage of the increasing sand concentrations. These samples can be used later to ascertain if the gel completely degrades. Keep in mind that if the breakers are designed for bottom-hole temperatures, which are much higher than ambient, the gel should not break on the surface and may never break until it undergoes bacterial degradation. Careful sampling and holding samples for a time can be extremely beneficial in troubleshooting problems that occur later in the well's life.

CHAPTER 7

QUALITY CONTROL OF OIL-BASED FLUIDS (1994)

The very first fracturing treatments were conducted with viscosified hydrocarbons. These fracture treatments were conducted using World War II surplus napalm and gasoline. The treatments were quite small. They consisted of gelling up the napalm in measuring tanks of cement trucks on location and adding propping agent through mixing tubs on the ground.

Obviously, these treatments were dangerous, and very quickly the industry moved toward the use of other means of viscosification of standard oils and for some time to the use of refined viscous oils. The refined viscous oils can still be used for fracturing, although to my knowledge there is little or no use of it today. They exhibit a great deal of friction because of their high viscosity and Newtonian character, and they are only applicable in shallow wells or wells that are completed down casing. This technique was widely used in the 1950s and 1960s, but to my knowledge it has little or no use today because of the high cost of the refined oils and certainly because of, in most cases, no availability.

Oil-based fluids as a last resort

I utilize a slide in fracturing schools that simply states that there are no formations that require oil-based fracturing fluids. Over the past few years, we have begun to recognize that many formations that were considered water sensitive, were not water sensitive in relationship to hydraulic fracturing, but were indeed sensitive to unbroken gel. With the advent of controlled release or encapsulated breaker systems, these formations that formerly were thought to be terribly water sensitive are being successfully stimulated with straight water-based systems. Examples of such formations in the United States are the Frontier formation in Wyoming, the Olmus in south Texas, the Morrow sandstone in New Mexico, Oklahoma, and Texas; and many more formations too numerous to mention. Although there may exist somewhere in the world a formation that requires oilbased fluids to stimulate, you need to recognize that these so-called "damaging fluids" relate to tests conducted in laboratories concerning regained permeability when water-based fluids are flowed through pore spaces in formations. Many of the formations we are talking about are very low permeability, and penetration outward from the fracture face of such fluids into the matrix is minimal at best. There have been many studies conducted indicating that you can accept quite high reduction in permeability many inches from the fracture face and still achieve excellent fracture stimulation as long as you have a: conductive propped fracture across the productive interval.

The strongest reason I have for not utilizing oil-based fluids is their potential danger. Although, as later noted in the chapter, I do utilize and recommend in some cases water external oil-based emulsions, I never recommend the polyemulsion fluid for large, multiple-tank treatments. I believe when a fracturing treatment starts to exceed three or four tanks, you need to consider the potential hazard on site. In my career I have had the unfortunate opportunity to see people burned alive. They had been unfortunate enough to be around when flash fires occurred during pumping of hydrocarbons. If there is any way possible not to utilize hydrocarbons, particularly when large volumes are involved, I tend to stay away from those particular fluids.

I am opposed to oil-based fracturing fluids because of the tremendous lack of research when compared to water-based fluids. I would estimate that 98% of all work in hydraulic fracturing fluids is done on water-based systems.

Since such a small portion of the fracturing work is oil-based, particularly gelled oils, then field personnel, both engineering and operational personnel, are typically not used to field testing and implementing these jobs. Secondarily, there is a distinct lack of safety equipment, such things as spark arresters and good firefighting equipment available on-site.

Another very good reason for not utilizing oil-based fluids is their poor proppant transport mechanisms as well as poor proppant suspending capabilities. Additionally, they are certainly not clean fracturing fluids, and the typical mechanism for breakdown of these fluids is simply changing the acidity or basicity of the fluids so that gelation is negated. It is not uncommon to have these materials flow back and regel once they come back to surface. The potential for damage from these fluids, as well as their inherent high-solid content does not make them a good choice for hydraulic fracturing.

If I haven't already convinced you not to use oil-based fluids, I will continue with my final argument that relates to leak off control with these systems. Service companies have not developed any fluid loss additives specifically for oil-based fluids within the last 20 years. This is primarily due to the low market. The only fluid loss additives specifically developed for oil-based fracturing fluid is an instantaneous degrading agent or breaker for the fluids. I am referring to Adomite Mark II. Adomite Mark II can't be used as a fluid-loss control agent for the standard oil gel systems used in the industry, which ate aluminum phosphate esters. The only fluid loss additives typically used are silica flour, starches, and other materials that were built and designed for use in water-based systems. Obviously, the silica-flour bridging agents can function in the same manner as they do in the water-based fluids, but starches and many other materials, such as bicarbonate and lime, are used as a combination fluid loss additive and degrading agent, which at very best, are very poor leak off control systems.

Therefore, there is very little technology put forth in the development of oilbased fracturing fluids, and I apologize for people who may have done a great deal of work after this book was written on a particular fluid. Along with very little work on these fluids, there also is a lack of expertise in the field in handling and pumping these fluids in a safe manner.

Safety first for oil-based fluids

If I still haven't convinced you against using oil-based fluids, I will be somewhat hypocritical and recommend the use of one type of oil-based fluid for a particular application. You need to understand the tremendous amount of safety procedures that need to be taken if you choose to pump these fluids. The ever present danger of fire and combustion needs to be considered when rigging up, and precautions should be taken right down to the treatment's final phases. Routine problems, such as a broken suction hose or a leak in a high-pressure line can be life-threatening when pumping hydrocarbons. The equipment set-up generally differs when using flammable fluids. Put more distance between the wellhead and the equipment, if at all possible, when pumping oil-based fluids.

All of the suction hoses between the pressurizing or blending unit and the manifold trailer and those hoses between the manifold trailer and the pumping units must be covered with tarpaulin-type material to protect personnel from spraying or leaking hydrocarbons.

I always like to ask students in my classes why I insist on wrapping discharge hoses with tarpaulin material, both on oil-based jobs and acid jobs. This is a good question because most of the hoses used are termed "bullet proof' hoses and have a 1,000 psi working pressure. The answer is that these hoses are subject to severe pressure pulsation if, for some reason, the suction valves on the high-pressure pumps don't hold completely on the discharge strokes of these piston pumps. We have physically measured pulsations well over 1,000 psi on suction hoses between the blender and the truck. It would behove personnel who seem to think it is terribly safe to stand around on the discharge site of the blender to perhaps think again.

On one particular occasion, I noted an individual who straddled a discharge hose throughout an entire treatment. When I explained the problems that go with discharge hoses, particularly while pumping abrasives and bridging materials, the individual turned pale at the potential problem. There was a definite potential for him to be speaking in quite a high voice from doing such an ignorant thing as sitting in the vicinity of high-pressure pumping equipment in this manner. If anyone is going to utilize large-volume gelled oil treatments, you should insist on having spark arresters which, as I stated earlier, are probably not available from most of the service companies.

There can be absolutely no smoking in the vicinity of the treatment. We, of course, recommend that there be no smoking on the location, regardless of the fluid pumped. Whatever you are pumping, a great deal of diesel (fuel for the pumping units) is on the location. It should be noted one of the most dangerous fires I ever saw on location was on a water-based fracturing treatment where a pump truck was leaking hydraulic oil that was spraying into the breather. At a critical point in the treatment, that hydraulic oil caught fire and a large fire was burning, which was being directed at the fuel tank on a pump truck. Thanks to some heroic efforts from on-site safety personnel and the service company, we were able to get the fire out before an explosion occurred.

Strangely enough, most of the fires and the resulting loss of life on oil-based treatment locations have not taken place during the actual treatment. These fires occur during rig-up or during the hydrocarbon viscosifying phase.

Spills and leaks must be minimized

One particular fire resulted because a portion of a valve was left out of a tank. When an employee knocked off the cap, a tank of high-gravity hydrocarbon leaked onto the location. Service company representatives who were gelling up several tanks of oil took the precautions of not smoking and of waiting some time for the area to clear. But they used a forklift, and its exhaust caught the location on fire. Several people died, and some were burned severely.

Spillage of flammable materials is very serious. The most dangerous hydrocarbon tank is not the full one, but the one that is empty and filled with fumes. When fracturing with hydrocarbons, the service company and the oil company should meet to prevent problems and to control the situation, if combustion occurs. There is nothing-equipment, job, or well-that is more important than human life. Of course, the most important thing to do is to keep a fire from starting.

No sparks or smokes allowed

Special brass hammers are available that give off minimal sparks. These should always be used around flammable fluids. A good practice is to instruct service company representatives to have all people at the location turn in cigarette lighters, cigarettes, and cigars. This eliminates any chance of someone's lighting up and causing a disaster. I came close to dying as a result of a fire on location, when I was standing on some tanks where we had gelled some 60 gravity condensate.

I looked back and noted the gentleman who was strapping the tanks leaning over the hatch of the tank. Without thinking, he was pulling a cigarette out of his pocket and reaching for his lighter. A properly placed drop kick negated perhaps one of the largest fires in the history of the oilfield. That individual had to be carried off of the frac tanks. When he realized what he was doing and what would have occurred if he had lit that lighter, he was shaken. Remember it is always important to take all smoking materials away from individuals if hazardous materials are on site.

Soap-type gelled oils

After refined oils, the next generation of oil-based fluids were viscosified, using soap-type materials. These fracturing fluids were commonly a mixture of oil-based fluids, caustic and toll-oil fatty acids. Service companies had various names for these. Dowell called it "PetroGel," Halliburton called it "Visofrac." Although excellent fracturing fluids, these systems suffered from having very high friction pressure, and thus have been basically replaced by the aluminum phosphate esters or polyemulsion fluids that we will discuss later.

High-gravity oils plus friction reducer

Another type of oil-based fracturing fluid that still is used on occasion today consists of high-gravity oil, crudes, diesel, or condensate combined with an oil based friction reducer. The base fluid is typically light crude or condensate or diesel, and the treatment is pumped at fairly high rate, which is accomplished through using a polyisodecalmethacrylate oil based friction reducer. Due to the low viscosity of this fluid, it is not an efficient sand-carrier medium and is probably only applicable when you want to remove near wellbore skin damage and break through damage, through drilling, and so forth. To my knowledge, very little usage of this type of treatment is conducted in the industry today.

It should be noted that if this type of treatment is proposed, the service company should not batch-mix the friction reducer because it is very shear sensitive. This is one of the rare occasions when I insist on adding the friction reducer on the fly. For the concentrate of the friction reducer, which is an emulsion, you must use a nonemulsifying surfactant as a breaker to allow the solubility of the friction reducer to be enhanced. Other service companies use a nonemulsified version of the friction reducer and add it on the fly at concentrations of about 10 gal./ 1,000. The concentrated emulsion is typically run at 2-3 gal./1,000.

Quality control on this type of treatment is very simple; you make sure the friction reducer is added properly and make sure that the hydrocarbons pumped are compatible with formation hydrocarbons. Oddly, one of the worst things that you can pump into a well is dead oil that is produced out of the well itself. If you are going to use an oil-based fluid, which is different from the oil being produced, then compatibility tests should be conducted. On this particular type of treatment, the previously mentioned Adomite Mark II can be utilized as a fluid loss additive. It should be noted that this type of treatment is very much a banking-type fluid, and it will be very difficult to pump high concentrations of proppant.

Other precautions involve testing of the pumped hydrocarbon with produced waters from the formation to make sure an emulsion doesn't occur and checking the crude itself for paraffin and sludge problems.

Water external emulsion-fracturing fluids

Still commonly utilized in the United States and somewhat overseas are the polyemulsion fluids developed in the early 1970s. This water (and sometimes acid) external emulsion was developed as a very economical means for creating a highly effective viscous fracturing fluid. The fluid uses two-thirds oil and one third water. It was particularly economical when the price of oil was extremely low. It is not so reasonable today, since the price of oil is hovering around $20 a barrel. The economies are good if you are relatively certain you can recover your load, so the polyemulsion fluids still find use in many areas. Common service company names for this produce include Super-emulsifrac*, Super Oil "E"*, and Super Frac Emulsion®.

Normally, you have two-thirds of your tanks on location filled with oil and one-third filled with water. You make a weak-based gel containing guar gum or hydroxypropylguar. This based gel should be tested, just as you would test a linear gel, for viscosity development and degradation with external breaker. The emulsifier is batch-mixed into the water. It is practical to batch-mix the emulsion, for it will remain stable for extended periods. As previously mentioned, I do not recommend doing large volume oil-based treatments. The polyemulsion treatments have become very popular again in West Texas and other parts of the country because of their inexpensive nature and their excellent proppant-carrying properties and fluid loss control. We do, however, recommend switching away from the polyemulsion-type treatment when tank volumes exceed four tanks on location. Batch-mixing larger treatments would require occasional mixing of the emulsion to make sure it is remaining stable and could cause, I believe, some serious problems from the safety and handling of such large quantities of oil.

Polyemulsion fluids have a two-fold breaker mechanism. First of all, breakers must be added to degrade the guar or other viscosifiers used as a stabilizer for the emulsion. Second, the emulsifiers tend to absorb out on the formation itself. These emulsifiers are usually cationic, so they absorb out on sandstone. Some criticize cationic emulsifiers for their tendency to leave a formation oil wet. Newer systems contain anionic emulsifiers that do not oil-wet but that still plate-out. Nevertheless, the breaking mechanism, with the plating-out of the emulsifiers and the internal breakers placed in the gel, functions very well.

Polyemulsions are a highly successful fracturing fluid across the United States. Quality control techniques for this fluid are similar to those for a linear gel system. Polyemulsion fracturing have had a rebirth over the last four to five years, and some people have tried to utilize it even down tubulars. Friction pressures for this fluid are very high, and it is only recommended for use down casing and for relatively small treatments.

One of the most common problems with this particular emulsion have occurred because the oil company inadvertently supplied crude oil for mixing of the emulsion that had already been treated with nonemulsifiers. This causes a very unstable emulsion system, and it can be very embarrassing. Again, it will come down to pilot-testing of fluids on site, which will be discussed in much more detail in Chapter 10. The only disadvantages of using this kind of fluid are relatively high-friction pressure and low-temperature stability (i.e., it cannot typically be used for long-duration treatments at temperatures much above 240°F).

Oil gels using aluminum phosphate esters

The last and most common oil-based fracturing fluid, and the one we will spend the most time discussing, utilizes aluminum phosphate esters. Aluminum phosphate esters were first used as fracturing fluids in 1971 or 1972. These systems closely resemble the napalm utilized by U.S. armed forces. The reaction that gels the oil is between an aluminum phosphate ester and, typically, sodium aluminate. Other bases, however, have been used. But since it is an oil system, the classic terminology for acids and bases does not relate.

These oil gels go by many names, anything from My-T-Oil®, and YF-GO® to Maxi Oil®. All have many similarities. The biggest problem is their critical nature in gelling.

Gelling oils can be exciting

Most of us who have been around the oilfield for any time and who have done many of these jobs have been extremely embarrassed by getting on the location and being unable, for one reason or another, to make an oil gel. This is particularly the case with crude oils. There is always a critical mixture in which you utilize the aluminum phosphate ester in proportion to the sodium aluminate. Typical loadings in the field are approximately 8-10 gals/1,000 of the aluminum phosphate ester to 1-3 gals/1,000 of the sodium aluminate or other base. This varies from crude to crude and sometimes from tank to tank on a location.

If you undertreat, you get no viscosity. If you over treat with the sodium aluminate, you get a degradation of the gel just as if you were breaking it. This is seen on-site as foaming material floating on the oil. If this occurs, you will have serious problems unless you can vigorously mix the tank, get the viscosifier and/or activator back in solution, and change your proportions. One troubling thing that relates to quality in crude selection is the relative heterogeneity of crudes from the same formation, even from offset wells. Sometimes even on 10-acre spacings, the crudes coming from wells side-by-side have different properties. These differences can affect the amount of gellant used.

One thing that causes most of the problems is the treating chemicals used on the wells. Crude additives such as paraffin inhibitors and scale inhibitors also act as degrading or breaking agents for the oil gel. To prevent these kinds of problems, you must exercise strict quality control on crude selection. In many cases, people simply select kerosene or diesel as the oil-based fracturing fluid. This does not always solve the problem. ·

Clean-burning additives common in kerosene or diesels can cause severe problems with gelling or gel stabilizing. Probably no other fluid requires more precise quality control tank-by-tank by the service company than do the oil gel systems.

There have been jobs in areas where a lot of oil-based fracturing had been done. The service company arrived on the location, gelled up the crude, and discovered it broke immediately. At that point they determined that they were gelling up load oil from another fracturing treatment that contained breaker. On other ·occasions, the service company gelled up the crude, and it became too viscous to pump. What they found was they had, in fact, a large amount of additives in the load oil left over from previous treatments.

The breaking mechanism for these types of oil gels is often an over-treatment in one direction or another. A gel is a neutral solution. The activator is a base. If too much activator is added, the gel breaks and becomes very fluid. If too little activator is used, the gel will never form. If surfactants are present, usually the gel will not form. By adding lime or sodium bicarbonate to the oil gel, you over treat it and a degradation occurs.

Another problem with high-gravity condensate is service companies tend to overtreat the materials (i.e., they run too much of the base gellant). I have seen 8 gal aluminum phosphate ester/1,000 gal oil gels that were unpumpable from the tank because they were too viscous. Typically, the gel looks fine as soon as you finish gelling. If any time passes between when you start to pump and when you finish, the gel becomes impossible to pump. Again, this problem can be eliminated by quality control and by checking the oil gel itself.

Oil gelling is difficult and often hazardous. With the present state of the art, there is no better system available than that using aluminum phosphate esters. Other oil gel systems suffer even more problems. Presently, some service companies are supplying continuous-mix oil gelling agents. In fact, in Table 8-1 a list of commonly available oil gel systems is shown. These continuous-mix oil gel systems do indicate some progress on the part of the service companies. Chemical suppliers are supplying chemicals that can gel fairly rapidly. As you might gather,

I am not in favor of utilizing these systems, since I am not in favor of running oil gel systems and not in favor of continuous mix. There are situations (such as offshore) where there is no choice from the standpoint of batch or continuous mix, where people have felt very strongly that oil gel should be utilized and that these systems have indeed been pumped. I would strongly recommend that any oil gel systems that is utilized be batch-mixed and pilot-tested in the manner of intense quality control, such as shown in Chapter 10.

Table 8-1 Oil Gel Systems-Aluminum Phosphate Esters

Four Types	
• Batch-mix, low-temp stability	150-180°F
• Batch-mix, moderate-temp stability	225°F
• Batch-mix, high-temp stability	300°F
• Continuous-mix, low-temp stability	200°F

Vigilant quality control is required

The oil company must insist that either its chemist or its engineer go to the location and take samples from each tank after the tanks have been agitated. Pilot tests should be run to make sure that the proper gel concentration is used and that the gels are neither too thin nor too thick. To give you some idea of the tremendous trauma that the service company went through in pumping or trying to pump oil-based gel systems in the early 1980s, one of the service companies came up with a very temperature-stable oil gel. This system had unique properties that gave it almost a constant viscosity regardless of temperature. Other companies did not have this technology. They were forced to utilize high concentrations of gelling agent and low concentrations of activator. This. was done because if they mixed the high-gel concentrations and a proportionate amount of activator, then the fluids were not pumpable out of the tanks. They added additional activator on the fly. This created some very exciting fracturing treatments and coincidentally many job failures. Today, most of the service companies do have fairly temperatures table oil gel systems; but as stated earlier, most of these are not run on any large treatments or on any large percentage of the fracturing treatments conducted in the United States.

Sequence of components and agitation are critical

The proper procedure for mixing an oil gel with aluminum phosphate ester is to add 8-10 gals. of aluminum phosphate ester/1,000 gal oil uniformly as the frac tank is rolled. Once the predetermined amount is in the frac tank, start adding the catalyst, sodium aluminate, or other base material to the gel carefully and very slowly as you monitor the viscosity. As you approach the predetermined level of the activator, you should see a viscosity rise. With any visible viscosity rise, stop the activator and wait to be sure you don't overtreat; when you overtreat, the fluid becomes too viscous. Continue to monitor the viscosity. If it approaches or equals the level anticipated from lab testing, do not add any more activator. It is extremely important that you do not add too much activator. This would totally break the gel.

Many times in this situation, there is insufficient agitation in the tank for proper mixing. To give you some idea of problems that can occur, let's develop some scenarios. First of all, if you mix the aluminum phosphate ester and the activator directly together, you will create a precipitate. This coagulated material floats to the top, and you won't get any gel, no matter how long you agitate. If you get too high a ratio of the gelling agent (i.e., the aluminum phosphate ester compared to the sodium aluminate and activator), no gel will form. If you get too high a proportion of the activator to the gellant, the gel will break and float. The point is that you need to get adequate dispersion of the aluminum phosphate ester in the tank and then uniformly disperse the activator.

Most often, fracturing tanks utilized in the field have a suction valve on the bottom and a roll line going into the front of the tank that runs to the back. You get excellent agitation when you roll the tank with the blender at around 20 bbl/min. If you do not have a roll line on your tank, then run a suction hose or a discharge hose from the blender over the top of the tank to make sure proper agitation is achieved.

Without a roll line, you are basically sucking out of the bottom of one tank back into the bottom of another without real mixing action. If this is the case, you will probably end up with no gel at all. You and the service company need to check for a roll line before gelling tanks. This sort of thing should be on your quality control list on oil gel jobs when you select your tank supplier, whether it's your service company or a local contractor.

Troubleshooting for gelation problems

If all of the gellant and activator have been added and no gel has formed, consider circulating the fluid at the highest rate possible for 15-20 minutes. At that point, have your service company select samples from the tub and add more gelling agent and/or activator. Adding the proper proportion of chemicals is an art. Your service company engineer should be helpful in this situation. Many times you have to supplement the fluid with one additive or another. You probably got the proportions wrong originally, so some precipitation or floating out of the additive occurred.

Another possibility is that the oil was contaminated. Many times there are residual surfactants or treating chemicals in the bottom of a frac tank that are stirred into and thus contaminate the oil during the actual mixing process. On many occasions, I planned to run 8 gallons of the gelling agent/1,000 gal oil and ended up adding 12 to get an acceptable viscosity. Check with your service company representative to be sure minimum acceptable viscosity is achieved before pumping the gel downhole.

Once the gel is of the proper consistency, take samples from each tank, set them aside, and monitor them for any further viscosity growth. It is not unusual, particularly when there have been gelling problems, for viscosity to increase to the point where you cannot pump the gel out of the tank. If this occurs, dilute this gel with diesel or another crude oil at the location. If you have multiple tanks on the job, open the valve of the problem gel tank and try to suck fairly large quantities out of that tank while pinching down on other tanks so the thick gel can be utilized. Surprisingly, these highly viscous oil gels do not exert extraordinarily high pressure when pumped, particularly if they are diluted with weaker gel systems.

Breaker pilot tests-a necessity

Breaker mechanisms are another area of concern with this type of oil gel systems. The major reason there is a problem with breaker mechanisms in gelled oils relates to our dealing in oil chemistry instead of water chemistry. The typical breakers for oil gels have not evolved much. The first breakers were great scientific discoveries. Adomite Mark II®, one of the first fluid loss additives for oil gel, was an excellent breaker. The problem with it was that, if enough was used to degrade, it degraded the gel immediately; if too little was used, it did not break the gel at all.

The second generation of oil gel breakers are the bases. These breakers work fairly well. The problem with them is that the concentration varies drastically with the kind of oil utilized. A very dangerous assumption for a service company to make is -that it will use the standard breaker concentration for crude oil that it uses for a diesel. Most of the testing and breaker recommendations are carried out with standard base diesel or kerosene solutions.

Many of the oils utilized in the field are raw hydrocarbons (i.e., crude oil) and vary greatly in pH. I am assuming that you can refer to such a thing when it is not an aqueous medium. Often times, very little breaker is required. When much greater concentrations are needed, laboratory tests must be run prior to pumping the fluid downhole. Many wells have been plugged because of insufficient breaker concentration in the gelled oils.

Typically, a serious problem occurs when wells with bottom hole temperatures of less than 120°F are treated. This is insufficient heat for the gel alone to degrade adequately. Most of the breakers mentioned are extremely critical in their ability to degrade at low temperatures. Some new breaking techniques have been devised, but these may need some water present in the breaker itself and/ or water in the oil.

Contaminants in crude oils can greatly affect the breaking mechanism, too. These breakers are also very critical when it comes to potential contamination from treating chemicals. The contamination problem is even more serious than the gelling problem.

Homogeneous sample from each tank is required

The procedures for gelling the oil and identifying the right amount of breaker are difficult at times because of the sensitive nature of some crude oils. This work having been completed, however, may not be the entire picture. On many cases, I have gone to a field location to gel up after considerable laboratory work was conducted. Before adding the gellant and/or activator to the crude oil, I agitated the tank, took a sample, and ran a shake-jar test. In doing this, I found that much more gellant or a different gallant-activator ratio was required.

Often this problem occurs because we do not really know the sources of the crude oils. Crudes coming from different wells or from varying locations where different treatment chemicals are used can cause many problems. Not only have the expenses of extra gellant and activator been added, but you may also have drastically changed the requirement for the breaker. The breaker designed in the laboratory with uncontaminated oil requires much more or much less breaker than the actual field treatment. The point is that you need to have a totally homogeneous oil samples that represent what is actually going to be pumped down the hole. Samples to be laboratory tested should be taken only after the tanks are full and agitated.

Other than the specialty breakers that require water, weak acids are used as breakers for low temperatures by some service companies. These systems also are very critical. Their successful use greatly depends on the crude oil, diesel condensate, or other oil with which they are mixed. The point made time and time again whether the fracturing fluid is water-based, oil-based, methanol, or whatever is that samples from the frac tanks should be tested. This testing ensures adequate gelation (i.e., viscosity is achieved) and that the proper breaker in the correct concentration is used to avoid a disastrous situation during the fracturing treatment. Once again, we will go into the absolute necessity of pilot-testing onsite (i.e., intense quality control), which is a very important part of success on any fracturing treatment. This is certainly the case for oil-based gels.

Success is possible

With all of the drawbacks I have mentioned and with all of the negative things, you can still successfully pump a gelled oil treatment. It just requires a great deal of effort and a great deal of care from the standpoint of safety. Again, I would reiterate that you should do everything possible to be absolutely sure that you need to run this type of treatment before conducting any large volume, oil-based fracturing treatments.

Table 8-2 Correction Factors for Use with Low-Density Carrier Fluids and Texas Nuclear Densimeters

Carrier Fluid, Specific Gravity U	**lb./gal**	**1**	**2**	**3**	**4**	**5**	**6**	**7**	**8**	**9**	**10**
0.70	5.842	0.784	0.862	0.933	0.999	1.060	1.116	1.169	1.218	1.264	1.307
0.71	5.925	0.794	0.871	0.942	1.007	1.068	1.124	1.177	1.226	1.271	1.314
0.72	6.008	0.803	0.880	0.951	1.016	1.046	1.132	1.184	1.233	1.278	1.321
0.73	6.092	0.813	0.889	0.959	1.024	1.084	1.140	1.992	1.240	1.286	1.328
0.74	6.175	0.823	0.898	0.968	1.033	1.093	1.148	1.200	1.248	1.293	1.335
0.75	6.259	0.832	0.908	0.977	1.041	1.100	1.156	1.207	1.255	1.300	1.342
0.76	6.342	0.842	0.917	0.986	1.050	1.109	1.163	1.215	1.262	1.307	1.349
0.77	6.426	0.851	0.926	0.995	1.058	1.117	1.171	1.222	1.270	1.314	1.356
0.78	6.509	0.861	0.935	1.003	1.066	1.125	1.179	1.230	1.277	1.321	1.362
0.79	6.593	0.871	0.944	1.012	1.075	1.133	1.187	1.237	1.284	1.328	1.369
0.80	6.676	0.880	0.953	1.021	1.083	1.141	1.195	1.124	1.292	1.335	13.376
0.82	6.843	0.899									1.397
0.83	6.926	0.909	0.981	1.047	1.101	1.116	1.218	1.268	1.314	1.357	1.397
0.84	7.010	0.918									1.404
1.00	8.345	1.071	1.137	1.197	1.253	1.304	1.352	1.387	1.438	1.477	1.514

Carrier fluids of < 0.8 specific gravity invalidate use of the pounds-of-sand-added (PSA) mode on specific gravity calibrates from 0.8 and greater since the PSA zero control cannot be adjusted properly. The above conversion chart will aid in determining PSA in terms of specific gravity readout. Note, however, that below 0.8 no output will appear on the recorder.

Pressure anomalies may be due to excess breaker

Watch for certain conditions on a gelled oil job. During the treatment, if you see a rapid pressure increase not related directly to sand concentrations or potential screen-out, check with your service company representative. See if the concentration of the breaker that he is adding has been increased. Too much breaker, due to the critical nature of the breakers in the system, gives rapid and complete break of the gel, so you no longer reduce friction. If that is the case, the service company needs to stop the breaker and restart it at a lower concentration. Obviously, this sort of thing should not occur. Flash-breaking [4]occurs because not enough time is spent evaluating the needed breaker concentration during the job. The service company personnel simply were not watching their business as they added the breakers on the fly.

[4] In an instant, the gel goes from being quite viscous to being very fluid.

Densimeter calibration

Of course, oil does not have the same specific gravity as water. This is important when operating a radioactive densimeter. Many service company densimeters use specific gravity. Some service companies are unable to plot pounds per gallon on their strip charts. But you can calculate this with a specific gravity readout and get comprehensive information on sand concentrations.

See Table 8-2. Most of the gelled oils do not have the transport capabilities of the high-temperature, water-based fluids. High-sand concentrations, however, have been pumped with gelled oils at moderate rates in very tight formations. Properly run, quality controlled oil-based fracturing fluids can yield excellent stimulation.

CHAPTER 7

QUALITY CONTROLLING FOAM FRACTURING FLUIDS (1994)

In the early days of foam fracturing, continuing through the 1970s and into the early 1980s, one of the most frightening experiences for an oil company representative was his first foam fracturing treatment. Normally, this individual would be used to seeing readouts of all the different additives and rates, listed as clean and dirty or slurry volumes, and so forth. On many of the early foam fracturing treatments, all that the company representative had was a readout of slurry rate of the fluids pumped. He was then told to believe that the nitrogen or CO, rate was being monitored by another individual. It did not take a rocket scientist to know that when pressures were varying greatly during the treatments and he could see visually many of the nitrogen or C02 trucks going down, that these rates were varying. He had nothing to quantify this in real time. The majority of these treatments were run in a shoddy manner, and very little quality control was conducted. In fact, it is surprising that foam fracturing succeeded as well as it did with all the negativism that was held in the oil community toward it in its early days. One of the significant early problems of foam fracturing was that two-thirds of the money from the treatment did not go to the conventional fracturing service companies; instead, it went to a nitrogen supplier. Today, as we have moved toward all of the service companies supplying nitrogen and $C0_2$, this is not the case. If anything, I think the pendulum has swung in the opposite direction with some service companies overselling foam as the answer to all problems.

Running a foam fracturing treatment can be fairly straightforward or terribly complicated. Many foam frac treatments are designed so that there is no real possibility at conducting the treatment with the equipment on-site. Well-meaning engineers with little knowledge of the equipment capabilities design foam fracturing treatments well beyond the capabilities of the equipment on-site. Examples of these are companies who design treatments and insist on a constant foam quality with a constant rate throughout the treatment.

If proppant concentrations are changed five or more times during the treatment, these treatments are almost impossible to conduct if you insist on constant foam quality and a constant slurry rate. In the latter part of the chapter, I will describe a more practical means of conducting foam fracturing treatments.

The theory of foam

First of all, let's discuss foam fracturing theory. Foam fracturing utilizes foam or foam bubbles as the transport-and-support medium for placement and suspension of proppants. A fairly new process, foam fracturing was introduced to the oilfield in the early 1970s. Some major oilfield service companies did a great deal of research and determined foam fracturing was not feasible. It took brave people to go out and actually conduct a few treatments. We were extremely lucky on early treatments to place proppant in light of the fluid, gas-metering, and sand-handling capabilities available.

Stable foam operates in an explicit range with nonstabilized, water-based fluids. Stabilized water-based foams are simply fluids that contain viscosifiers such as guar, hydroxypropylguar, and in some cases· xanthan or other gelling agents. Figure 9-1 illustrates that a volume of nitrogen, approximately 60-90%, properly dispersed in water and containing a foaming agent, yields a stable foam. The foam's stability and viscosity increase as the quality increases from 60-90%; then the foam reverts to a mist. For an ordinary foam fracturing treatment, this indicates that the major criterion is to stay within a stable foam range during the treatment. Typical fracs are designed for what is termed 70, 75, or 80 quality. This actually means that 70%, 75%, or 80% of the fracturing fluid is gas. Figure 9-2 illustrates viscosity-versus-foam quality for stabilized foams. As you can see, you can have stable fracturing fluids below the 60-quality range shown in Figure 9-1. Virtually no foam fracturing treatments today utilize straight water and foamer because of the inherent problems therein.

We have learned a great deal about foam fracturing primarily due to a great deal of research conducted by the service companies and other consortiums. We have learned that you can enhance the viscosity of foam fracturing fluids by utilizing cross-linked foam. Extensive studies by the service companies indicate a three-to-four-fold increase in downhole viscosity by cross-linking the foam. Not only does this enhance the viscosity and proppant transport, but it also greatly enhances half-life, which will be discussed in following paragraphs. I recommend very strongly that if one is going to utilize a foam fracturing fluid that cross-linked foam systems should be utilized wherever possible.

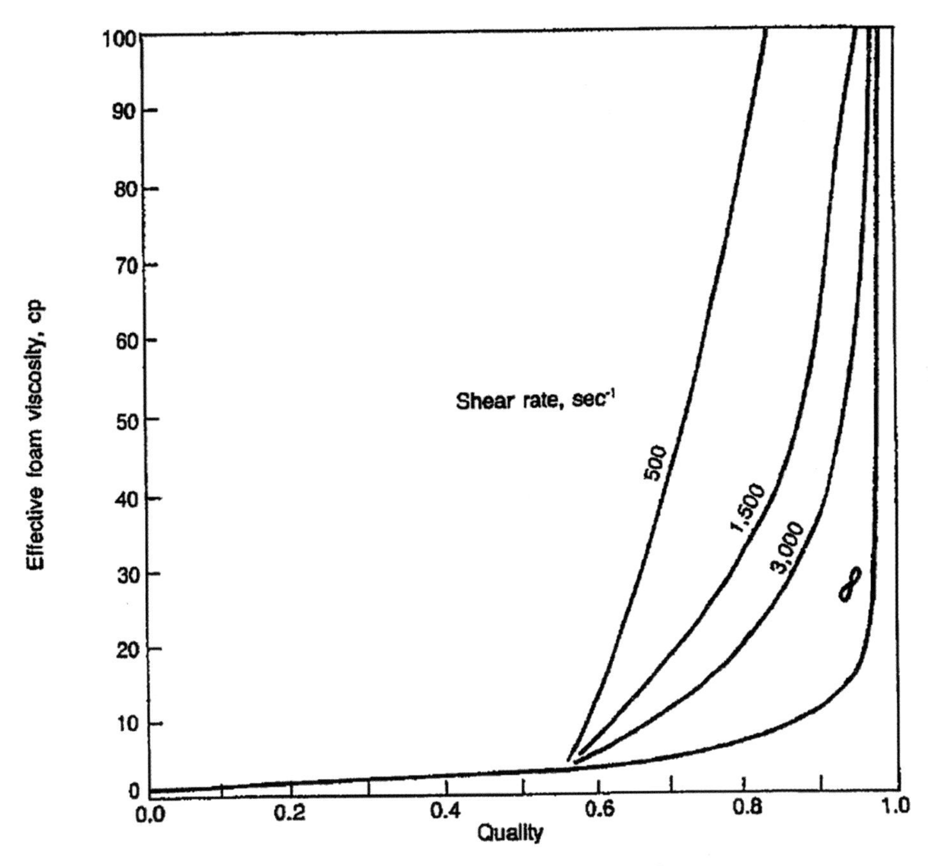

Figure 9-1 Foam viscosity versus quality

Half-life

Stable bubbles are created by turbulence when liquid and gas are mixed. Standing together, these bubbles make a foam that will not break out for a measured period called drain time or half-life. This is the time the foam takes to break out under atmospheric conditions (60°F, 14.7 psia), when half of the water used in preparing the foam mixture returns to its original state. A foam quality of 70-80%, with a good quality foamer without stabilizers, generally yields a half-life of 3-4 minutes. Many fracturing treatments have been conducted in this manner. If you add stabilizers, which are usually just viscosifiers or gelling agents, half-lives in the range of 20-30 minutes are· feasible and easily obtained. Much longer half-lives are seen under pressure.

Half-life tests are therefore something that oilfield service company chemists or engineers normally check before conducting a treatment. A simple test can be carried out in a Waring blender with the additives, stabilizers, foamers, and other ingredients to be used on the treatment. This test is particularly advisable if any additives not previously run are added to the treatment. Many additives can drastically affect the quality and the half-life of the foam. When one cross-links the foams, almost infinite half-lives are seen until the breaking agents actually break the cross-link and finally the polymer.

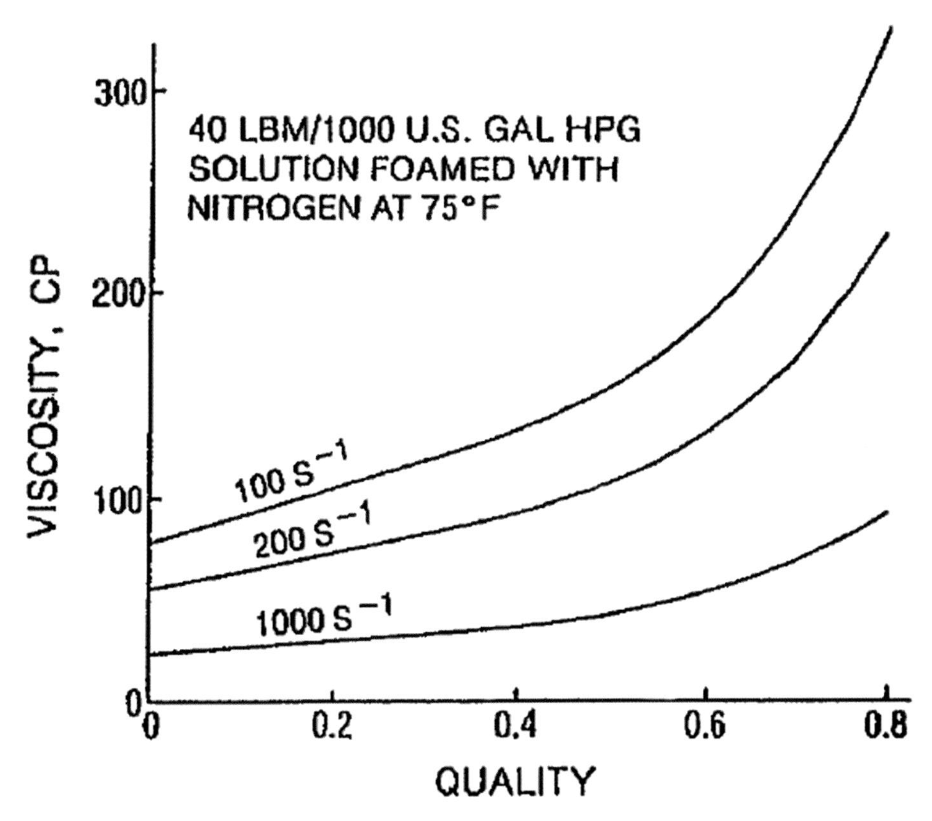

Figure 9-2 Effect of quality and shear rate on the viscosity of a 40-lbm/1,000-gal HPG solution foamed with N2 at 75°F.

Quality control is more difficult for foam

Obviously, quality control for a foam fracturing treatment, whether nitrogen or carbon dioxide, is perhaps more difficult than for gelled oil, gelled water, or crosslinked fracturing fluid treatments. Two and sometimes three totally different mediums are added together, and their concentrations are extremely critical. Variations from these concentrations can be disastrous for proppant carrying and proppant suspension in the fracture. Good planning and persistent quality control must be conducted both before and during a foam fracturing treatment.

Table 9-1 gives a treatment program for a constant clean fluid side and constant nitrogen-rate foam frac treatment. Table 9-2 illustrates a constant downhole-rate foam frac treatment. Table 9-3 illustrates a constant internal phase foam fracturing treatment. In this case, the rate of nitrogen and/or C02 is dropped proportionately to the volume of proppant utilized. This is based on a theory that the proppant becomes part of the internal phase. In my opinion, the viscosity and proportionality of the internal phase may or may not be a function of the sand, but you are able to utilize this technique to allow much higher proppant concentrations to be pumped. Even with decreased foam quality with foam stabilizers, you do have a good proppant transport mechanism, particularly if the foam is cross-linked. For the inexperienced engineer, Tables 9-1, 9-2, and 9-3 can be complex and confusing.

In fact, they are a precise means of controlling a treatment. With this kind of prejob preparation, you know precisely what to pump. I don't think it takes long to see that Table 9-1 would be a much easier type treatment to conduct, whereby one would maintain a constant dean rate, a constant nitrogen or C02 rate, and just increase the rate as the proppant concentrations are increased. Obviously, this type of treatment is impossible to conduct if you are at maximum pressure just prior to starting sand. Because of this event occurring, (i.e., maximum pressure and being unable to do increasing slurry rate) Table 9-3 or constant internal phase technique was developed.

Table 9-1 Increasing Slurry Rate Foam Frac Pump Schedule

Foam	**Liquid**	**Proppant**			**Slurry Volume**		**Rate**				
Volumes (gals)	**Volumes (gals)**	Foam (ppg)	Liquid (ppg)	Total (lbs)	Foam (gals)	Blend (gals)	N_2 (scfm)	Liquid (bpm)	Sand (bpm)	Total (bpm))	Time Min-Sec
35,000	10,500	0.0	0.0	0	35,000	10,500	13,820	4.5	0.0	15.0	55:33
25,000	7,500	1.0	3.3	25,000	26,130	8,630	13,820	4.5	0.7	15.7	39:37
30,000	9,000	2.0	6.7	60,000	37,712	11,712	13,820	4.5	1.4	16.4	47:29
12,500	3,750	3.0	10.0	37,500	14,195	5,442	13,820	4.5	2.0	17.0	19:53
10,000	3,000	4.0	13.3	40,000	11,808	4,808	13,820	4.5	2.7	17.4	15:53
7,500	2,250	5.0	16.7	37,500	9,195	3,945	13,820	4.5	3.4	18.4	11:54
7,500	2,250	6.0	20.0	45,000	9,534	4,284	13,820	4.5	4.06	19.06	11:54
1,800	540	0.0	0.0	0	1,800	540	13,820	8.56	0.0	19.06	2:52

Table 9-2 Constant Slurry Rate Foam Frac Pump Schedule

Foam Volumes (gals)	Liquid Volumes (gals)	Proppant Foam (ppg)	Proppant Liquid (ppg)	Proppant Total (lbs)	Slurry Volume Foam (gals)	Slurry Volume Blend (gals)	Rate N_2 (scfm)	Rate Liquid (bpm)	Rate Sand (bpm)	Rate Total (bpm))	Time Min-Sec
35,000	10,500	0.0	0.0	0	35,000	10,500	13,820	4.5	0.0	15.0	55:33
25,000	7,500	1.0	3.3	25,000	26,130	8,630	13,225	4.3	0.06	15.0	41:28
30,000	9,000	2.0	6.7	60,000	37,712	11,712	12,700	4.1	1.25	15.0	51:55
12,500	3,750	3.0	10.0	37,500	14,195	5,442	12,200	3.96	1.8	15.0	22:32
10,000	3,000	4.0	13.3	40,000	11,808	4,808	11,700	3.81	2.3	15.0	18:44
7,500	2,250	5.0	16.7	37,500	9,195	3,945	11,223	6.67	2.8	15.0	14:35
7,500	2,250	6.0	20.0	45,000	9,534	4,284	10,860	3.55	3.2	15.0	15:08
1,800	540	0.0	0.0	0	1,800	540	13,820	8.56	0.0	15.0	2:51

Table 9-3 Constant Internal-Phase 70 Quality Cross-linked Foam

Foam Volumes (gals)	Liquid Volumes (gals)	Proppant Foam (ppg)	Proppant Liquid (ppg)	Proppant Total (lbs)	Slurry Volume Foam (gals)	Slurry Volume Blend (gals)	Rate CO_2 (bpm)	Rate Liquid (bpm)	Rate Sand (bpm)	Rate Total (bpm))
40,000	12,00	0.0	0.0	0	40,000	12,000	21.0	9.0	0.0	30
2,500	450	2.0	5.3	5,000	2,728	1,178	18.6	9.0	2.4	30
4,000	1,867	4.0	8.6	16,000	4,721	2,597	16.0	9.0	5.0	30
6,000	2,760	6.0	13.04	36,000	7,642	4,402	13.8	9.0	7.2	30
8,000	4,853	8.0	13.20	64,000	10,918	7,771	11.8	9.0	9.2	30
10,000	6,733	10.0	14.85	100,000	14,560	19,109	9.8	9.0	11.2	30
15,000	10,901	12.0	16.5	180,000	23,208	504	8	9.0	12.8	30
1,800	540	0.0	0.0	0	1,800	540	21.0*	9.0	0.0	30

*If practical, prefer to flush with no $C0_2$

Design

To understand better what goes into the design of a foam fracturing treatment, refer to Table 9-4. Here the procedures are set forth for foam frac design. By following these procedures, you can determine the amounts of liquid and gas required for a certain foam quality. Particular characteristics of a typical well are given in Table 9-5, and the problem is solved on the completed form. Some of the data in Table 9-5 are obtained by utilizing several curves from a nitrogen-foam frac design manual.

Table 9-4 Foam Fracture Design Worksheet [5]

BHTP	Bottom hole treating pressure, psi	Ph	Hydrostatic pressure, psi
BN_2	Nitrogen volume factor, scf/bbl.	Phf	Foam hydrostatic pressure, psi
Cs	Sand concentration, ppg	Ppf	Perforation friction, psi
Df	Depth of formation, ft	Pw	Wellhead pressure, psi
Dp	Depth to packer, ft	QN	Nitrogen injection rate, bbl./min
Fg	Foam gradient, psf ·	qN	Nitrogen injection rate, scfm
FG	Fra.c gradient, psf	Qs	Sand injection rate, lb/min
FQ	Foam quality	Qt	Total foam injection rate, bbl./min
hhp	Hydraulic horsepower, hhp	Qw	Water injection rate, bbl./min
ISIP	Instant shutin pressure, psi	Tbh	Bottom hole temperature, op
Pf	Pipe friction, psi	Ts	Surface temperature, °F

Given Data

1. Dp = __________ ft.
2. Df = __________ ft.
3. Tubing - __________ in., __________ lb./ft.
4. Casing- __________ in., __________ lb./ft.
5. Perforation diameter __________ in. Number of perfs __________
6. ISIP = __________ psi
7. FG = (ISIP + Ph)/Df
   FG = (__________+__________) / __________= __________ = __________psi/ft.
8. Ts= __________ °F
9. Tbh = __________ °F

Assumed Data (at bottom hole conditions)

1. FQ = ____________
2. Qt = ____________ bbl./min.
3. Cs = ____________ ppg

Flow Rate Calculations

1. BHTP = FG x Df = __________ x __________ = __________ psi
2. BN_2 = __________ scf/bbl/ of space (at BHTP and Tbh)
3. QN_2= GQxQt = ____________ x ____________ = ____________ bbl./min.
4. $qN_2 = QN_2 \times BN_2$ = ____________x ____________ = ____________ scfm
5. Qw = Qt - QN_2 = ____________ - ______________ = ____________ bbl./min.
6. qN_{2}/ Qw = ____________ / ____________ = ____________ scf/bbl. of liquid

Surface Pressure Calculations

1. ΔPf = __________ psi/100 ft. (from service company tables)
2. Pf = 0.01 x Dp x Δ = 0.01 x ____________ x ____________ = ____________ psi

[5] Adapted from Nowsco Services Technical Manual, © 1980, Nowsco Services, USA.

3. Ppd = ____________ psi
4. Fg (at BHTP, Tbh, and FQ) = ___________ psi/ft (from Table IV)
5. Phf = Df x Fg = ___________ x ____________ = ____________ psi
6. Pw = BHTP - Phf + Pf + Ppf

   Pw= __________ - __________ + _________ +__________ = __________ psi

Hydraulic Horsepower Calculation

hhp = 0.0245 x Qw x Pw = 0.0245 x _________ x _________ = _________ hhp

Surface Foam Quality and Rate Calculations

1. BN_2 = _____________ scf/bbl. of space
2. $QN_2 = qN_2 / BN_2$ = ____________ / ____________ = ____________ bbl./min.
3. $Qt = Qw + QN_2$ = ____________ + ____________ = ____________ bbl./min.
4. $FQ = QN_2 / Qt$ = ____________ / ____________ = ____________

Sand Concentration Calculation

1. Qs = 42 x Qt x Cs (at bottom hold conditions)

   Qs = 42 x ____________ + ____________ = ____________ lb./min.
2. Cs = Qs / (42 x Qw) (at blender

   Cs = ____________ / (___________ x ____________) = lb./gal.

Table 9-5 Completed Foam Fracture Design Worksheet

BHTP	Bottom hole treating pressure, psi	Ph	Hydrostatic pressure, psi
BN_2	Nitrogen volume factor, scf/bbl.	Phf	Foam hydrostatic pressure, psi
Cs	Sand concentration, ppg	Ppf	Perforation friction, psi
Df	Depth of formation, ft	Pw	Wellhead pressure, psi
Dp	Depth to packer, ft	QN	Nitrogen injection rate, bbl./min
Fg	Foam gradient, psf ·	qN	Nitrogen injection rate, scfm
FG	Fra.c gradient, psf	Qs	Sand injection rate, lb/min
FQ	Foam quality	Qt	Total foam injection rate, bbl./min
hhp	Hydraulic horsepower, hhp	Qw	Water injection rate, bbl./min
ISIP	Instant shutin pressure, psi	Tbh	Bottom hole temperature, op
Pf	Pipe friction, psi	Ts	Surface temperature, °F

Given Data

1. Dp = 4,000 ft.
2. Df = 4,100 ft.
3. Tubing – 2 3/8 in., 4.7 lb./ft.
4. Casing- 4 ½ in., 9.5 lb./ft.
5. Perforation diameter 0.35 in. Number of perfs 40
6. ISIP = 1,200 psi
7. FG = (ISIP + Ph)/Df

   FG = (1,200 + 1,775) / 14,100 = 0.73 psi/ft.

8. Ts= 80 °F
9. Tbh = 120 °F

Assumed Data (at bottom hole conditions)

1. FQ = 0.75
2. Qt = 12 bbl./min.
3. Cs = 2 ppg

Flow Rate Calculations

1. BHTP = FG x Df = 0.73 x 4,100 = 2,933 psi
2. BN_2 = 961 scf/bbl/ of space (at BHTP and Tbh)
3. QN_2= GQxQt = 0.75 x 12 = 9 bbl./min.
4. $qN_2 = QN_2 \times BN_2$ = 9 x 961 = 8,649 scfm
5. Qw = Qt - QN_2 = 12 - 9 = 3 bbl./min.
6. $qN_{2/}$Qw = 8,649 / 3 = 2,833 scf/bbl. of liquid

Surface Pressure Calculations

1. ΔPf = 76 psi/100 ft. (from service company tables)
2. Pf = 0.01 x Dp x Δ = 0.01 x 4,000 x 76 = 3,040 psi
3. Ppd = 4 psi
4. Fg (at BHTP, Tbh, and FQ) = 0.1729 psi/ft (from Table IV)
5. Phf = Df x Fg = 4,100 x 0.1729 = 709 psi
6. Pw = BHTP - Phf + Pf + Ppf

   Pw= 2,993 - 709 + 3,040 + 4 = 5,328 psi

Hydraulic Horsepower Calculation

hhp = 0.0245 x Qw x Pw = 0.0245 x 3 x 5,328 = 392 hhp

Surface Foam Quality and Rate Calculations

1. BN_2 = 1,608 scf/bbl. of space
2. $QN_2 = qN_2 / BN_2$ = 8,649 / 1,608 = 5.38 bbl./min.
3. Qt = Qw + $QN_{2=}$ 3 + 5.38 = 8.38 bbl./min.
4. FQ = QN_2/ Qt = 5.38 / 8.38 = 0.64

Sand Concentration Calculation

a. Qs = 42 x Qt x Cs (at bottom hold conditions)

   Qs = 42 x 12 + 2 = 1,008 lb./min.

b. Cs = Qs / (42 x Qw) (at blender

   Cs = 1,008 / 42 x 3 = 8 lb./gal.

Calculating required quantities of nitrogen under bottom hole conditions is laborious and requires a great deal of work or a computer program. If the amounts of water and nitrogen available must be changed frequently because of a constant slurry rate, obviously you can become quite confused on a treatment. This is true because the actual downhole pressures during a treatment can change due to an increasing bottom hole treating pressure. If, in fact, this does occur, surface changes must be made to accommodate for the pressure change, which implies a different confining pressure upon the compressible gas.

Consider carefully the inherent problems when you try to make several changes in flow rate during a treatment. Changing the pump rate on one or two pumps, which is typical on the fluid side, is not a big problem. However, most modern nitrogen treatments utilize a large number of trucks. The treatment shown in Figure 9-3 utilized about 30 trucks. Coordinating and controlling this many units while varying nitrogen rates becomes a logistical nightmare. The less confusion, the greater the chances for a successful treatment.

Guidelines

Step-by-step instructions follow for the observing engineer to monitor and quality control a foam fracturing treatment.

a. Follow procedures for evaluating viscosity, temperature, pH, and other base fluids characteristics (See Chapters 6-8). Usually, you will utilize either a 2D-40 lb. gelled fracturing fluid on water or a low-strength aluminum phosphate gel if the fluid is oil based. Quality controlling the viscosity of this fluid is just as important as it is on a linear or a cross-linked fracturing treatment. Determine the volume of base fluid present before the job and again on its completion. Also monitor fluid usage during the treatment to be sure foam quality is maintained.

Figure 9-3 Nitrogen foam treatment

c. Follow intense quality control procedures, whereby the fluids are tested at insitu conditions for viscosity maintenance and breakdown of the fluid. When utilizing C02 as a fluid, it is important to buffer the fluid down to 3.5 to 4 pH so that the breaker conditions are similar to those of the fracturing fluid. If cross-linked foam is used, you should, of course, cross-link the fluid and test it under cross-linked conditions with a B2 bob as discussed in Chapter 10 for intense quality control.
d. Consult with the service company about how it will obtain the required sand concentrations downhole. Previously, some service companies were utilizing sand concentrators, which was a very dangerous and work-intensive device. Fluids were concentrated and clean fluid was sent back to tanks while high concentration sand fluids were sent downhole under high-pressure conditions. Today, all service companies can mix up to 22 lbs./gallon at the blender tub, allowing for dilution of sand for a constant slurry-rate treatment, thereby achieving about 8 lbs./gallon downhole. With constant internal phase treatments, you should be able to achieve 12 lbs./gallon and higher downhole concentrations by modifying the gaseous phase to accommodate the sand. important criteria, which were fully discussed in the chapter on Rig-Up, relate to maintaining adequate velocity to be sure that sand does not pack off in the pumps. It is an important consideration to be sure that one has adequate flow rates on high-rate treatments, such as having sufficient suction hoses. For foam frac treatments, where the clean fluid rate or slurry fluid rate can become quite low, you should use smaller hoses such as 3 in. between the blender and the pump truck to make sure that the velocity is high enough to carry the sand from the blender to the pump truck without packing off or causing serious problems due to slugging of sand into the triplex pumps.
e. Consult the service company engineer about his monitoring techniques on the clean and dirty sides of the blender as well as his monitoring capabilities for the nitrogen or C02.The clean side flow absolutely must be monitored. This flow is the actual fluid that will be foamed and the fluid with which you calculate the foam quality. On the dirty side, you are measuring a mixture of the fluid and proppant. Obviously, proppant will not foam. You also need an inline nitrogen or C02 flowmeter. Several types of these in-line systems are available in the industry. One meter is a turbine that is temperature and pressure compensated. There are also mass flowmeters that do not require temperature or pressure compensation. Properly calibrated, any of these meters will suffice. Conventional turbine meters work well for C02.
f. Check with the service company to be absolutely sure all flow rates are monitored easily from a digital readout. These rates also must be strip-charted so you can follow the records during the treatment and see trends. Work closely with your service company representative in calculating anticipated treating pressure.

   In foam fracturing, pressure tells you more about the quality of the fluids downhole than it does in any other treatment. If you are treating at a lower pressure than anticipated, you are most likely pumping a low-quality foam (i.e., one with a low nitrogen-to-fluid ratio). If you are pumping at a much higher than expected pressure, then the opposite is probably the case.
g. During the treatment, one service company representative should be communicating with the personnel metering the foamer downstream of the pressurizing pump on the blender. Not adding the foamer, of course, negates any chance for a successful treatment. Some service companies have a separate pump that pumps into a discharge manifold or pumps downstream of the blender pump. Other companies actually pump from a tank on the blender into the suction of the discharge centrifugal pump. A running count of volumes pumped as well as pumping rates should be checked regularly to be sure the proper amount of foamer is in the fluid.

Quality control guidelines for fluid loss and other solid additives added continuously to linear and cross-linked fluids need to be followed for foaming fluids. There is one exception for the foaming fluids. Obviously, if you plan to run 25lbs. of a fluid loss additive per 1,000 gals. fluid in the foam, you need to run a lot of this additive-more than that -required for the foam quality. For instance, if you run a 75 quality foam, you will be running 100 lbs./1000 gal of the clean side fluid to obtain 25 lbs. downhole. This holds for proppant or any other additives. The foamer is an exception, since the foamer concentrations are for the clean fluid only.

Carbon dioxide foam fracturing

Stable foam fracturing may be accomplished when carbon dioxide is the gaseous component with much the same results as the more conventional nitrogenbased foams. There are similarities as well as differences between the two systems. This section presents information pertinent for the design and implementation of carbon dioxide foamed-fracturing systems with specific quality control considerations.

Comparison of nitrogen and carbon dioxide

The most notable difference between carbon dioxide and nitrogen is that the gas is pumped in its liquid phase and allowed to vaporize downhole in the carbon dioxide systems. This is in direct contrast with nitrogen foams, where the gas is vaporized before mixing with the liquid component and a stable foam is generated on the surface. A basic similarity for the two systems is that in each the components are blended to yield a predetermined gas-to-liquid ratio or "quality" at bottom hole treating conditions. This ratio is critical to the success of foamed treatments and must be considered with the peculiarities of the carbon dioxide foam process in mind. Table 9-6 illustrates comparative properties of C02 and N2.

Composition

The carbon dioxide foam system is comprised of abase liquid (usually water), liquid carbon dioxide, a foaming agent, and an emulsifier. The base liquid is gelled, and the various surfactants and proppants are introduced at the blender. The mixture is subsequently emulsified with the liquid carbon dioxide downstream from the high-pressure pumps. Even though the majority of the mixture's

Table 9-6 Nitrogen versus Carbon Dioxide

Feature	N_2	CO_2
Relative Density	Low	High
Solubility	Low	High
Surface Tension	No effect	Lower
Compatibility	Good	Limited
Reactivity	Inert	Acidic

volume is liquid C02, which has a low viscosity, the fact that the liquids are emulsified imparts high viscosity to the system. There has been a great deal of confusion within the industry relating to terminology. Although you can see definite change in phase when you reach the critical temperature of CO, where the emulsion does change into a foam, they effectively perform the same function, one with liquid C02 as the internal phase and the other with gaseous CO, as the internal phase. C02 emulsions have severe pumping friction due to the high density of C02 (i.e., that akin to water).

Treatment design

The foam quality of a carbon dioxide foam treatment should be calculated with reservoir temperature and bottom hole treating pressure in mind. Usually, "treatments are designed for ultimate foam quality in the 0.65-0.70 range. This recommendation is based on such considerations as optimum system performance, economics, and minimization of friction loss in tubulars. In terms of fluid performance and friction loss values, the CO_2-water emulsion behaves like the polyemulsion systems. In those fluids, variations from the normal 2:3-1:3 mixture result in drastic changes in fluid performance and associated friction loss values. The same may be expected with the carbon dioxide foam systems. Field experience shows that 0.65-0.70 quality carbon dioxide foams are optimum at the previous mentioned consideration. After running C02 energized and CO, foam treatments for many years, one of the important discoveries was that the use of the constant internal-phase technique was a very valuable one for CO, foam treatments or just simply C02 energized treatments. Once you approach 70 quality CO, foam, you start to achieve very high friction pressures. Very good empirical work has shown that friction pressures can go up exponentially as sand concentrations rise in CO, foam systems. I very much recommend that you utilize constant internal phase or something akin to modified constant internal phase systems where C02 foams are involved, particularly when you are pumping down deep tubulars. On high sand concentration treatments, where one starts out the treatment at 70 quality foam, you may have less than 40 percent C02 in the system due to the addition of the sand at fairly high concentrations. For C02 energized systems, even at 30%, you tend to decrease foam quality due to friction pressure seen in tubular goods. CO, causes so much friction pressure compared to nitrogen because of its high density. The fact that it creates an emulsion is secondary to the effect of relative density even in gaseous state.

Determining carbon dioxide requirements

A step-by-step procedure for determining carbon dioxide requirements for desired foam quality in the fracture is presented below. The example illustrates the steps necessary to design a treatment with the following parameters:

- 0. 70 quality foam
- 200°F BHT
- 6,000 psia bottom hole treating pressure
- 20 bbl./min. total rate (foam)

1. Determine the desired foam quality. The 0.70 quality is recommended but not requisite; 0.70 quality is used in this example.

2. To yield the quality determined in item 1, a certain quantity of carbon dioxide must be mixed with the base fluid to yield the necessary standard cubic feet of carbon dioxide per barrel of fluid. This quantity, in standard cubic feet per cubic foot, is calculated as follows:

$$V_1 = \frac{(35.3741\ P_2)V_2}{z_2 T_2}$$

Where:

V_1 = volume at standard conditions of temperature and pressure (14.7 psia, 60°F), scf
P_2 = bottom hole treating pressure, psia
V_2 = volume at treating conditions of temperature and pressure, psia
(This is a decimal equivalent of the desired foam quality and may be viewed as "relative volume.")
z_2 = compressibility factor for carbon dioxide at treating conditions[6]
T_2 = reservoir temperature, °R (°F + 450)

So: $V_1 = \frac{(35.3741\ (6{,}000)(0.7)}{(0.76)(200 + 460)}$

V_1 = 296.19 scf/cu ft

3. Convert item 2 value to standard cubic feet per barrel of base liquid:

$$\frac{V_1}{1 - FQ} \text{ x } 5.6146 = \text{scf/bbl}$$

Where:

V_1 = scf/cu ft (from item 2)
FQ = foam quality

Therefore: $\frac{296.19}{0.3}$ x 5.6146 = 5,543 scf/bbl

4. Compensate for the solubility of carbon dioxide in the base fluid.t Add the water solubility to the value obtained in item 3.
Solubility= 190 scf/bbl.

5,543 + 190 = 5,733 scf/bbl.

5. Convert the item 4 to the liquid carbon dioxide rate to barrels per minute at the appropriate discharge pressure and temperature.
   a. Determine the temperature of the liquid carbon dioxide at the discharge pressure and the vessel pressure. [7]Assume the vessel pressure is 260 psi for the example.+
   6,000 psi x 260 psi = 22.17°F
   b. Determine the density of the liquid carbon dioxide at the discharge temperature.
   22° = 58.64 lb./cu ft

---

6 Angus, McReynolds in Nowsco Services C02 Manual, © 1984, Nowsco, USA, Fig. 4. +McReynolds, op. cit., Fig. 6
7 McReynolds, op. cit., Table VII.
+McReyoolds, op. cit., Table V.
+McReyoolds, op. cit., Table VI.
§McReynolds. The pressure-temperature relationship of liquid carbon dioxide is illustrated by the saturated liquid curve in Figure 9.4.

c. Determine liquid carbon dioxide after correcting for density.

Assume: 1 lb. =8.61 scf @ 0°F

Density @ 0°F = 61.95 lb./cu ft

$$8.61 \quad \frac{58.64 \text{ lb./cu ft}}{61.95 \text{ lb./cu ft}} = 329.25 \text{ lb./bbl}$$

d. Determine the pounds per barrel of liquid carbon dioxide at the discharge temperature.

Assume 1 cu ft = 0.1781 bbl.

$$\frac{\text{lb./cu ft}}{\text{lb./cu ft}} = \frac{58.64}{0.1781} = 329.25 \text{ lb./bbl.}$$

e. Determine the standard cubic feet per pound of liquid $C0_{2.}$

1.25 b./bbl. x 8 scf / 149 lb. = 2,683 scf/bbl.

f. Convert standard cubic feet of carbon dioxide per barrel of base fluid (from item 4) to barrels of carbon dioxide per barrel of base fluid.

$$\frac{5{,}733}{2{,}683} = 2{,}136 \text{ bbl. liquid } CO_2 \text{ / bbl. base fluid}$$

PVT relationships

Liquid carbon dioxide exists between -70°F and 87.8°F and at pressures between 60 and 1,056.3 psi, respectively. If the pressure is reduced, carbon dioxide converts to either a vapor or a solid. The minimum pressure required to maintain a liquid state is a function of temperature. § This pressure-temperature relationship explains the need to maintain at least 60 psi on treating lines at all times. The problem of maintaining pressure is obviated on flow backs because of the significantly higher temperatures of the carbon dioxide then. The transition from liquid to vapor is also illustrated in Figure 9-4. Once the critical temperature is exceeded, it is impossible to maintain a liquid phase, regardless of pressure. The transition from vapor-liquid to a completely vapor phase, from -70 to 87.9°F, is a function of pressure. This relationship is defined by the saturated vapor curve in Figure 9-4. In oilfield applications, the transition to a complete vapor state is generally reached as the C02-treating fluid mixture nears the formation or when the mixture just slightly penetrates the formation; The probable phase behavior of carbon dioxide during a treatment is illustrated in Figure 9-4.

Solubility

The solubility of carbon dioxide in fluids is generally high, especially when compared with another gas such as nitrogen. This aspect of carbon dioxide's physical properties can be beneficial in well stimulation treatments. Carbon dioxide's solubility in fluids depends on the liquid in question, on pressure, and on temperature.

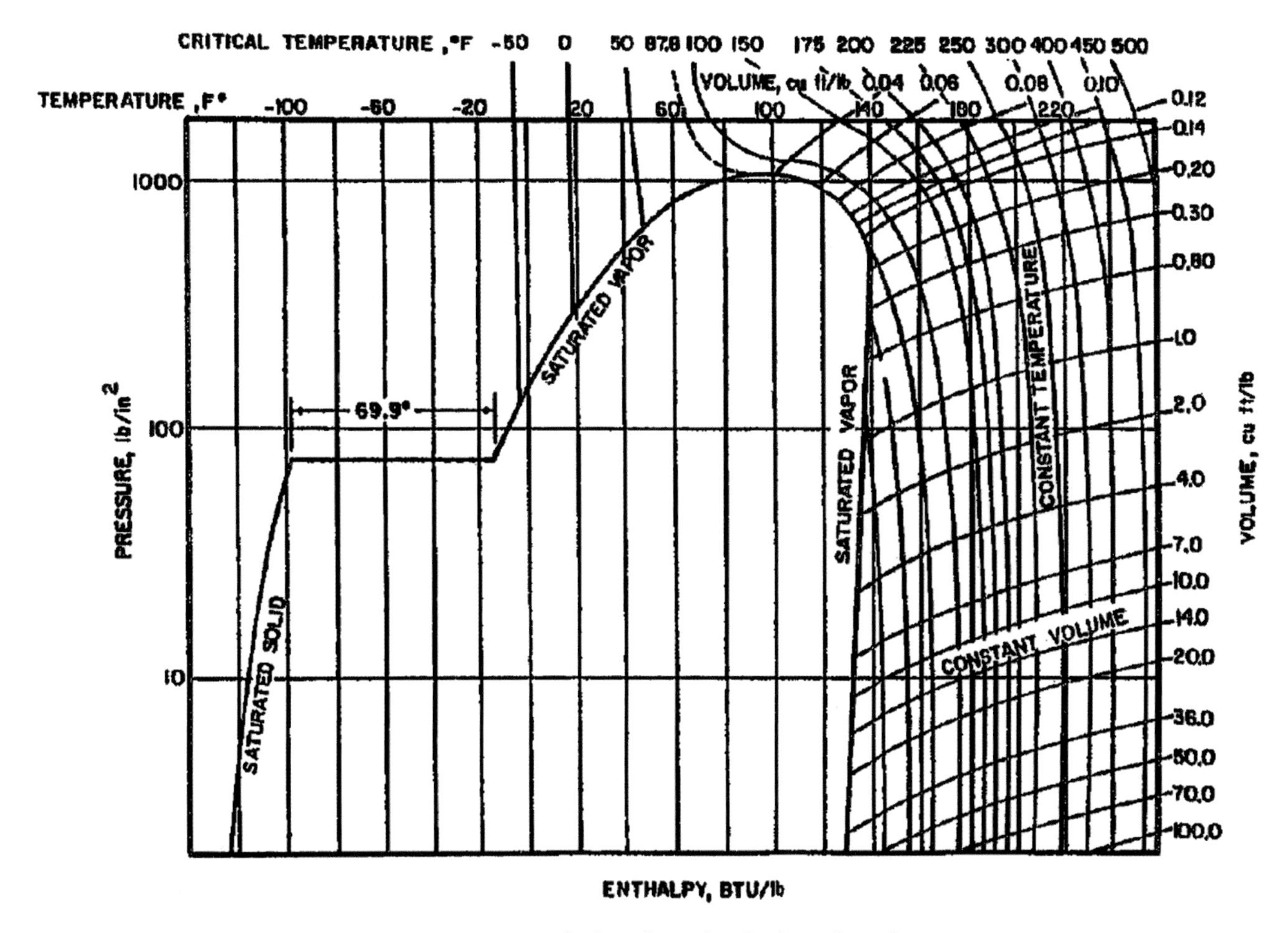

Figure 9-4 Enthalpy chart for Carbon dioxide.

Carbon dioxide's ability to dissolve in fresh water is illustrated by Figure 9-5. Note that at any given pressure, the amount of carbon dioxide to go into solution decreases with an increase in temperature. Conversely, at any given temperature the solubility increases with pressure.

Somewhat less carbon dioxide goes into solution in light brines than in fresh water. Carbon dioxide solubility decreases with increased salinity or total dissolved solids. Table 9-7 shows the correction factors to compensate carbon dioxide solubility in fresh water for the presence of dissolved solids in brines.

Multiply the appropriate correction factor by the solubility of carbon dioxide (in standard cubic feet) in water (in barrels). Carbon dioxide dissolves readily in most crudes. As in aqueous fluids, solubility is a function of density. As a result, more carbon dioxide will dissolve in lighter, higher-gravity crudes than in denser, lower-gravity oils. An interesting corollary to the C02-crudes interaction is that once a given amount of gas has dissolved in a crude, little more gas will go into solution, regardless of increasing pressure.

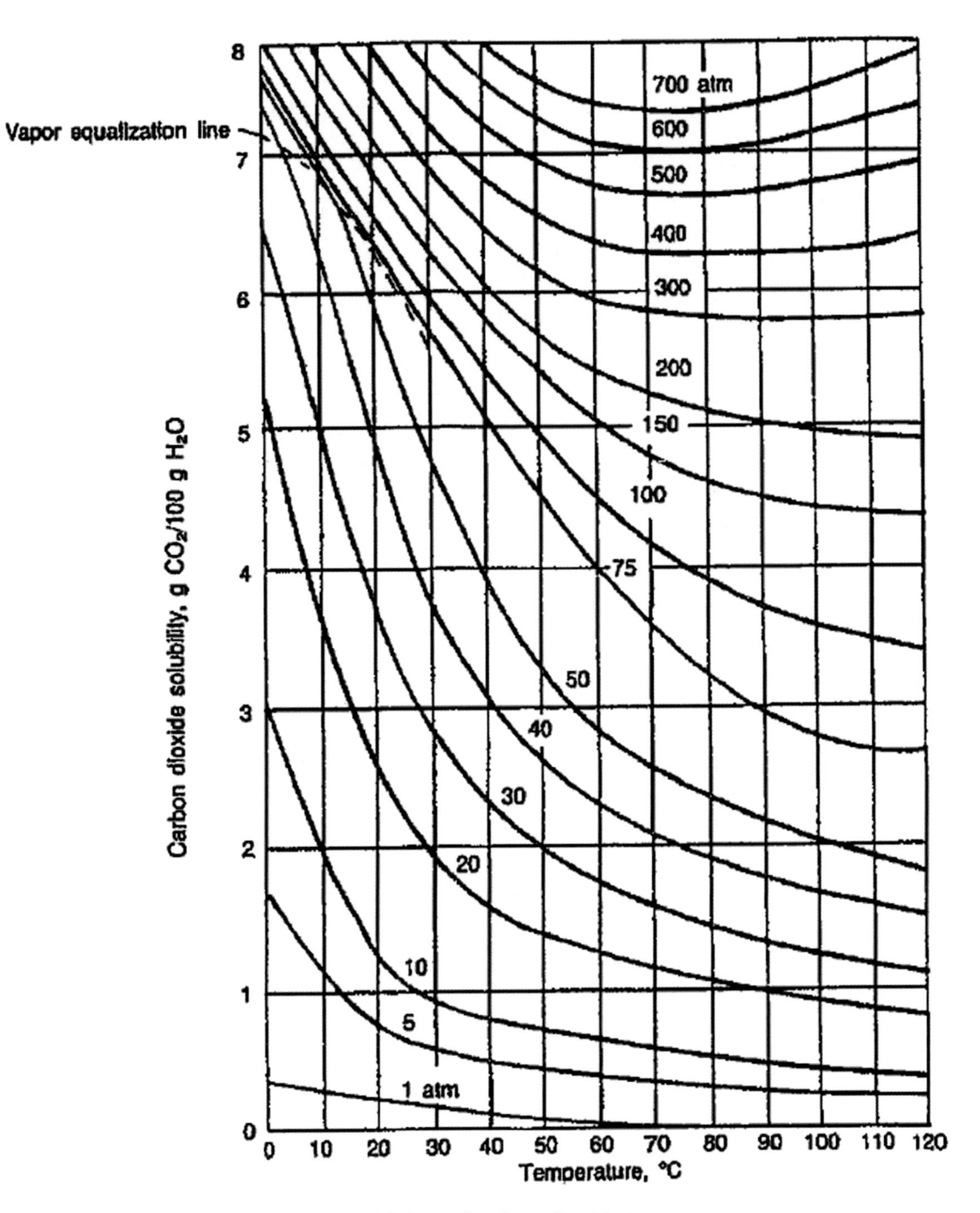

Figure 9-5 Solubility of carbon dioxide in water.

Table 9-7 Carbon dioxide solubility in standard cubic feet barrels for various solutions.

Solubility of Carbon Dioxide scf/bbl.	Pressure, psi			
	100	**1,000**	**2,000**	**4,000**
Fresh water (100°F)	20	152	174	191
Salt water (100,000 ppm, 100°F)	13	108	127	139
Salt wliter (260,000 ppm, 100°F)	6	53	63	69
Crude oil, 38°API gravity (85°F)	45	1,025	1,075	1,075
Crude oil, 20°API gravity (120°F)	35	415	700	700

Quality control procedures

Once you are satisfied that the service company has followed the correct procedures in the treatment design, quality control is quite similar to that for a nitrogen-foam treatment. The following points need to be added to the nitrogen foam treatment quality control procedures.

Be absolutely sure that primary and standby arrangements are made for monitoring the carbon dioxide liquid as it is pumped. A common procedure is to put a low-pressure flowmeter between the carbon dioxide transports or storage units and the high-pressure pumps and to put a high-pressure flowmeter downstream of these units. Unfortunately, these flowmeters, particularly if they are turbine units, are often destroyed when the carbon dioxide is vented to the atmosphere to cool down the pumping units. Additionally, if the service company is using a foamer and an emulsifier, monitor both during the treatment. Some service companies utilize a combination foamer-emulsifier.

Unique properties

A carbon dioxide foam fracturing treatment, or just a treatment using $C0_2$ as an energizing medium, is interesting from the standpoint of the cool down procedures used on the high-pressure pumps. Of course, carbon dioxide is a unique compound. It exists in a liquid form below 88°F, assuming that sufficient pressure is maintained to keep it from vaporizing. Above 88°F, carbon dioxide is a gas, regardless of the pressure. Carbon dioxide can also be a solid with which most of us are familiar (dry ice) when pressure is lowered very rapidly and a low-temperature environment is maintained. These unique properties make carbon dioxide a valuable compound as well as one that can be dangerous for personnel on location.

Rig-ups

A typical rig-up on a treatment involves one of two set-ups by service companies. Some service companies have carbon dioxide-carrying tankers that may be pressurized with vaporized C02 or nitrogen gas. This pressure, usually in the 300-350 lb. range, forces the carbon dioxide into a liquid state and on to the high pressure pump.

Other service companies place gear or centrifugal booster pumps between the $C0_2$ transports and the high-pressure pumps. The liquid is taken into the centrifugals, and it is boosted approximately 100 psi so the carbon dioxide remains liquid.

Cool-down

Regardless of which is used, a procedure must be followed to cool down the pumping units. These carbon dioxide pumping units are generally conventional triplex pumps or intensifiers. Cool-down is accomplished so the fluid, prior to pumping, remains in a liquid state and gassing out of the pumps does not occur. This simple procedure involves circulating liquid carbon dioxide through the pumps and venting it or circulating the carbon dioxide back to the transport so all of the equipment to pump the C02 is frosted up, indicating the fluid is in a liquid state.

Problems

Sometimes during a venting stage, the carbon dioxide may vent off at a very high rate, destroying turbine meters by over-spinning the turbines. This, of course, can only occur with a turbine meter and with people inexperienced in running carbon dioxide treatments. A little preplanning by the service company operators prevents these occurrences.

If you destroy a flowmeter and do not have spare parts, I recommend not conducting a carbon dioxide treatment because you cannot closely monitor the carbon dioxide rate. Many service companies utilize stroke counters on their pumps. Although good for back-up, this technique should never be used as a primary means of monitoring pump rate. Obviously, a pump will tum over and register pump rate even if it does not have fluid in it.

Erroneous pump readout has happened many times and has caused job failures on numerous occasions. The pump stroke counters work well if you understand the efficiency of the pump and if sufficient booster pressure is applied pumping through only liquid, not gasified fluid.

Once the cool down procedure is completed, conduct the treatment like a nitrogen-foam frac. As mentioned previously in this chapter, the only difference other than the cool down procedure is the use of a foamer or an emulsifier combination chemical or two chemicals, a foamer, and an emulsifier. Proper ratios must be maintained during the treatment; otherwise, a stable foam is not achieved downhole.

If for some reason a shut-down occurs when fracturing with carbon dioxide foam, the high-pressure pumps could ice up, leaving ice plugs in the pump. If this happens, there is an excellent possibility when the pumps restart that fluid ends will be pumped off, and the probability of personal injury will increase. This risk can be negated by keeping the high-pressure pumps turning over and by continuing to vent the carbon dioxide while the equipment is shut down. Ice plugs present the most potential for significant danger that can occur on a carbon dioxide treatment.

Another possible problem is a rupture on either the low-pressure or the high pressure side of the $C0_2$ line. If this should occur without sufficient valves to control the flow of the carbon dioxide, the location might be flooded with pure carbon dioxide; people may become disoriented and could suffocate. Personnel have died because they could not find the way out of a cloud of carbon dioxide. Warn your employees of this possibility before the job. They should always have a clear path set up to leave the location without being able to see. Treatments like this can be made much safer, of course, by using remote equipment.

Always double-check the rig-ups, particularly on the low-pressure side, when utilizing carbon dioxide to determine if the service company is using proper high pressure rated hosing or steel iron. Simple bulletproof hoses are commonly used between the blender and the pump trucks on ordinary fracture treatments. Typically, these hoses are unsafe for day-to-day use with $C0_2$. The service company supervisor should see that this does not occur, but it does not hurt to have the oil company representative double-checks for such situations. People in a hurry sometimes overlook safety.

Binary or Tri-Phase Energized or Foam Systems

In the late 1980s, Halliburton was issued a patent whereby it was the only company that had the rights to pump C02 at concentrations greater than 50%. Any company that pumped C02 at concentrations greater than 50% on a job at that particular time was required to pay a sizable royalty fee to Halliburton for the treatment. This put other service companies· in a bad competitive position and put Halliburton in a positive position where customers wanted to pump $C0_2$. Because of this patent, a considerable amount of laboratory work was conducted looking at mixtures of nitrogen and $C0_2$ to circumvent the Halliburton patent. Publications have been presented showing synergistic effects on blends of nitrogen and $C0_2$.

Coincidentally, other publications have been presented showing no real benefit between pumping $C0_2$ artd nitrogen singularly. It is beyond the scope of this book to get into the argument about the benefits of binary fluids or simply fluids that are a mixture of linear or cross-linked gels combined with various proportions of nitrogen and $C0_2$.

It has been my experience that these are very difficult treatments to conduct from the standpoint of quality control and assuring what is being pumped. However, there are successful treatments being pumped utilizing these mixtures. At the present time, due to some other legal circumstances, Dowell, as well as Halliburton, now has the right to pump high concentrations of $C0_2$.

I believe that anything that you can do to simplify a treatment should be done. I am opposed to the use of energized or foam treatments where formations are normal or geopressured. I have seen no real evidence in the industry of enhanced production due to the use of the foam over nonfoam fluids where formations were normally pressured. There were instances in the past where this appeared to be the case, but it has now been shown that these problems with normal, conventional fracturing fluids were due to the cleanup of the fracturing gel, filter cake, and so forth. I am in favor of foams and particularly cross-linked foams where moderate under pressure exists. To try to conduct a treatment with more than one energizing mode greatly complicates the treatment; and if at all possible, I would prefer to run a straight foamed nitrogen or foamed C02 treatment. I certainly have empathy for the service companies in this situation, and there is some data published indicating that there is some synergism that exists when pumping multiple-phase fluids.

CHAPTER 7

INTENSE QUALITY CONTROL (1994)

A great deal of the emphasis in this book has been built around quality control of stimulation treatments. As has been the case throughout my career, and I hope will continue to be, I have continued to learn and hopefully to change as new technology and new things develop. One of the most startling things in relationship to understanding fracturing treatments was some work in relationship to taking conventional quality control to another level that started approximately in 1987.

Because of the hard times that had come upon the oilfield service industry, there was a great deal of viscosity-testing equipment available at very inexpensive prices, particularly between 1986 and 1987. While working for the Gas Research Institute, we decided to purchase high-pressure, high-temperature Fann 50 viscometers and shock mount them in a van to allow us to do on-site field testing. We had no idea where this would lead us. In fact, we were quite concerned with potential repercussions of this kind of field testing. However, what we found, I believe, led to a revolution in enhancement of on-site quality of hydraulic fracturing fluids.

Throughout the industry, there has always been very common assumptions made by those in charge of quality control. These assumptions were made prior to a lot of the aforementioned on-site testing. Listed are four basic assumptions that were very common, if not universal, within our industry prior to 1987.

1. There was a correlation between downhole in-situ viscosity and the measured surface viscosity of the base fluid at ambient conditions.
2. The addition of cross-linking agents, retarders, activators, breakers and other additives, when mixed in a precise manner, will yield viscosity profiles similar to those measured in the laboratory.
3. The chemicals and products available in field are identical to those tested in the laboratory.
4. Small variations in the quantities of cross-linkers, buffers, and solvent salts will have minimal effect on downhole viscosity.

These early assumptions were wrong. We found out through our on-site testing procedures that wide variations (sometimes orders of magnitude) in downhole viscosity are possible with small variations in cross-linker loadings, buffers, salts, and shear and temperature history of the fluid. Most individuals who had been familiar with testing would have agreed very quickly with the shear and temperature history comments, but the other comments were indeed startling and revolutionary within our industry. We now know that if we are going to quantify the viscosity of the frac fluid that measurement or testing should be done at the wellsite. We have found little or no relationship between laboratory-generated viscosity (conventional) and the actual viscosity of cross-link fluids at in-situ conditions.

With the writing of this book, the API committee, which has been working for some 12 years, is coming out with guidelines for testing of cross-link fluids. I am not being critical of a lot of the testing procedures utilized by the service companies because there have not been universal guidelines such as API procedures set up for testing of sand, cement, and so forth. At the present time, testing procedures for cross-link fluids in the laboratory itself vary dramatically from company to company. Perhaps with the new API guidelines, there can be more of a correlation between field measurements and laboratory measurements.

At the same time, what we are going to describe under intense quality control may or may not be actual downhole viscosities at in-situ conditions, but it is a technique to verify the relative viscosity characteristics of the fracturing fluids onsite at all in-situ conditions-except pressure.

At this time, for those who may only be reading this chapter, we should perhaps define conventional quality control versus intense quality control. Conventional quality control, talked about so much in separate chapters, involves all of the pre and post-job inventories, the measuring of viscosities, pH, contaminants, cross-link time, and so forth. It also involves the evaluation of the pumping equipment, treating iron, valves, etc. What is involved in intense quality control, on the other hand, is taking these testing procedures one step further and evaluating the fracturing fluid on-site at the in-situ conditions of temperature and shear.

What is involved in intense quality control is, in effect, pilot testing of on-site fluids at in-situ conditions. Basically, we perform conventional quality control, doing all of the things mentioned in the previous chapters of the book. We then evaluate the fracturing fluids at bottom hole conditions of temperature and shear. What we have found (and what continually amazes us) is that dramatic differences exist job to job when we test the actual chemicals on-site with the fluid on-site, as compared to testing in the laboratory with laboratory chemicals and sometimes even the same fluid that has been taken from the site. We want to reiterate the absolute importance of testing in the field. At the present time, oilfield service companies are rapidly putting into the field on-site quality control vans. There are individuals who feel strongly that they can go out and get water from the site and bring it into the laboratory and test the day before the job. On numerous occasions, even taking such precautions as these does not allow for unforeseen contaminants that perhaps were not apparent because the tanks were not properly mixed or stirred or because of changes that have occurred in the fluids on-site. Perhaps the most dramatic examples have been in relationship to the way the fluids were mixed and when and how particular salts were added to the fluids.

Historical perspective

When the first Fann 50's were installed in what was termed "the GRI Rheology van," we were concerned about legal implications and potential problems with the service companies. A verbal arrangement was made with the service companies concerning testing and what to do with the van on-site. Basically, the arrangement was that we would test their fluids, work with them in modifying them if they were not up to specifications, and then during the treatment we would catch samples of the fluid and evaluate the service company's performance. Our original specifications, which were reported in SPE paper 17715, involved having the service company deliver 60% of pre-job viscosity during the treatment, and we would term the treatment a success from the standpoint of fluid preparation performance. In the process of presenting this paper, we monitored 14 treatments in the field and evaluated major service company performances in relationship to the aforementioned specifications. The most startling thing that occurred in the publication of the paper was that 14 treatments were conducted and not one ended up as successful under the 60% specifications.

Figure 10-1 shows the original GRI Rheology Van. In this van, we worked with the service company, testing their fluids on site, modifying them and then evaluating the fluids and several samples caught during the job. Listed below are several examples of field performance in the testing of fracturing fluids using the GRI Rheology van. Figure 1Q-2 is an illustration of shear stress versus time for a well that was treated in south central Mississippi. As can be seen from Figure 10-3, bottom hole temperature was 330°F. Figure 1Q-2 illustrates prejob optimization of fracturing fluids in cooperation with the service company. This particular treatment was designed to be pumped for 5 hours. The original fluid as set up by the service company is shown in Figure 10-2 at curve #1. Working with the service company, we came up with fluid #2, which gave us approximately 1,000 cps after 70 minutes. and, incidentally, remained fairly stable for the duration of the treatment. Fluid #1, 3, and 4 were falling off dramatically in stability at the bottom hole temperature of 330°F. What was done on-site were variations in the particular buffer system and cross-linker loadings. We have found, now that we have been doing this procedure for more than six years, that this is a very common occurrence.

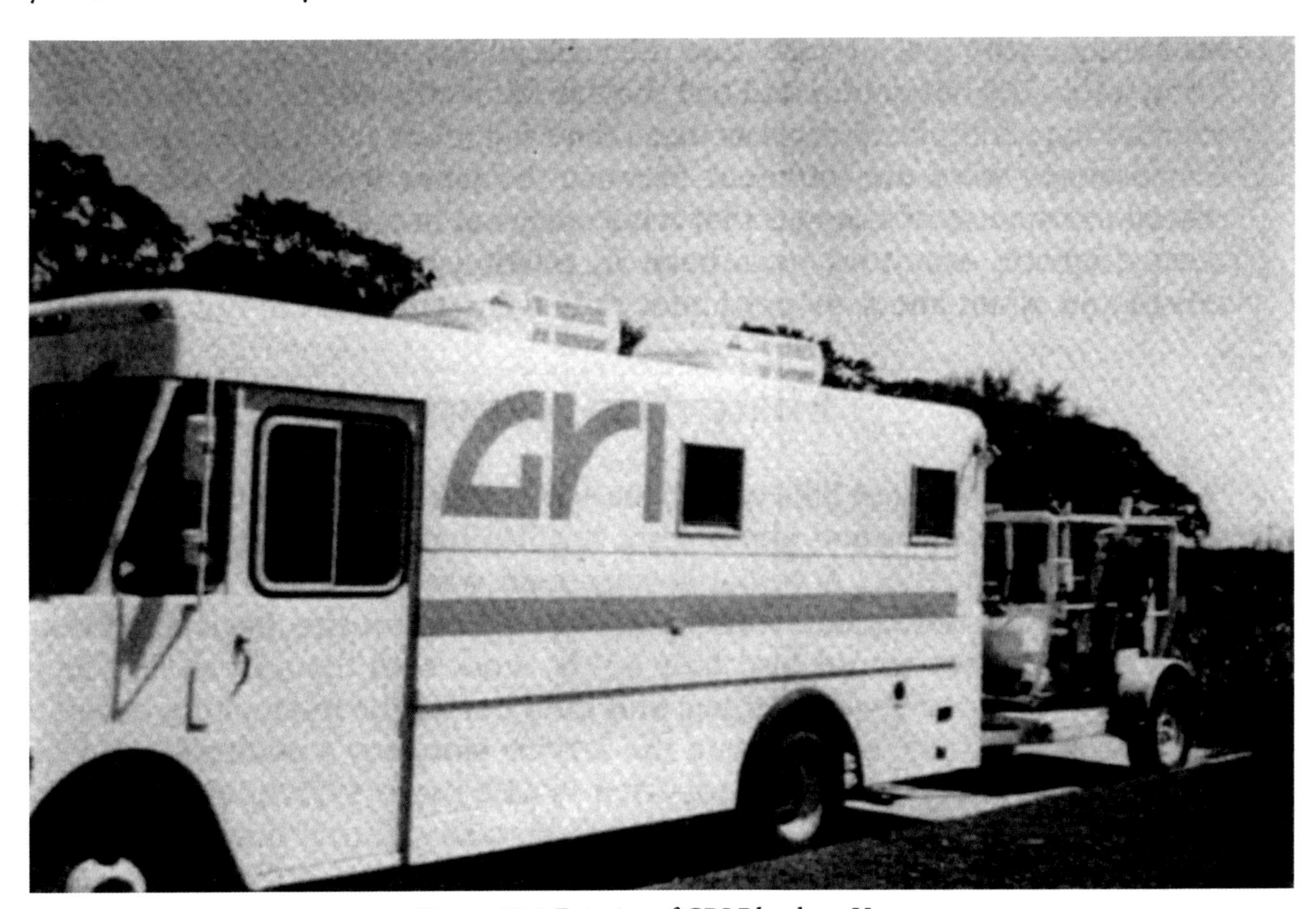

Figure 10-1 Exterior of GRI Rheology Van.

There has been a lot of controversy and confrontation between the service company and testing people on-site. What we find so often is that the fluid that is brought to site, which is set up per service company specifications, has no stability whatsoever when tested at in-situ conditions. Often this relates to the breaker concentrations, but it may be due to contaminants in the water, salt mixtures, or simply shelf-degraded or contaminated cross-linkers, buffers, and so forth. No fracturing fluid should be pumped into the ground without testing this fluid on site at in-situ conditions of temperature and shear. To make the assumption that the fluid is going to perform as it did in the laboratory in glass beakers with distilled or city water is ludicrous.

Furthermore, it is almost unheard of or unthinkable for an independent or major oil company to pump a cement job on a long string without doing pilot testing of cement under job conditions of shear and temperature. Because of the lack of such testing, a tremendous volume of wells are plugged up with unbroken frac gel or conversely, and there are large numbers of treatments that fail because of poor proppant transport, proppant suspension, or perhaps early screen-out that was blamed on formation factors rather than poor quality fluids.

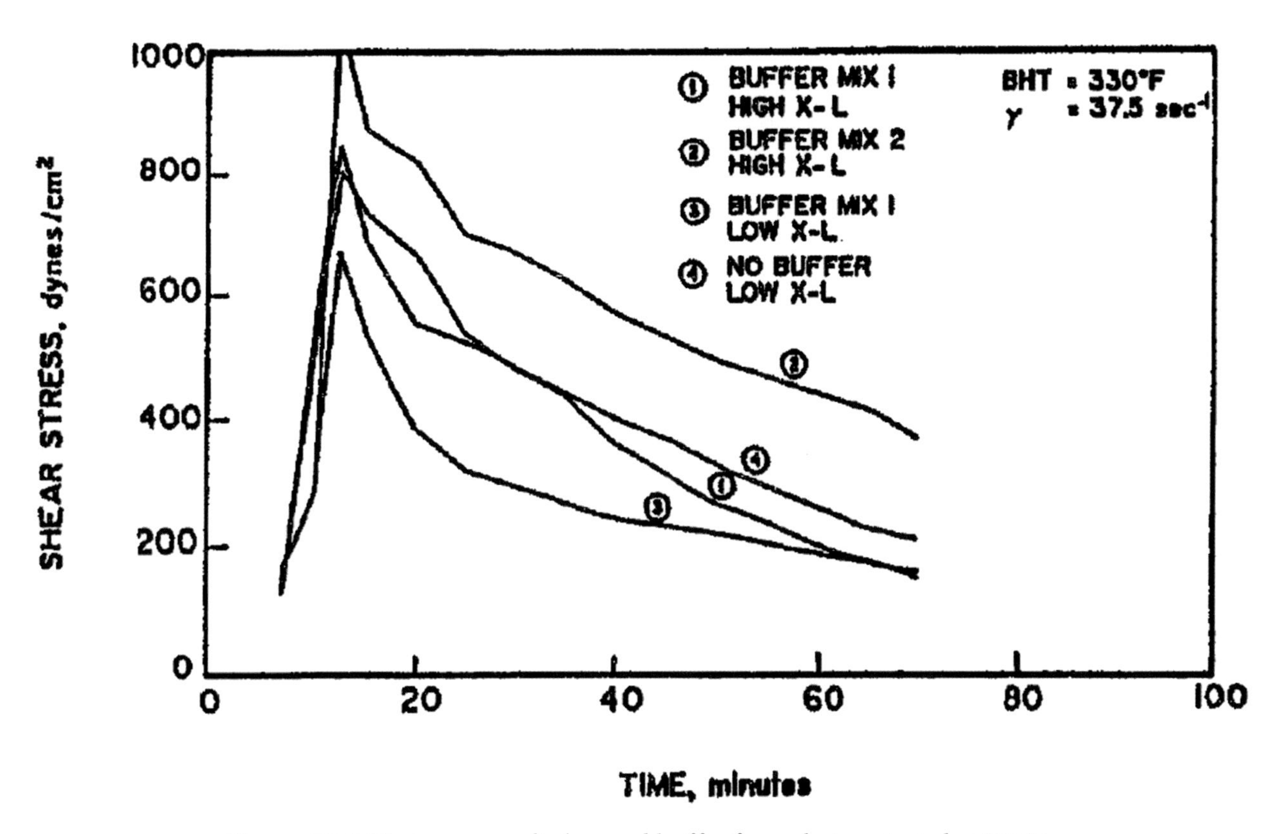

Figure 10-2 Various cross-linker and buffer formulations tested at 330°F.

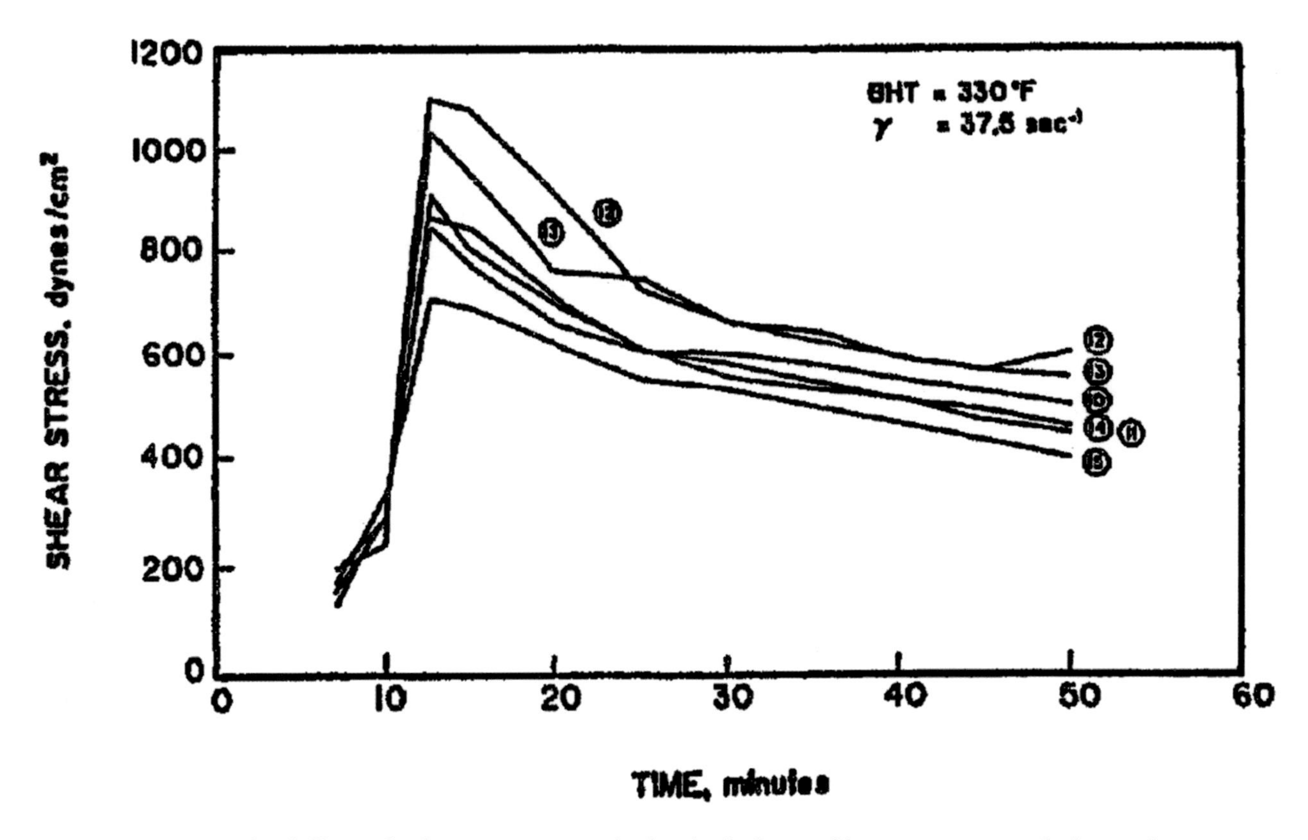

Figure 10-3 Eight different high-temperature gels that had identical base viscosity and identical cross-linker concentration but yielded different viscosities at bottomhole temperature conditions.

Figure 10-3 illustrates the same job as Figure 10-2, but it simply shows eight different frac tanks tested at identical conditions of temperature and shear. What is most interesting about this figure is that all of the tanks shown have identical base gel viscosities and identical cross-linker concentrations. Figure 10-5 gives values of shear stress. Viscosity is defined as shear stress + shear rate x 100; therefore, at a shear stress of 800 dines per square centimeter, we have an approximate viscosity here of 2,000 cps. This figure is interesting because in the early stages of the treatment we are seeing viscosity variations of as much as 1,000 cps, and later art in the treatment we are still seeing fairly large differences in viscosity for tanks that had identical base gel viscosity and identical cross-linker concentrations. As we continued with this work, we found this not to be an unusual case. Based upon our under. standing about what was going on, we found that this variation in viscosity was due to the mixing of salts and their inherent effect upon final viscosity measurements and the way the viscosities were measured. It should be noted that all of these tests were recorded in duplicate, negating problems with measurement.

Figure 10-4 is another treatment where we measured several different tanks at bottom hole temperature and also measured a composite viscosity for all the tanks, which consisted of a mixture of all the fluid on-site and measured that viscosity for purposes of inputting n' and k' data into on-site computers. This figure shows again some 300--400 cps difference for identical gel concentrations, viscosities, · and cross-linker loadings.

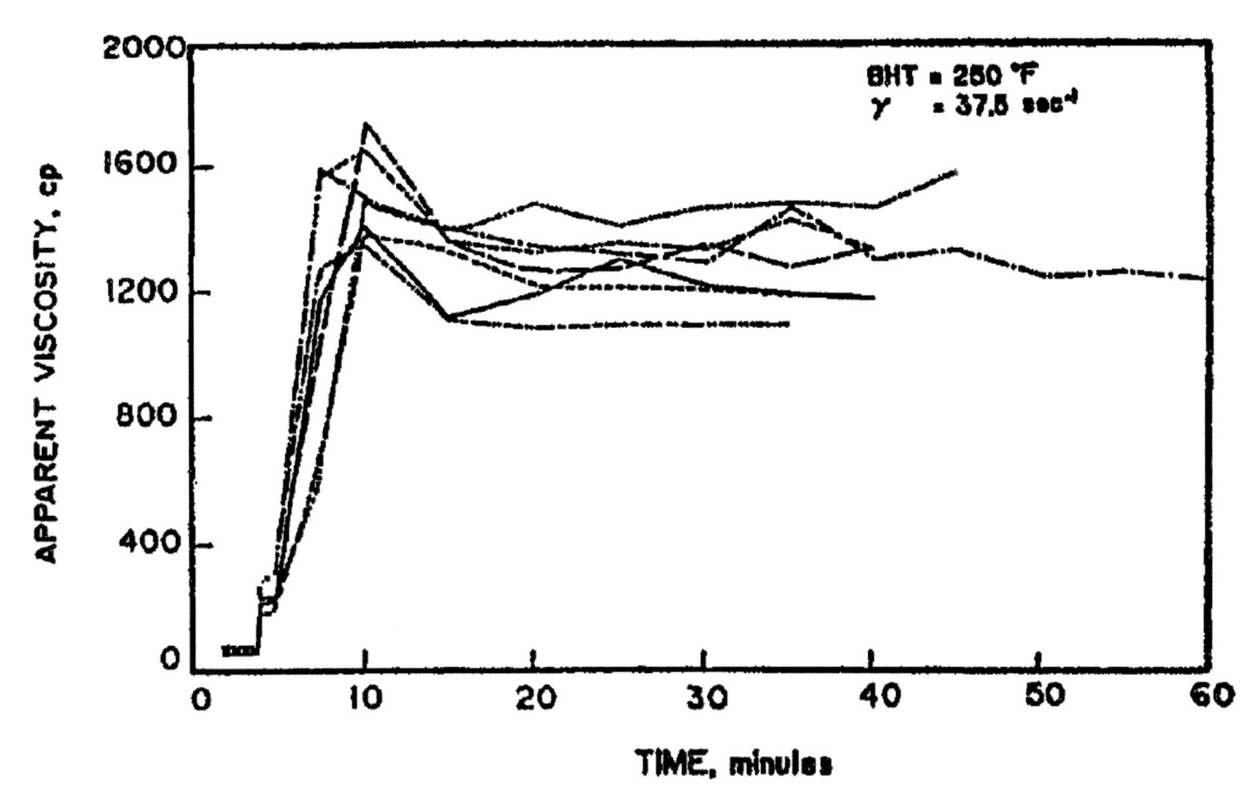

Figure 10-4 Apparent viscosity versus time of six different tanks of gel and a com. posite run of all the tanks at 250°F.

Figure 10-5 is the first example of before and after viscosity measurement as compared to Figure 10-4. If you will note approximately 1,200 cps viscosity reading for the composite run in Figure 10-4 and look at Figure 10-5 for three different samples caught during the job, you will see that it was only on the dotted line later in the treatment that we were approaching the prejob composite run as is illustrated by the top line in Figure 10-5. On this job, by utilizing pre- and post-job inventory, we learned that the service company was running the cross-linker at approximately 15% high. This was further confirmed on-site as the service company ran out of cross-linker in the 12 ppg stage of this treatment.

It should be noted that this fracture treatment was not a failure from the standpoint of stimulation and placing proppant. It just did not achieve viscosities previously measured prior to the treatment. We had discovered with Figures 10-4 and 10-5 that there was an extreme sensitivity to cross-linker concentrations in this particular fracturing fluid. Fifteen percent variation is certainly not uncommon for cross-linker loadings during the treatment. It should be noted that many individuals have made what I consider to be very dangerous decisions on-site adjusting cross-linker loading based on visual observation of a cross-linker at ambient conditions. Many individuals will look at a cross-link sample during the job and tell the service company to vary cross-linker based upon whether a sample is wet, dry, chunky, or whatever. I am very much in favor of catching multiple cross-link samples during a job, but I only use that to be assured that the crosslinker is indeed being added. I base my evaluation of the performance of the crosslink system on information obtained by testing it at bottom hole conditions.

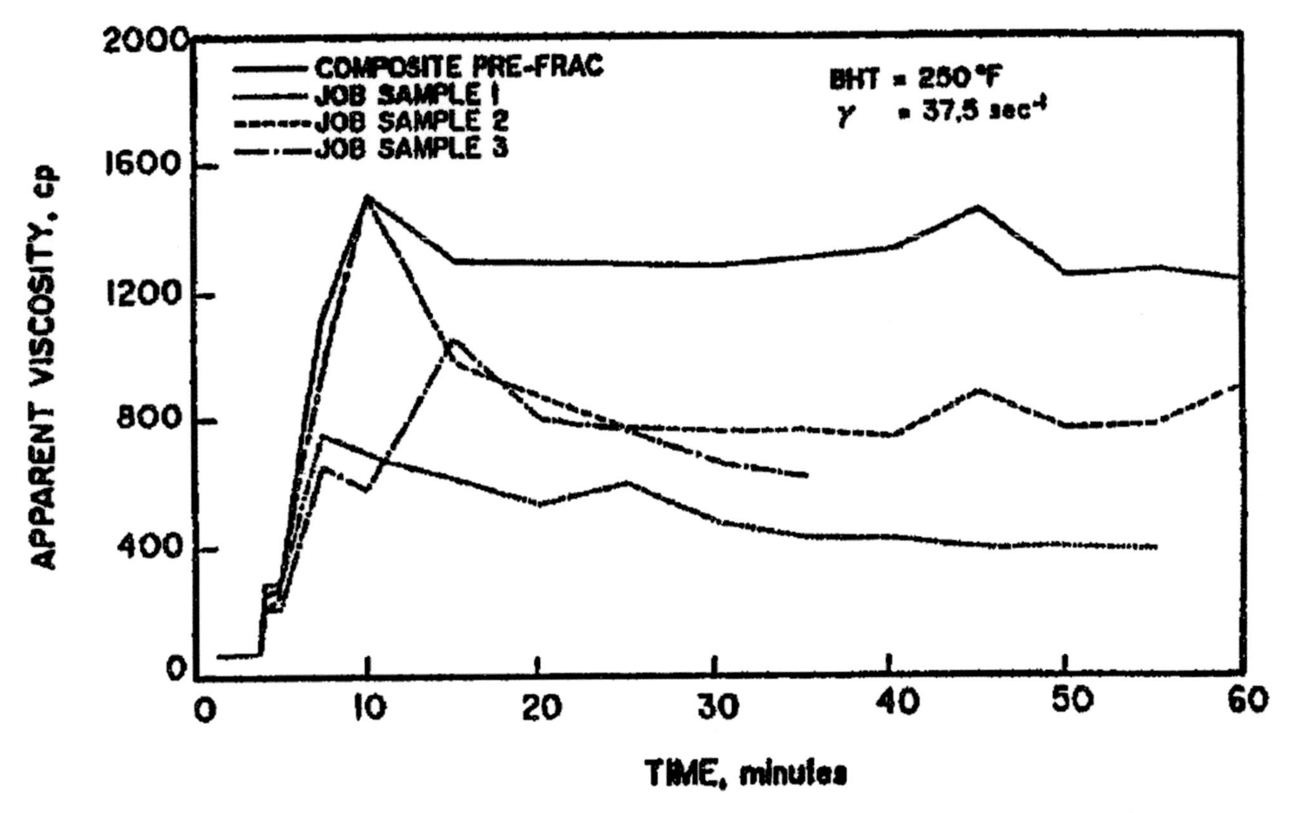

Figure 10-5 Viscosity versus time profile indicating samples caught during the job, noted as samples 1, 2, 3 compared with a composite pre-frac test shown on the plot. This illustrates much lower viscosity achieved during the treatment due to over cross-linking.

The testing of cross-link fluids at in-situ conditions of temperature is very .important for delayed cross-link systems that are activated by downhole temperature, which may look very wet and under cross-linked on the surface due to various heating techniques used by the service companies. High-temperature borate systems will look terrible at ambient conditions because of the higher concentrations of borate required at elevated temperatures.

Figure 10-6 is another "before" of a before and after demonstration of shear stress versus time at 205"F. Figure 10-6 illustrates viscosity profiles for a well in the Rocky Mountains utilizing a delayed cross-link fracturing fluid. These wide variations in viscosity noted in Figure 10-6 were a direct result of poor mixing procedures in the frac tanks and illustrates vividly the wide difference in viscosity that can occur when precise mixing and quality control are not followed.

As it turns out, Figure 10-6 was perhaps the good news on this particular treatment. When we note Figure 10-7 (examples of samples caught during the treatment), you see tremendously lower viscosities. This was an interesting treatment because much of our monitoring equipment showed that the additives were

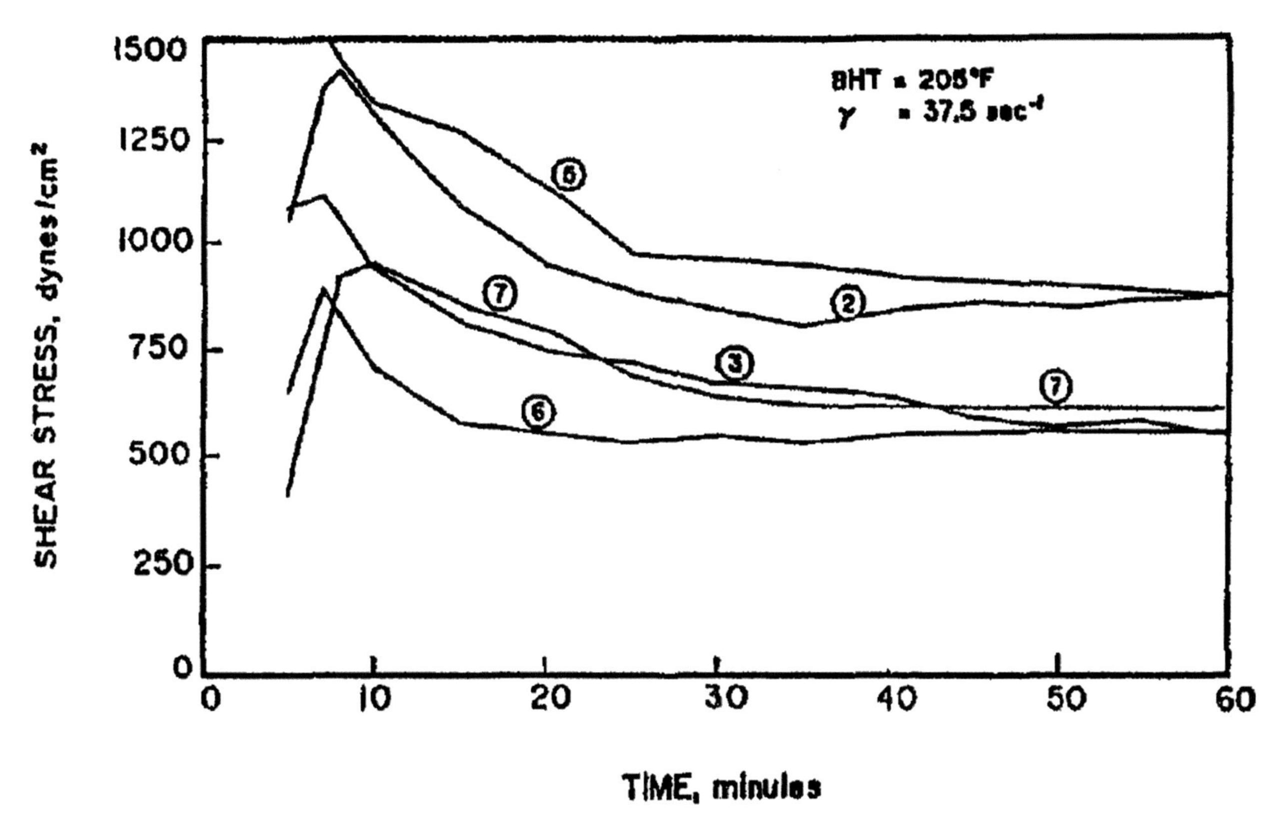

Figure 10-6 Viscosity versus time plot of six different frac tanks of gel having identical base gel viscosities and identical cross-linker concentrations.

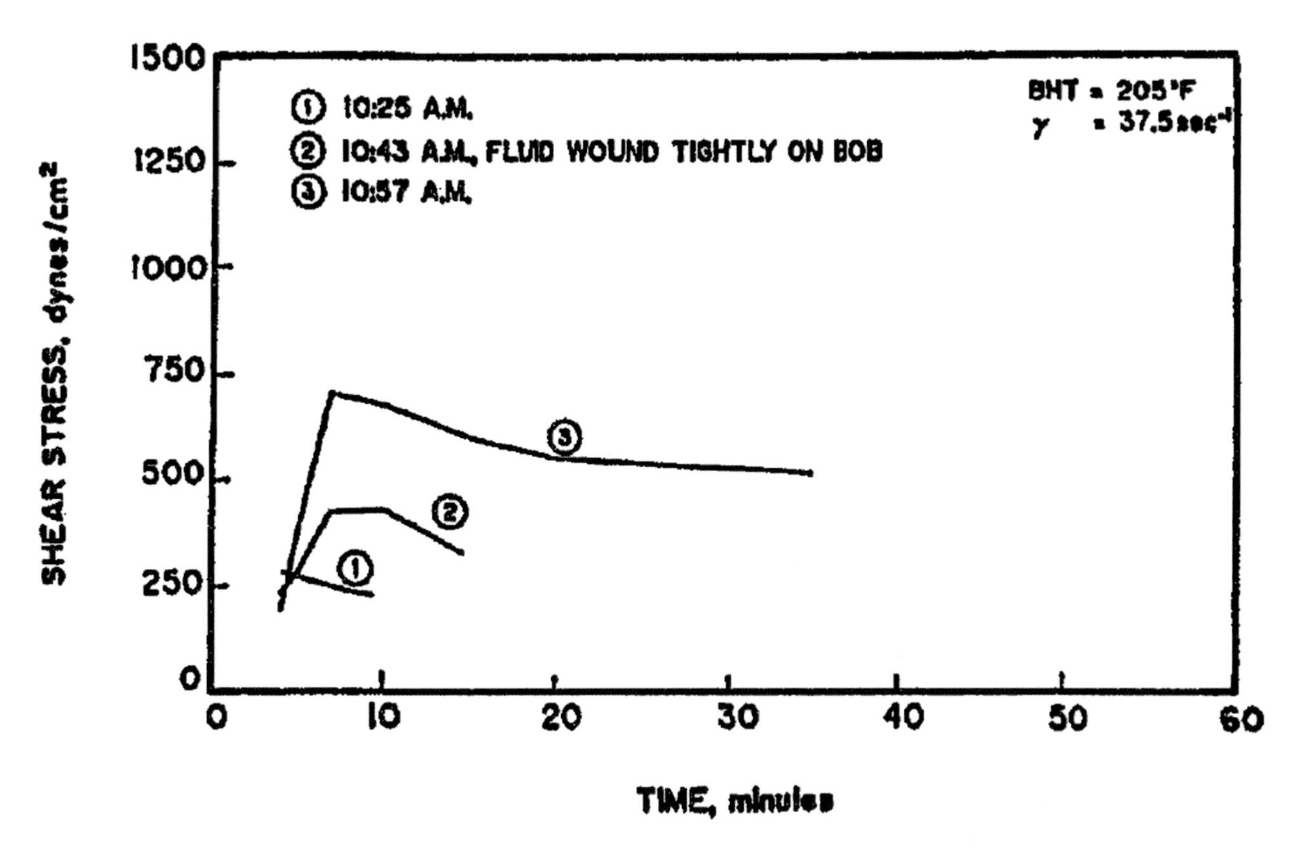

Figure 10-7 Viscosity versus time profiles of samples caught during the treatment compared to pre-job tests illustrated in Figure 10-6.

all being added properly. This treatment screened out fairly early on, and many of us were mystified because of the lack of performance of the fracturing fluid and assumed that we had high leak-off downhole or complex fracture geometry (etc.). I discovered, however, that the problem had nothing to do with these factors. The problem on the treatment was due to a break in a cross-linker line downstream from the monitoring skids where the cross-linker was, in fact, being blown into the air. This was reported to me by an on-site technician. This took quite a bit of communication and searching to find the problem. I was told that the service company was very diligent in working on the problem and got the problem worked out by late in the pad. If this bit of investigation had not occurred, we would have had another example (I believe) of a fracture treatment that failed due to unknown reasons. In this particular case, there was little doubt about the problem. I might add that this is not a case of a service company doing a poor treatment, and they were certainly diligent in their efforts to solve the problem. One of the things I stress in communications with service companies on fracture treatments is that they report problems immediately and that we work through them. Several of the operating companies I work with are perfectly willing to pay the cost of fluids and the cost of personnel on-site for an unforeseen and unexpected problem if the service company points it out immediately. By utilizing intense quality control procedures, many of these problems can be noted very early, even if the service company personnel have not spotted the problem themselves.

Figure 10-8 illustrates the first treatment where we utilized the GRI Rheology van for a low-temperature application. As seen in Figure 1D-8, viscosity profiles of several different tanks were anywhere from 900 down to perhaps 350 cps. Once again, these were all the same base gel concentration, the same apparent viscosity, and had the identical cross-linker loadings. At this point in the testing, this had become a common occurrence; and once we had run duplicate tests for all of these viscosity profiles, we proceeded with further testing. The line shown in Figure 10-8, which goes out much further than the others, is a composite of all the gels. Since we were going to pull out of the tanks simultaneously, we did not see this as a particular problem in the treatment. It was just noted that once again we saw wide variations in viscosity, which in this case we felt were due to variations in mixtures of buffers, salts, and so forth.

What we did find out with Figure 10-9 was something very startling and something that has dramatically changed our ways of quality control testing in field operations. Since we had the time, we decided to go ahead and test the breaker concentrations. Oil this particular job, we tested what was a normal breaker loading of .2#/1000 gals. of a persulfate breaker, which certainly should have been adequate to break the gel at 150°F. It should be noted from Figure 10-9 that when we added the breaker, we measured the highest viscosity that we had seen to date. We immediately put on additional tests, running 50% more, and finally up to more than 10 times the normal loading of breaker that would be pumped on this treatment. As it turned out on this particular treatment, the breaker, which had only been shipped to the service company the day before and was white and free flowing, was more than 90% shelf-degraded. Once again, if the service company had done the normal procedure, which was to follow guidelines for breaker loading and had been diligent in adding these, the well would have been permanently plugged.

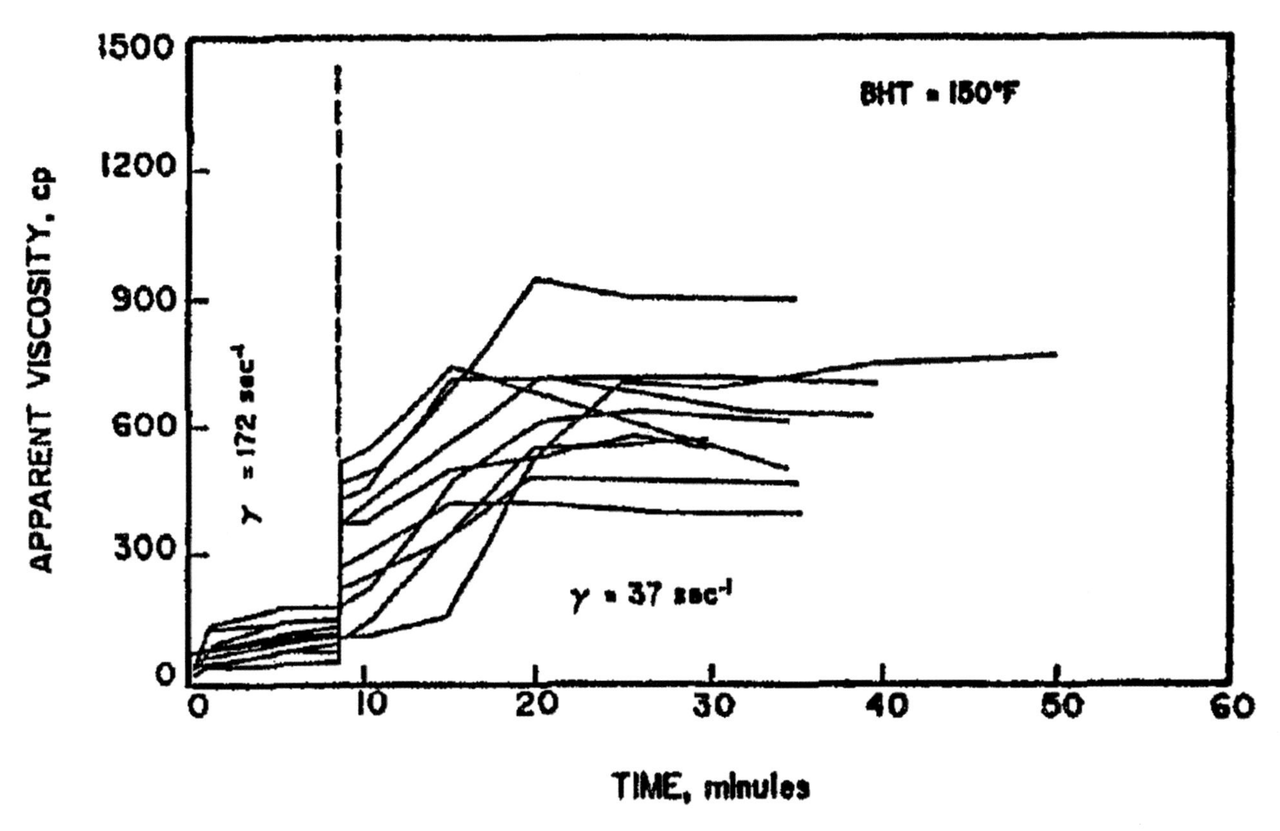

Figure 10-8 Viscosity versus time profiles at 150°F of several different tanks having identical base gel viscosities and identical cross-linker concentrations.

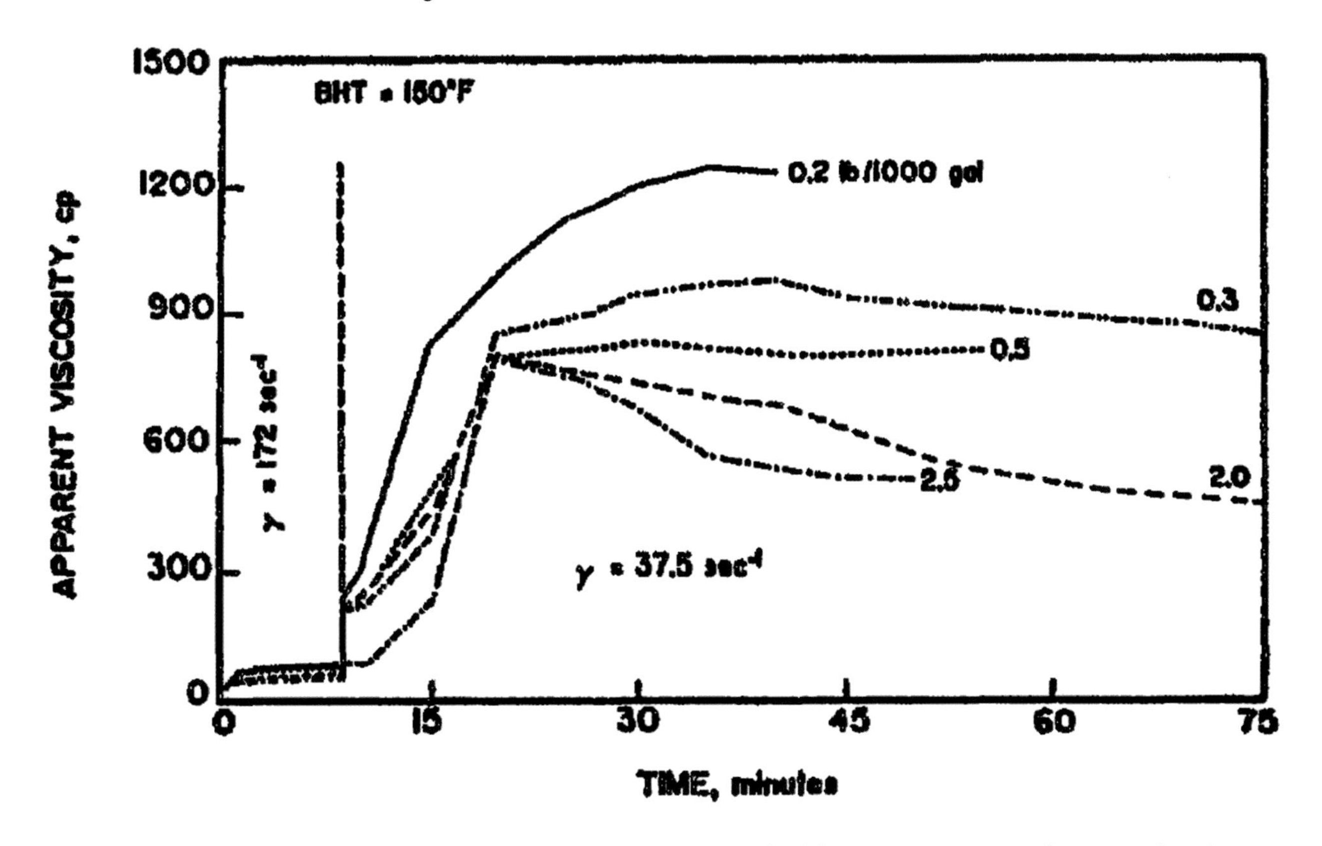

Figure 10-9 Viscosity versus time profile of composite runs of gel from Figure 10-8 with various breaker concentrations. This illustrates for this particular job some shelf-degraded breaker that was on-site.

I will state once again a comment that seems to be somewhat revolutionary in our industry. This is that fracturing gels, without proper amounts of internal breakers at temperatures less than 280°F, will never degrade. Perhaps the most important thing that can be done on a treatment is the addition of breakers in order to assure that the fluids will break back to water. For very low-pressure wells, even low-concentration linear gels will plug a well permanently. Recently, I was involved in a project where a service company inadvertently substituted oxidizer for an enzyme and plugged 11 wells. The fracturing fluid used was foam with only 20 lbs. of gelling agent, but bottom-hole pressures in this particular area are less than 100 psi. Subsequent clean-up treatments have proven that gel plugging was the major factor in poor well performance.

After doing all of this testing with very expensive Fann 50 viscometers, I realized that the same kind of testing could be done with much less expensive, conventional testing equipment. Figure 10-10 illustrates a conventional Fann 35 viscometer. In this picture is also shown a B-2 bob, which allows testing of crosslink. fluids. Additionally, shown in the picture is a heat cup that is applicable for heating the fluids up to 200°F and, in this case, a digital thermometer is also shown.

Obviously, the standard steel thermometers used with these heat cups are adequate. We have been able to test fracturing fluids successfully at temperatures approaching 200°. By utilizing a B-2 bob at 100 rpm on either a Fann, Chandler, or other viscometer, you are testing the fluids at 37 reciprocal seconds shear. This turns out to be an excellent procedure as this shear rate is in the middle of average fracture shear, which most people feel is between 10 and 100 reciprocal seconds.

Figure 10-10 Fann 35 Viscometer with B2 bob, heat cup and digital thermometer.

I won't get into an argument about whether this is absolute viscosity or get into the argument about whether a B-2 or a B-5 bob should be used because of variable shear across the gap. In effect, this test gives relative viscosity and illustrates the stability of the fluids during job time and also conversely illustrates if the fluids will break back to water at a later time.

Figure 10-11 illustrates the benefits of intense quality control being used during a mini-frac for another high-temperature well, which was able to save a disastrous occurrence during the actual fracture treatment. This was an interesting treatment from the standpoint that the samples caught during the treatment and the performance of the service company, as far as monitoring equipment, looked very good. Right after the mini-frac, when we looked at the plots as shown in Figure 10-11, a very large controversy occurred between the service company and me. We were able, however, to ascertain from postjob inventory that the service company's cross-link monitoring equipment was off (low) by 15%. As exemplified earlier, being over 15% and under 15% in some cases can cause dramatic effects upon downhole viscosity. People have looked at this particular figure and stated, "Well, you would have caught this during the job and been able to change it." One of the problems with high-temperature testing as its done today is that it takes about 20-30 minutes to get up to temperature .and be tested. In 20-30 minutes at relatively high rates, you have pumped, in many cases, a fairly large percentage of the pad with unacceptable viscosities.

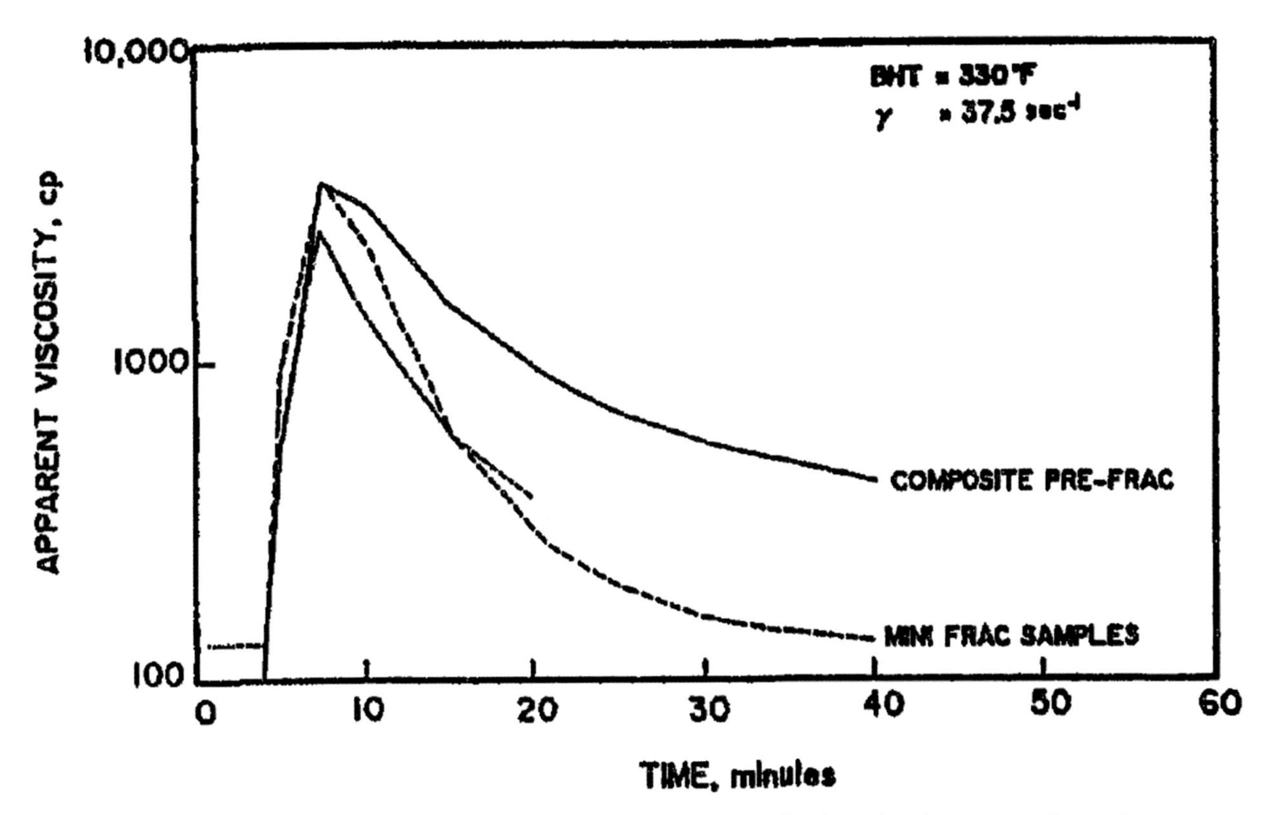

Figure 10-11 Viscosity versus time profile for fracturing fluids utilized on a minifrac. The highest viscosity curve were samples taken prior to the treatment. The other two curves illustrate samples caught during the treatment. The lower viscosity was due to undercross-linking.

Figure 10-12 illustrates another high-temperature well; this one is in Alabama. On this particular treatment, as can be seen from Figure 10-12, the service company did an excellent job of mixing their base gels and buffers, salts, and so forth; within experimental error. They had four identical frac tanks. This was another of our "before and after" tests. When looking at this, perhaps we could have had our first success with this treatment, but it was not to be.

Figure 10-13 is the "after" of this particular treatment. You can see in Figure 10-13 five different curves as well as the top curve, which is a composite of all of the curves shown in Figure 10-12. As can be seen, there were some very low viscosities measured. These low-viscosity values were taken very early in the treatment. As the job progressed, we saw gradually improving fracturing fluids.

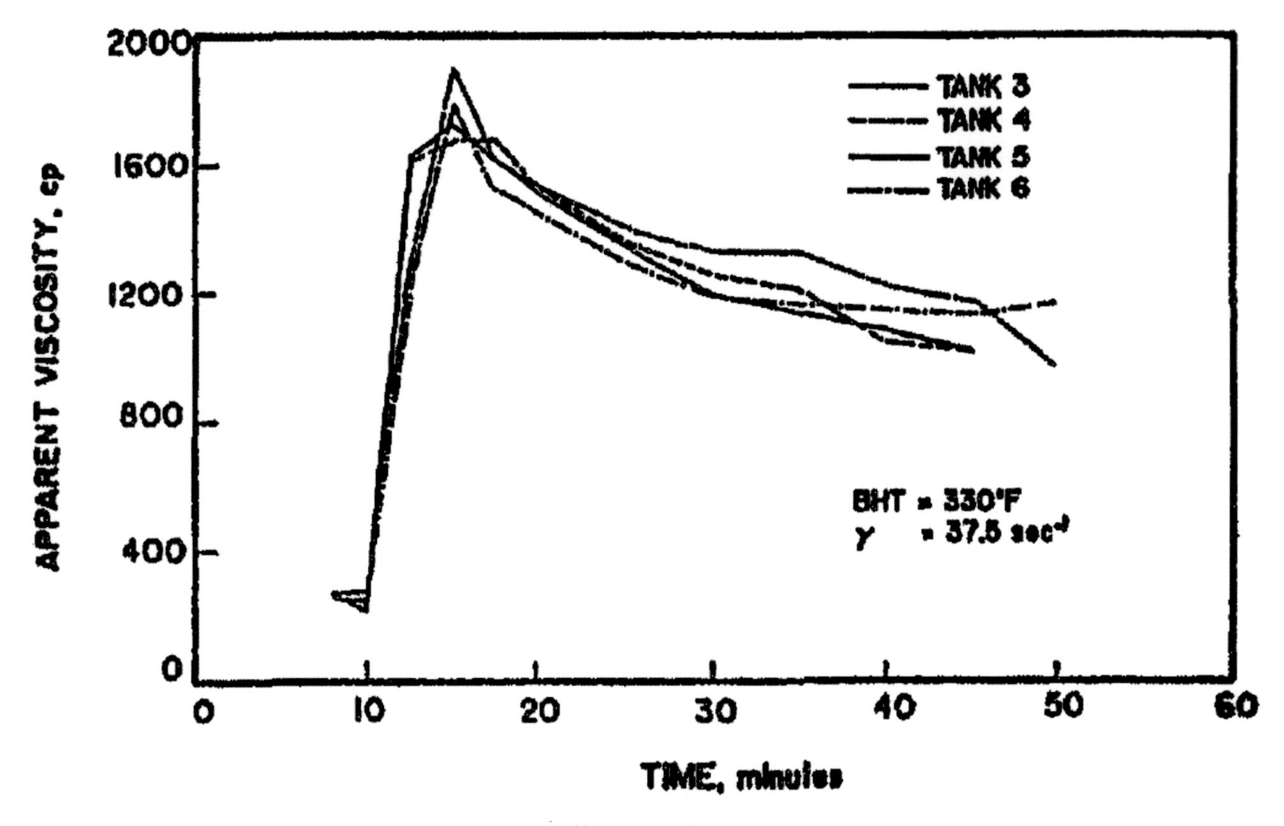

Figure 10-12 Example of four different tanks tested on location with identical base gel viscosities and identical cross-linker concentrations.

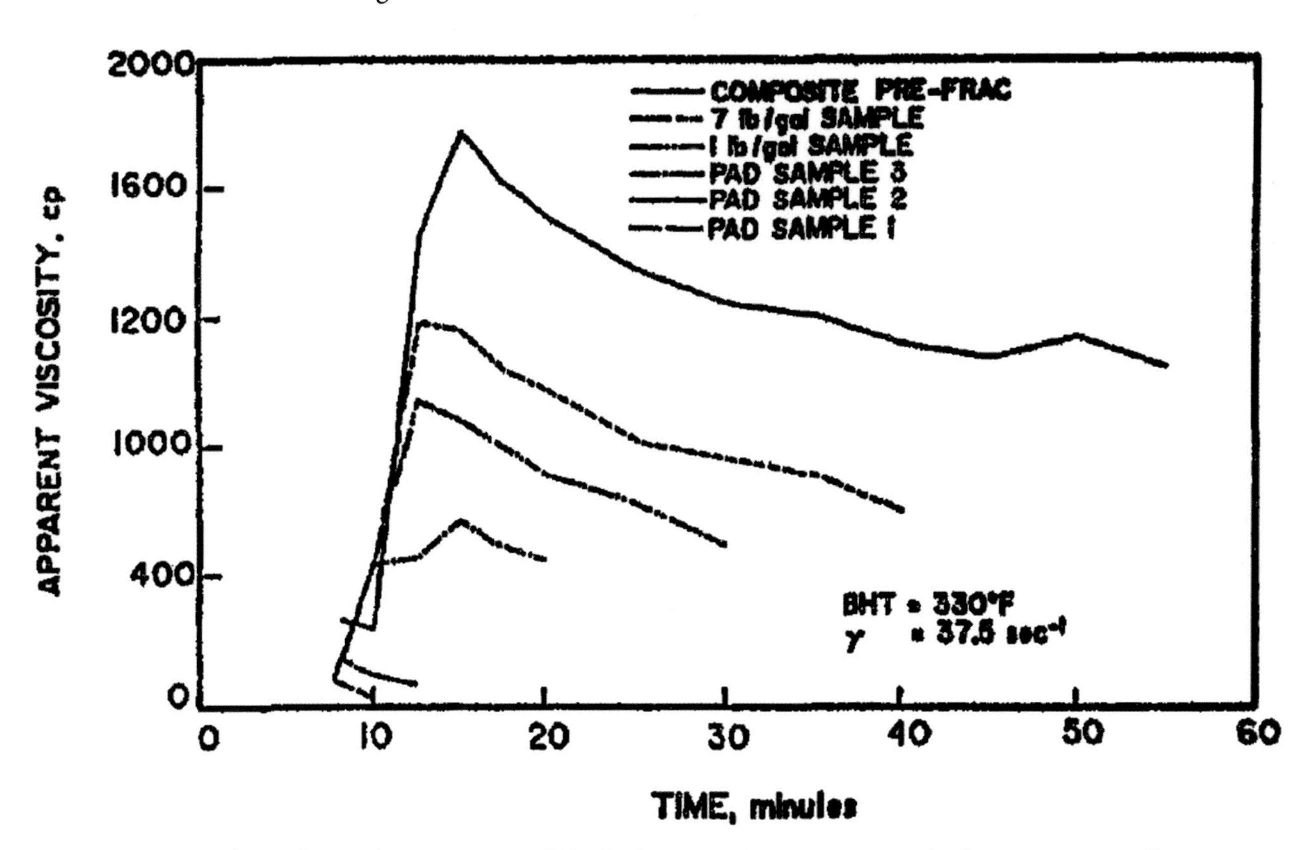

Figure 10-13 This is the real-time testing of fluids shown on Figure 10-14 with the upper curve illustrating the composite pre-frac testing. All the other curves illustrate samples caught during the treatment.

As noted in Figure 10-13, this treatment was a very interesting one as this was a 20,000 ft well with a bottomhole temperature of 330°F. On the day of the treatment there was a driving rainstorm and the treating pressures were approaching 18,000 psi. When we noted very early on in the treatment the very poor viscosities as shown from our testing, we immediately communicated with the service company, and they came back to us and told us that their straps on their crosslinker were fine and that perhaps we were not sampling fluids correctly. We countered with changing our sampling point; and, as can be seen in Figure 10-13, we were seeing gradually better viscosity as time went on. Concurrently, with the low viscosities, when 100 Mesh sand, used as a fluid loss additive in this particular treatment hit bottom, we saw a very large increase in pressure. This increase was due to lack of fracture width because we indeed did have low-viscosity fracturing fluids. At this point, we increased the pump rate on the treatment to the limit of the service company's capability. When the bauxite proppant used on this frac hit bottom in the initial low concentrations, we saw very large pressure spikes, indicating again that we had problems with fracture width due to poor fracturing fluid. This was very disconcerting because we had continually taken new samples, checked, and confirmed again that indeed the fluid was not as anticipated.

This particular fracture treatment is an illustration of what can result with well-meaning people doing the wrong thing. When we had proceeded toward the latter stages of the treatment, I asked the treater how much fracturing fluid we had left. We should have had approximately 70 bbls. All through the treatment, I had been talking to them about strapping their tanks versus pump rate, and these strap rates and volumes coincided very well. When I asked the treater how much fluid we had left, he became very quiet and somewhat withdrawn. At that point, I asked him what was wrong, and he indicated that instead of having about 70 bbls. of gel fluid left, we had nearly 500 bbls. I was very startled because the entire treatment was only about 2,000 bbls., and I asked him how that could be possible. What had occurred on this actual treatment was that a brand new employee of the service company had decided to help us out. Knowing that we were fairly close on volumes, he started very early in the treatment transferring water from an extra frac tank into our 50# cross-link gel. What we were measuring was indeed correct as, very early in the treatment, we were getting very watered-down fracturing fluid. This two-day employee of the service company had created a disaster in relationship to viscosity. I should state at this point that this fracturing treatment was successful in placing 330,000 lbs. of proppant. Although the treating pressures were erratic and there were a lot of indications of potential screen-out, we did get the job away and the well was stimulated. Obviously, however, we did not approach the expected viscosities.

Figures 10-14 and 10-15 illustrate some of our early testing with the Fann 35 viscometer, B-2 bob and heat cup. This procedure, which is now being used across the country as a technique for on-site job testing, allows both the service company and the operator to quantify the fracture fluid's stability for pump time and also

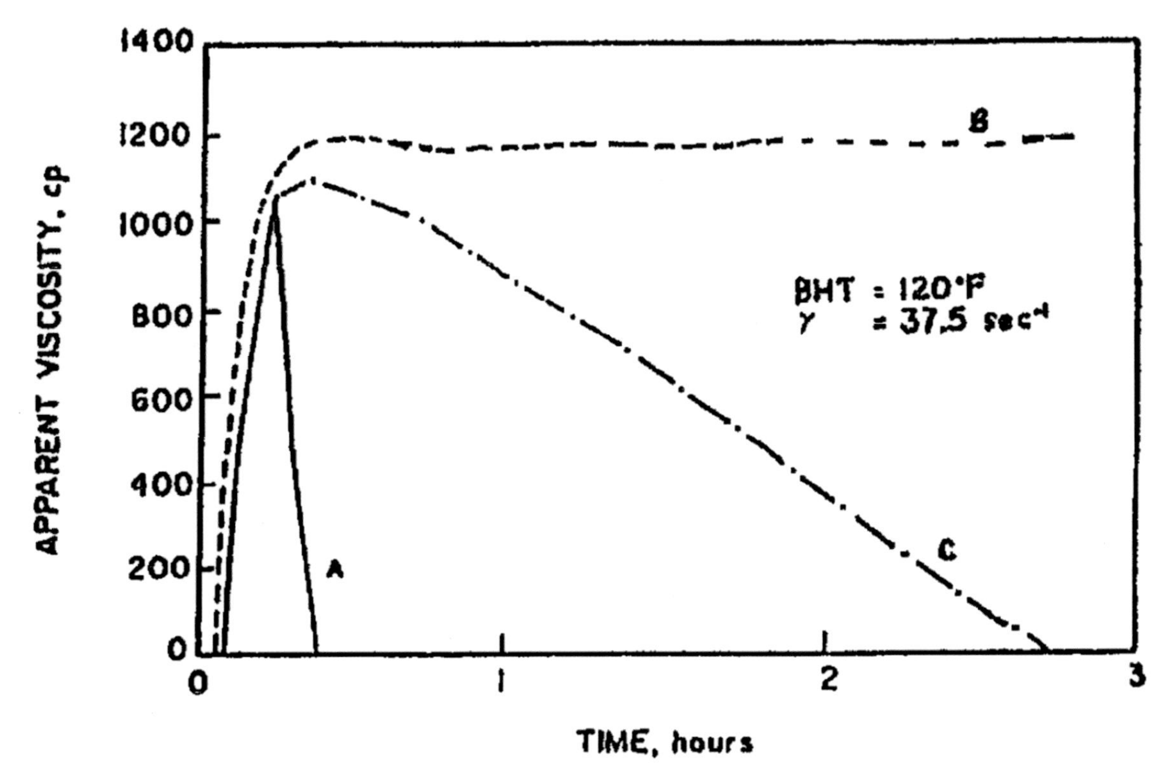

Figure 10-14 Viscosity versus time of a 30# Borate gel. Curve B is a 30# borate gel properly prepared without breaker. Curve A is a borate gel that was slightly less than 30# gel that we were unable to stabilize. Curve Cis a 30# Borate gel with proper amounts of catalyzed oxidizer breakers.

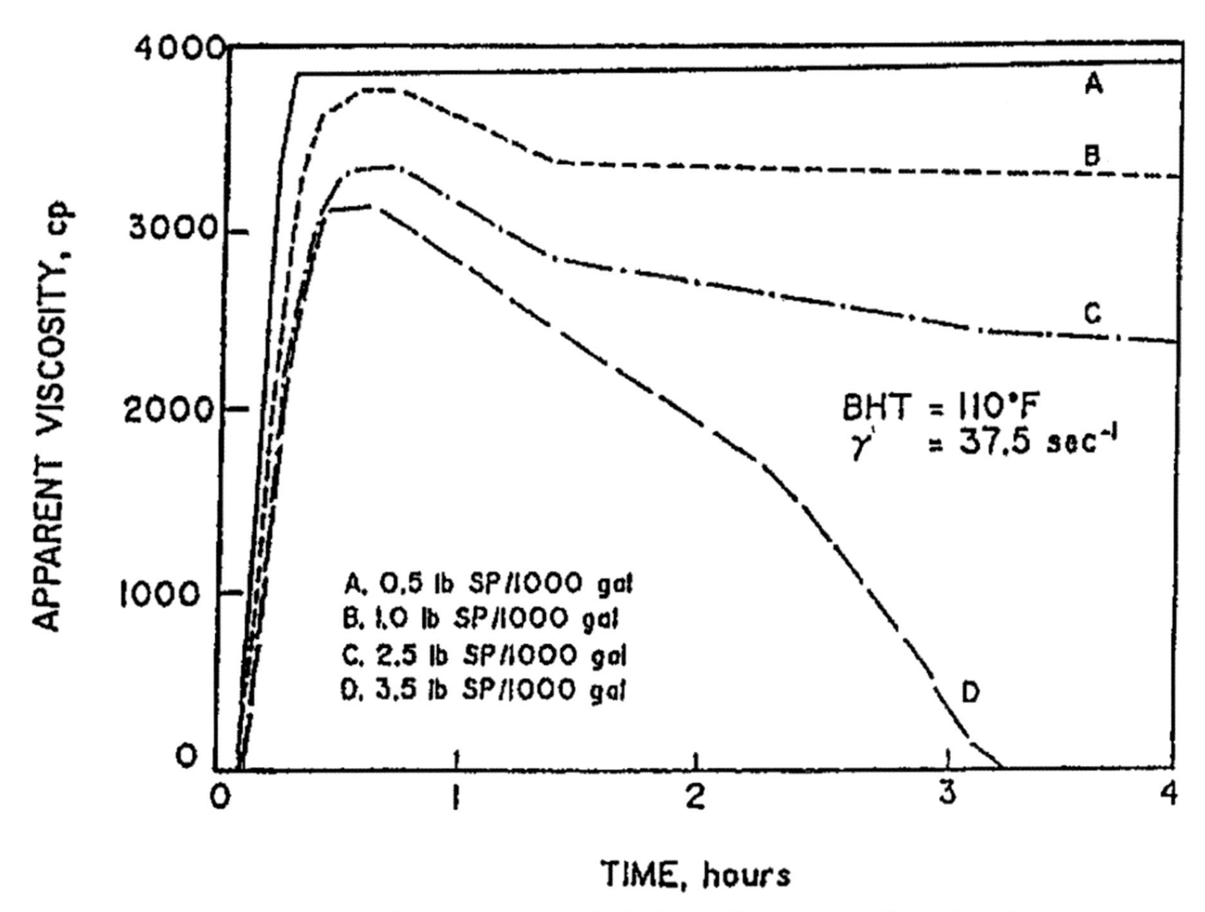

Figure 10-15 Viscosity versus time of 35# Borate cross-linked gel, illustrating different breaker concentrations.

to evaluate if the fluid will totally break prior to pumping anything into the ground. Figure 10-14 illustrates a viscosity-versus-time profile for a 30# borate cross-link. The curve A illustrates a phenomenon we have found many times in the field with conventional borates. If the concentration falls below 30#/1000 gal and you add any conventional breakers to the fracturing fluid, the stability of the fracturing fluid is minimal at best. Curve A actually illustrated a 27# gel on which many tests were conducted and no stability could be achieved. The fluid immediately broke when they got it to temperature. In communications with field people who were conducting the tests, we beefed up the gel by adding a slurry of polymer so that the concentration got above 30# and then we had a very stable fracturing fluid. This problem is caused by a phenomenon that relates to minimum acceptable polymer and crossclinker loading. There have been many attempts in the history of hydraulic fracturing to utilize very low gel concentrations to save money. From the standpoint of conventional borate systems, if you go much below 30#, the fracturing fluid is essentially unstable with conventional breaker systems. I want to define "conventional" versus "new generation" borate systems because many of the "high-temperature borates" have much higher borate concentrations that have more stability at lower gel loadings. I would, however, highly recommend that in-situ testing in the manner of intense quality control be followed even with higher borate concentrations for any gels below 30#, and obviously I recommend that for all gel systems.

At this point, there may be many of you reading this who have pumped many jobs with 20# or even 25# borate gels. We have never been able to test a conventional borate with breakers present that had any stability at all at bottom hole temperature. What this illustrates is that it is possible to pump proppant with very poor fluids. Obviously, the proppant transport, suspension mechanisms (etc.) are not what we would anticipate or require.

In Figure 10-14, curve B shows the fracturing fluid with insufficient breaker to degrade the fluid with time. With no breaker, fracturing fluids should thin down to a certain point and then maintain viscosity almost indefinitely. If you are seeing a degradation of viscosity with this testing procedure with time, it is an indication of over or undercross-linking of the fracturing fluid. Curve C, as shown in this report, contains an oxidizer and an oxidizer catalyst that controllably degrade the fluid to complete breakdown and, in this case, in something approaching 3 hours. This particular job had a pump time of approximately 1 hour, and as you can see, we had very good viscosity at bottom hole temperature for job duration. Again, I want to state that I always test at bottom hole static as I very strongly believe that very minimal cooling occurs as compared to some predictions within the industry based on parallel smooth wall plates. I feel with the tremendous amount of surface area available and the infinite heat sink available in the fractures where we are treating, that very little cooling actually occurs, even late in the treatment.

Figure 10-15 illustrates a 35# borate cross-link gel viscosity versus time. Curve A illustrates no breaker, curve B is a mixture of an oxidizer and a catalyst, curve C is a higher concentration of oxidizer and catalyst, and curve D is a proper amount of oxidizer and catalyst. It should be noted that neither curve A, B, or C will ever completely degrade. It is extremely important that in pilot-testing the fluids that you not only test for job time, but also that long-term tests need to be conducted to be sure the fluid breaks back to less than 3 cps at ambient temperature at 511 reciprocal seconds.

Another thing illustrated by Figure 10-15 is the difference between viscosities measured for a 30# versus 35# gel. Shown in Figure 10-14, 30# gel indicates an apparent viscosity of about 1,100 cps.

I recommend to my customers that on borate systems, whether conventional or "high-temperature borates," that they use a minimum 35# concentration and unless they are using the truly delayed borate for deep high-temperature wells that they not exceed 35#. The reason for this recommendation is that if one develops a true 40# borate system with immediate cross-link (particularly in casing jobs), it is quite easy to gel out and have a very embarrassing situation. I have never had a problem with a 35# gel from the standpoint of excessive friction pressure. Obviously, we will have friction pressure higher than a 30# system or a delayed system. The 35# gives adequate viscosity for fracturing, even at temperatures approaching 260°F for the high-temperature borates, but it will not create excessive friction pressure problems causing gel outs. The other reason for using 35# is to stay away from that minimum gel-loading problem that can occur. By utilizing 35# gel, if for whatever reason the service company ends up 2-3lbs. low, you still have an excellent fracturing fluid. If you try to pump 30# gel and miss low by 2-3 lbs., you have a fracturing fluid that most probably is not acceptable for hydraulic fracturing.

Summary

Intense Quality Control is a procedure that has been actively followed in the field for more than 6 years. We have documented a failure rate of over 70%. This failure rate relates to the proposed fracturing fluid's ability to transport proppant for job duration at in-situ conditions of temperature and shear and/or properly degrade post-treatment back to water. It is very uncommon for us to go to a site and actually pump what is on the proposal. Most commonly, we will vary breaker concentrations sometimes by order of a magnitude even on offset wells. We believe this problem relates to contaminants in the water, variations in salt content, and sometimes in variations of chemicals. I had a long, ongoing argument with one of the major service companies about shelf-life of some of their breakers. We found that the shelf-life of many breakers was quite long and acceptable, but we still saw wide variations in breaker concentrations required to break gels in the field.

What was shown by SPE paper 17715 and by this ongoing work is that it is imperative that no fracture treatment be conducted without testing the fluids on site at in-situ conditions. Testing fluids for fracturing with a bottom hole temperature of less than 200°F is relatively simple and straightforward and not costly. At the present time, testing fluids at higher temperatures is relatively expensive and requires some very competent technical people on-site. Presently, there are only three pieces of equipment available for testing on-site. This testing typically costs in the range of $10,000 per job. Secondarily, it requires competent, skilled rheologists or engineers to evaluate what is actually seen in the field. There is work going on evaluating some less-expensive, high-temperature viscometers, and I predict that eventually all fracturing treatments will incorporate high temperature testing systems on site. The procedure I presently use on treatments of high temperature is to require the service company to get samples of the actual frac fluid, chemicals, and so forth and to test them at the regional or research laboratories at bottom hole temperature conditions. I do feel that this is less acceptable than actual on-site testing; but at the present time, the lack of equipment and competent personnel dictates this procedure for most jobs.

CHAPTER 7

QUALITY CONTROL OF FRAC FLUIDS (2022)

I have decided to put all the quality control chapters together primarily because quality control is essentially the same for each of the fluids although with different parameters and protocol. Much has changed since 1994 but being assured that what one is pumping matches up to the design has not changed. Testing of oil-based fluids is quite like water base systems with more emphasis on safety as well as gravity of non-man-made hydrocarbons. I would hope that young minds who do not choose to study the history of oil-based stimulation do not reinvent the system.

I recall vividly arriving to the Nowsco laboratory at 7 AM one morning and finding two of the staff had been working all night to break an oil gel. They said they had tried everything but to no avail. They were well trained and followed my definition of a broken gel, which would would yield less than 2 centipoise at 300 rpm at room temperature. Not knowing what the problem was, I asked a few questions and, trying to be a good boss, let them down slowly. They had gelled up a crude less than 30 API and had applied the standard breaker systems. No matter what they did the final viscosity measured greater than 9 centipoise. I finally told them that we would not need refineries if they had got the viscosity to less than 9 centipoise. The breakers had reduced the viscosity of the crude to its original API gravity and refinery cracking would be required to get it any lower. But another reason not to use oil for fracturing.

Quality control of stabilized and non-stabilized foam is challenging to give an understatement. Early on in the development of CO2 foam there was high pressure testing equipment which measured capillary viscosity of the CO2 foam or crosslinked foam. To my knowledge there is no such equipment available today. Since there is no means of measuring the viscosity of foams, I am reluctant to recommend foams for conventional fracs. I do use non stabilized (ungelled) high quality foam for a replacement for water fracs where there is either no water available or extreme high costs are incurred. I also use 30 quality nitrogen foam for conventional fracs because of excellent data published by Dr. Phil Harris at Halliburton who illustrated that at 30 quality and below the nitrogen had little or no effect on viscosity of the crosslinked fluids utilized.

I will discuss the quality control of fracturing fluids in relation to changes or developments over the past 40 years.

QC: Water Based Fluids

Quality control on water based fracturing fluids came directly from the Gas Research work done in the early 80's into the early 90's. On-site evaluation of what was actually being pumped was a huge eye opener for our industry. There were excellent fracturing fluids that existed in the 80's as long as all additives were added exactly as done in the laboratory. The field evaluation of titanium and zirconium-based guar and guar derivatives essentially proved that the fluids were almost impossible to control because of their sensitivity to even minor changes in loadings. Although there was significant effort made to make these fluids work, they simply were extremely sensitive to critical additives. The invention of the name "Intense Quality Control" which involved testing the actual fluids to be pumped as well as the actual chemicals proved without a doubt that the industry needed to move to fracturing fluids that had a larger error factor. The move to Zirconium crosslinked CMHPG and later moves to using Borate as the dominate crosslink fluid for temperatures up to 300 degrees Fahrenheit created functional fracturing fluids that could be pumped successfully on most treatments. Mostly all service companies had field laboratories where water samples and isolated chemicals for the job were tested, typically the night before the job. All the major service companies had not only field labs but research facilities where improved chemicals and products were evaluated. Historically, chemical suppliers also had research facilities and submitted products to be tested by the service company. As stated earlier, these times, in my opinion, were some of the best times in our industry. We, in some cases, were not pumping the correct fluid but at the end of the day we knew what we had pumped, its stability and break properties and we knew what size proppant had been placed in the formation.

I conducted a very interesting and, to some degree, successful investigation on the usage of Nano surfactants. There are a significant number of formations where various forms of the nano-surfactant products enhanced production. Conversely there were many reservoirs where little or no effect was seen. Extensive research by suppliers and service companies was underway when our unfortunate drop in oil prices and the Covid 19 event occurred. I predict that there will be success with products once we truly understand how these products work.

Prior to the development of the Nano surfactants, I had for the most part recommended negating usage of typical water-based surfactants in slick water fracs. I held this recommendation for many years but like most general guidelines there is always an exception. For ultra-tight nano Darcy reservoirs, which produce high quantities of water, there is in some cases a tendency to form an emulsion which can inhibit production. For dry gas and oil wells with low water production, I can find no rational reason to pump conventional surfactants.

I came upon this decision not to run conventional surfactants from an evil cheap customer. The guy kept beating us down on price wanting to know what he could take out of the frac to cut his cost. The job was a simple, crosslinked gel treatment in a dry gas formation. I finally told him we could take the surfactant out but would not recommend doing so. He took this as a challenge and left it out. I was concerned that God would strike us down with lightening, as everyone knew we had to lower the surface tension to get the fluid to flow back. I was concerned enough to stay for the flowback of the well and was amazed that the well came back very strong. I believe that great understanding most times comes from what I call the 'Peanut Butter and Chocolate' accident creating Reese's cups. On considering the surfactant we were using, it was basically a soap which by its very nature plaits out very quickly and unlike some of the newer products had literally no function and perhaps was detrimental beyond the cost. Surfactant suppliers do not shoot me—we have come a long way since that time.

Sadly, today there are literally hundreds upon thousands of jobs where little or no quality control efforts are even attempted. The huge volumes being pumped, the drastic cost reductions, and the lack of field experience of service companies has led, in my opinion, to a situation where, in an age of massive data manipulation capabilities, we may be off by as much as 20 % in critical volumes such as fluid, proppant, and chemicals. With no oversight of sand sizing, we then will not know how much we pumped or what was placed in the reservoir. Massive data to optimize results will be unsuccessful if when our data is not representative of what is pumped.

QC: Oil Based Fluids

To state the obvious, I am very much in opposition to the use of oil-based fluids for stimulation. Over the years I have supervised hundreds of gelled hydrocarbon treatments. I had an occasion to illustrate graphically what the personnel were dealing with during a safety meeting in western Oklahoma. As was the case then, we pumped whatever the customer required. In this case it was 60,000 gallons of gelled diesel. The crew was fairly experienced and got through the batch mixing process safely. Prior to the job we all gathered for a safety meeting. After the safety officer for the service company had finished his talk on safety, I addressed the crew and asked everyone if they knew what we were pumping. Immediately one of the crew stated we were pumping My-T-Oil which was the Halliburton name for the gelled hydrocarbon utilized on the job. I immediately told the crew that we were in fact pumping Napalm and subsequently explained that Napalm was in fact gelled diesel. I seemed to have everyone's attention so carried on with how many of the safety meeting remembered the movie *Forest Gump*? I then had the most attentive safety meeting crew ever. I then asked if they remembered the scene where Forest went back in the jungle rescuing his companions and finally was carrying his best friend Bubba. Everyone remembered and I asked if they recalled the phantom jets coming in behind Forest and Bubba dropping Napalm containers. Again, everyone remembered, so I asked if they knew how much Napalm was in the containers which caused the massive fireball which almost engulfed Forest and Bubba. There was no response, so I told them less than 100 gallons. From the back of the gathered crew, one gentleman said very loudly "We're all going to die!

I have more true scary stories, but the above paragraphs illustrate the major reason I am opposed to any form of hydrocarbon as the major component in fracturing.

QC: Foam Fracturing Fluids

The original 1994 chapter covers QC of foam fracturing fluids pretty well and if one is so disposed there are manual calculations one can follow to design your own CO_2 and N2 volumes. As is the case with cursive writing, no one in our world today is doing anything other than using programs and algorithms already present in service company computers. In some coal seam areas and some isolated locations, foam fracturing is still being conducted, but it is rare in the overall picture. CO_2 fracturing has almost disappeared from the US. In the following paragraphs I will discuss my basic rational for minimizing the use of energized fluids.

While spending five years of my life as Engineering Manager for Nowsco (Nitrogen Oil Well Services company), I learned a great deal about foam fracturing. In looking back on those times, I am amazed at our ability to place proppant with non-stabilized foam and very rudimentary control of nitrogen volume.

We also pumped a very large number of CO_2 foams in the Lobo Wilcox and in the cotton valley sands. During that timeframe and continuing into the 90's there was a massive number of foam fracturing treatments pumped particularly in the Rocky Mountains and East Texas. There were, in fact, companies that would pump nothing other than foam fluids due to the obvious, they believed, damage caused to reservoirs by water. My company eventually re-fraced hundreds of these wells with non-energized perfect proppant transport borate crosslinked gels. Many of the early vertical re-fracs were on wells that were initially stimulated with CO_2 foam.

A CO_2 foam frac is an amazing thing to watch. One has to vent CO_2 vapor until the CO_2 turns to liquid. A significant number of CO2 and N2 jobs are shown at the end of the chapter. One can watch all of the iron and fluid ends on the pumps frosting over as the venting process ensues. Typically, the venting process goes through a vertical pipe yielding a huge cloud of vaporized CO_2 over the location. As soon as everyone is convinced that the CO_2 is coming out of the vent pipe is liquid, the valve is closed on the vent line and the liquid CO_2 is diverted to the downhole piping. Simultaneously the water or gelled water pumping is initiated, and the fracture treatment is underway. Without a doubt CO_2 fracs are exciting and using standard intuitive reasoning, this would be almost the perfect fracture fluid.

Nitrogen foams are a great deal less exciting. The liquid nitrogen is vaporized after going though special cryogenic pumps. The nitrogen foams are therefore a mixture of nitrogen gas and fluid, while the CO_2 is pumped as a liquid emulsion down casing or tubulars. There are, I believe, multiple reasons not to run both CO2 and nitrogen foam and are listed below:

1. The execution of CO_2 foams is inherently dangerous. If for some reason there is a shutdown of the frac, one must open the vent lines and once again cool down the entire iron and fluid ends. Initiating pumping without venting will result in attempting to pump solid dry ice which will result in explosive breakup of fluid ends and treating iron.

   I was on a job in West Virginia where we were pumping gelled methanol and CO2. This was a fluid I had developed and was on the job, having traveled from the research lab. This job was late 60's or early 70's and the crew was not familiar with pumping CO_2 or conducting foam fracs. I had personnel from Halliburton mechanical research running a remote blender but otherwise had no real experienced personnel on location. This was a time before frac vans so all supervision was done in open air. After going over the job with the site supervisor and pump truck operators' multiple times, the supervisor turned to me, handed me the sound power equipment, and said I was in charge. This was highly unusual for a person from research to run a field operation, but I was young and took on the job with enthusiasm. I once again gathered all the pump truck operators and told them multiple times that, if there was an unintended shutdown, we must not start pumping again without going through the cooldown process.

   Throughout my career, I have had many times where direct supervision was not followed but on this occasion my life and everyone who was on location were lucky to have survived. The explosion of iron that occurred when about 2/3 of the way through the job a leak occurred, and a momentary shutdown followed by an immediate restart of the job proved the inability to pump solid dry ice. The supervisor who had handed me the control of the job, saw the leak, called for shutdown, and then immediately signaled for the pumps to come back online. I was shouting not to bring the pumps back on, but to no avail, and five fluid ends were destroyed, and high velocity shrapnel was flying all over location. The good Lord was watching over us that day as no one was injured and we were able to get the well shut in.

In today's world of better communication and better control, this event may not have occurred. I do believe that with remote control of pumps combined with multiple safety devices these types of job can be conducted.

2. I am not a fan of gelled or crosslinked fluids containing CO_2. Due to the acidity from CO_2, one cannot have a perfect proppant transport fluid. The presence of massive amounts of carbonic acid degrades the gel very quickly, particularly at elevated temperatures, i.e., 200 F+. The other reason for my dislike of the fluid in conventional fracturing of permeable reservoirs is due to the inability to quantify the viscosity of the mixture versus time and temperature. There simply is no equipment available to measure quantitatively the viscosity profile of viscous CO_2 fluid mixtures versus time and temperature. It seems counterintuitive to utilize a fluid you have no way of quantifying its viscosity
3. Nitrogen foam does not have the inherent danger of CO_2 and today is typically pumped by a separate nitrogen company working alongside the pumping company. When nitrogen foam was first initiated in the late 70's, individual nitrogen companies existed and worked with the service companies. As the popularity of foam increased all the major companies had their own nitrogen pumpers. Today there are a few companies that pump both nitrogen and fluid, but they are relatively small operations. My major problem with 60 + quality foam for conventional treatments is the lack of perfect proppant transport. Foam, particularly crosslinked foam, has the ability to carry high concentrations of proppant into the formation but the strong possibility of settling out of zone always pushes me toward 30 quality or lower nitrogen foam where the suspension quality is not denigrated or simply use aqueous fluids incorporating forced closure and then either swab or pump the well to retrieve the load. For operators who have under-pressured reservoirs the use of energized fluids has virtually always been pushed toward some type of energized fluid. I believe this has been driven primarily due to the fear of leaving evil water on the formation. It is my opinion that properly treated water will not damage the reservoir. By using a perfect proppant transport fluid, which by its very design, creates a superior pathway to produce hydrocarbons by being adjacent to the producing interval and not settling out of zone. The success of these fluids in conventional reservoirs for low pressure reservoirs is very dependent upon rapid flowback, i.e., taking advantage of inherent supercharge from the fracture treatment.
4. Although nitrogen foam has relatively low friction pressure, many times approaching gel friction values, you do lose hydrostatic resulting in higher treating pressures which increases ticket charges, not to mention the cost of the nitrogen.

CO_2 on the other hand creates a viscous liquid emulsion resulting in high treating pressures. Because of the high friction pressures, most CO_2 foams are pumped down casing. The hydrostatic is essentially equal to water, but friction pressure goes up dramatically as proppant is added to the mixture. According to Halliburton the proppant becomes part of the internal phase thereby effectively increasing foam quality and friction pressure.

To compensate for these high treating pressures, the CO_2 or nitrogen is decreased matching the volume of the proppant. This technique is termed "Constant Internal Phase" and is widely used particularly on CO_2 foam jobs. At the end of the chapter there are illustrations of constant internal rate, increasing slurry rate & constant internal phase treatments. It should be noted that a constant slurry rate treatment will probably require divine intervention to be pumped accurately, with constant changes in clean rate, slurry rate, and variations of chemical concentrations.

Based on a significant amount of success with non-stabilized nitrogen foam as substitutes for un-accessible water, high quality nitrogen non stabilized (no gel) foams have been utilized in the San Juan basin. CO_2 can be used similarly. There are areas where there simply is no water available and areas where location size and logistics negate the ability to transfer water. Before I am verbally attacked about no water available or tiny locations, what I am really expressing is the cost viability. I have conducted jobs in Appalachia where we pumped water 5 miles up a mountain and have conducted treatments on locations by spending enormous amounts of money building locations out of the side of mountains. Just about anything can be done but the cost can be prohibitive.

The good news is that 60 quality plus, non-stabilized foamed fluids can be used as a substitute for 100% slick water and many times, particularly with shallow wells, nitrogen is much cheaper than trucked water. These types of jobs have been very successful from a production standpoint appearing to mirror similar slick water volumes. It should be noted that conventional foaming agents are typically adversely affected by high salinity, similar to anionic friction reducers. As is the case with newly developed cationic and copolymers of acrylamides, there are available specialized foaming agents that can handle high salinity waters. These foamers are typically zwitterion products that function in virtually all types of salinity and even in acid.

I would state that these are my opinions based on more than 50 years in the fracturing side of the business, I have no doubt that monetary success has been achieved with foams. When we were discussing water sensitivity, I mentioned one area that could not be treated with water. On the foam side there are very low-pressure wells in which if treated with large volumes of water are simply overwhelmed and although there is no damage from water it might require months to get the water back and on production.

I was involved in a 10,000-foot T.V.D. well in Europe that after pumping nearly 3 million gallons of water and 8,000,000 pounds of proppant the operator finally admitted the interval had a bottom hole pressure gradient of .3 psi/foot. To achieve normal first flow of oil with 30 % load recovery would require very costly jetting with nitrogen, extensive swabbing or various other mechanism to recover the load. Had we know about the low bottom hole pressure, a radically different stimulation approach would have been proposed. Most probably the treatment of the lateral would have to be broken up into foam stages allowing the fluids to be recovered following each stage. There is also a possibility with the low pressure that the economics of recovering profitable hydrocarbons is questionable.

Stateside there are shallow low permeability wells, with less than water pressure gradient, that do require high volume slick water type treatments to achieve economic production. These wells, which show promise of economic production though pressure analysis, can be stimulated using high quality nitrogen foam. Typically, these non-stabilized nitrogen foams are a minimum of 70 quality and sometimes approaching 90 quality.

Intense Quality Control

I will not expand greatly on the old chapter 10 as it explains in detail that "Intense Quality Control" is basically testing the actual water and chemicals to be used on the job at bottom hole shear and temperature conditions. This procedure is built around conventional crosslinked fluids. The Gas Research Institute work behind this procedure dramatically changed our industry, even illustrating the need for more user-friendly fluids such as CMHPG versus HPG.

At the time of the writing of the second edition of my book some service companies built and put into the field vans not unlike the GRI rheology van. We found that although the vans were operationally sound that finding personnel to do the testing and maintain the equipment was virtually impossible. The person was required to be superhuman. They needed to be rheologists, chemists, electricians, mechanical Engineers, and not require sleep or have a family. Although high temperature fluid testing devices are present on offshore stimulation vessels, all of the work on land was moved back to local service company labs requiring water and chemical samples to be gathered and tested just prior to the job. For temperatures less than 200 degrees Fahrenheit, standard heat cups were carried in all the service company frac vans. These procedures were very rigorously followed until the movement to horizontal ultra-low permeability fracturing. Although there are sporadic jobs run conventionally, it is extremely difficult to find functional laboratories that can do high temperature fluid testing. Where available these labs are at service company regional labs or at their headquarters and requires sometimes days to gather fluids and samples for testing. The same problem holds for heat cups and B2 bobs for testing below 200F. We recently had to purchase equipment for a small service company and instruct the personnel how the tests were to be conducted. There are numerous reservoirs in the US and worldwide that will respond to conventional fluids. It is my fear late in my career that we are losing the ability to properly test fluids prior to their execution. I would agree that chem add monitoring has improved but the vast differences in waters utilized certainly requires testing the local water at in-situ conditions to be sure we are pumping treatments as planned.

I have stated many times in frac schools that all our generated data on fluids is wrong. This includes the friction data and all the viscosity data. The reason the FR data is wrong from service company manuals or electronic data is that the vast majority of all data was measured or is still being tested in smooth wall pipes and we do not frac down smooth wall tubing or casing, thereby ignoring wall roughness. Unless the actual water is tested with the actual chemicals, the viscosity can truly vary as much a magnitude. If you are using n' or K' data from your frac model, you should know that you are not running data from your water, but water from the headquarters of the service company. Also, perhaps even more important, the data does not include necessary degrading agents. If you want to input real data on a conventional frac, one must gather rheological data from the actual fluid and chemicals to be used.

Image A

Venting the energized line during a coiled tubing operation, utilizing liquid nitrogen and associated equipment, on a remote wellsite in Australia. Venting of all energized fluid in the lines must occur to prevent a pressure build up and possible equipment rupture from expansion of any gas left to warm to surrounding temperatures.

Image B

CO2 boost pump frosting over during pumping operations.

Image C

Aerial view of N2 frac job, with Nowsco Service and Western Company of North America.

Image D

Aerial view of energized CO2 foam hydraulic fracturing treatment in Eastern Canada – Saskatchewan province.
Source: Oil and Gas Product News

Image E

Dowell CO_2 frac job.

Image F

Combination N_2/fluid pumping unit; flow rate up to 180,000 SCFH with working pressure between 10,000 and 15,000 psi. On-board triplex and centrifugal pump, as well as mixing tank capabilities.
Source: CSP Technologies

Image G

Ely cutting open alcohol gelling agent bags, into the hopper of the remote blender.

Image H

Nowsco N2 pumping units with Western Company pump trucks.

Image I

INCREASING SLURRY RATE											
70 QUALITY CROSSLINKED FOAM											
		PROPPANT			SLURRY VOLUME		RATE				
Foam Volume (gals)	Liquid Volume (gals)	Foam (ppg)	Liquid (ppg)	Total Lbs.	Foam (gals)	Blend (gals)	N2 scfm	Liquid (bpm)	Sand (bpm)	Total (bpm)	Time Mins-Secs
40,000	12,000	0.0	0.00	0	40,000	10,500	13,820	4.50	0.00	15.00	55:33
2,500	950	2.0	5.30	5,000	2,728	8,630	13,820	4.50	0.70	15.70	39:37
4,000	1,867	4.0	8.60	16,000	4,721	11,712	13,820	4.50	1.40	16.40	47:29
6,000	2,760	6.0	13.04	36,000	7,642	5,445	13,820	4.50	2.00	17.00	19:53
8,000	4,853	8.0	13.20	64,000	10,918	4,808	13,820	4.50	2.70	17.40	15:53
10,000	6,733	10.0	14.85	100,000	14,560	3,945	13,820	4.50	3.40	18.40	11:54
15,000	10,901	12.0	16.50	180,000	23,208	4,284	13,820	4.50	4.06	19.06	0.50
1,800	540	0.0	0.00	0	1,800	540	13,820	8.56	0.00	19.06	2:92

FOAM FRAC PUMP SCHEDULE											
CONSTANT SLURRY RATE											
		PROPPANT			SLURRY VOLUME		RATE				
Foam Volume (gals)	Liquid Volume (gals)	Foam (ppg)	Liquid (ppg)	Total Lbs.	Foam (gals)	Blend (gals)	N2 scfm	Liquid (bpm)	Sand (bpm)	Total (bpm)	Time Mins-Secs
35,000	10,5000	0.0	0.0	0	35,000	10,500	13,820	4.50	0.00	15.0	55:33
25,500	7,500	1.0	3.3	25,000	26,130	8,630	13,225	4.30	0.60	15.0	41:28
30,000	9,000	2.0	6.7	60,000	37,712	11,712	12,700	4.10	1.25	15.0	51:55
12,500	3,750	3.0	10.0	37,500	14,195	5,445	12,200	3.96	1.80	15.0	22.32
10,000	3,000	4.0	13.3	40,000	11,808	4,808	11,700	3.81	2.30	15.0	18:44
7,500	2,250	5.0	16.7	37,500	9,195	3,945	11,223	3.67	2.80	15.0	14:35
7,500	2,250	6.0	20.0	45,000	9,534	4,284	10,860	3.55	3.20	15.0	15:08
1,800	540	0.0	0.0	0	1,800	540	13,820	4.50	0.00	15.0	2:51

Image J

CONSTANT INTERNAL PHASE										
70 QUALITY CROSSLINKED FOAM										
		PROPPANT			SLURRY VOLUME		RATE			
Foam Volume (gals)	Liquid Volume (gals)	Foam (ppg)	Liquid (ppg)	Total Lbs.	Foam (gals)	Blend (gals)	C02 (bmp)	Liquid (bpm)	Sand (bpm)	Total (bpm)
40,000	12,000	0.0	0.00	0	40,000	12,000	21.0	9.0	0.0	30
2,500	950	2.0	5.30	5,000	2,728	1,178	18.6	9.0	2.4	30
4,000	1,867	4.0	8.60	16,000	4,721	2,597	16.0	9.0	5.0	30
6,000	2,760	6.0	13.04	36,000	7,642	4,402	13.8	9.0	7.2	30
8,000	4,853	8.0	13.20	64,000	10,918	7,771	11.8	9.0	9.2	30
10,000	6,733	10.0	14.85	100,000	14,560	11,293	9.8	9.0	11.2	30
15,000	10,901	12.0	16.50	180,000	23,208	19,109	8.2	9.0	12.8	30
1,800	540	0.0	0.00	0	1,800	540	21.0**	9.0	0.0	30
**If practical, prefer to flush with no C02										

CHAPTER 8

FORCED CLOSURE (1994)

Perhaps one of the most serious mistakes made in reference to the technique of forced closure was its name. Although very early on we really did believe that we were causing the fracture to close by flowing the well back, we now know that if that occurs that it is only coincidental; and certainly at the flow rates that we use, it is not a dominating factor in fracture closure. Perhaps a better term would be "reverse gravel packing." We are attempting to pack off the perforations externally with proppant and negate near wellbore settling as well as flowing back unbroken fluid at a fairly aggressive rate out of the wellbore.

The concept of forced closure was developed as a response to a paper presented whereby the authors suggested that when you don't see fairly rapid indications of fracture closure then you should flow the well at a low rate to accelerate closure time. My particular response to that was that any waiting period could be detrimental and that I couldn't see anything wrong with flowing all wells back at a controlled rate.

Very early on, wells were produced at fairly low rates, in the range of 10-15 gals per minute. It should be noted that some particularly cautious operators used lower rates than this; namely, in the range of 5 gals per minute. There was a tremendous amount of concern with operators that flowing back wells when the gel is not broken will result in producing the sand out of the perforations into the wellbore. What we have found is that flowing the wells in this way actually reduces the amount of sand produced out of the formation; and, as an added bonus, it helps you to produce sand that is left in the wellbore from the underflush out of the wellbore, negating any cleanup operations post frac. We were taught for so long about the importance of completely breaking the gel (i.e. shutting in the well for long periods of time) to make sure that everything was completely broken. We also were taught that if we brought gel back that we would bring sand with it. There is a very simple and straightforward answer to combat this old false hypothesis. There are no wells being fractured with proppant at .5-1 bbl./minute. The sand bridges on the perforations, negating the ability to pass through and into the formation. The same thing occurs in reverse mode with forced-closure. You simply bridge at the perforations at nominal rates. Perhaps one of the most dramatic things that is seen in forced closure is when bottoms up occurs and you go from heavy sand, which is of course the under flushed proppant, to a relatively clean gel with no proppant. This clean viscous gel is the first fluid from the formation. It should be noted that we have started continuing the addition of the cross-linker on treatments until no sand exists on the surface (i.e. we are cross-linking part of the flush). This, we have found, assists us in removing sand from the wellbore at the typical forced-closure rate, which today is around 1 bbl./minute. If you flush with linear gel with some sand in it, it will be strung out and you will make sand for some time. With a cross-linked gel, you typically bring the sand up as a single plug and only have to deal with it for a very short period of time.

Fairly early on in the implementation of the forced-closure technique, we were very careful to monitor a plot of pressure vs. the square root of time, watching carefully for a deflection that we thought early on was fracture closure. It should be noted that the earliest applications of forced closure were done in 1986. In other words, this technique has been utilized on hundreds of wells for the last 7 years. While monitoring the first 20 or 30 treatments, and watching for an inflection of the square root of time plot, we came fairly quickly to a decision in reference to fracture closure. On most of these wells (see Figures 11-1 and 11-2), we saw a very early deflection in the square root of time plot. It was apparent that regardless of job size, permeability, or other pertinent factors relating to fracture closure that we saw this early deflection at almost an identical period of time on virtually every well. In relationship to volumes, we saw these deflections in less than 5 bbls. flowed back. Obviously, this did not indicate closure of the fracture, but it relates to a packed proppant pack whereby the flow regime that had previously been one of flowing through an open fracture now change to flowing through a packed proppant pack at the perforations.

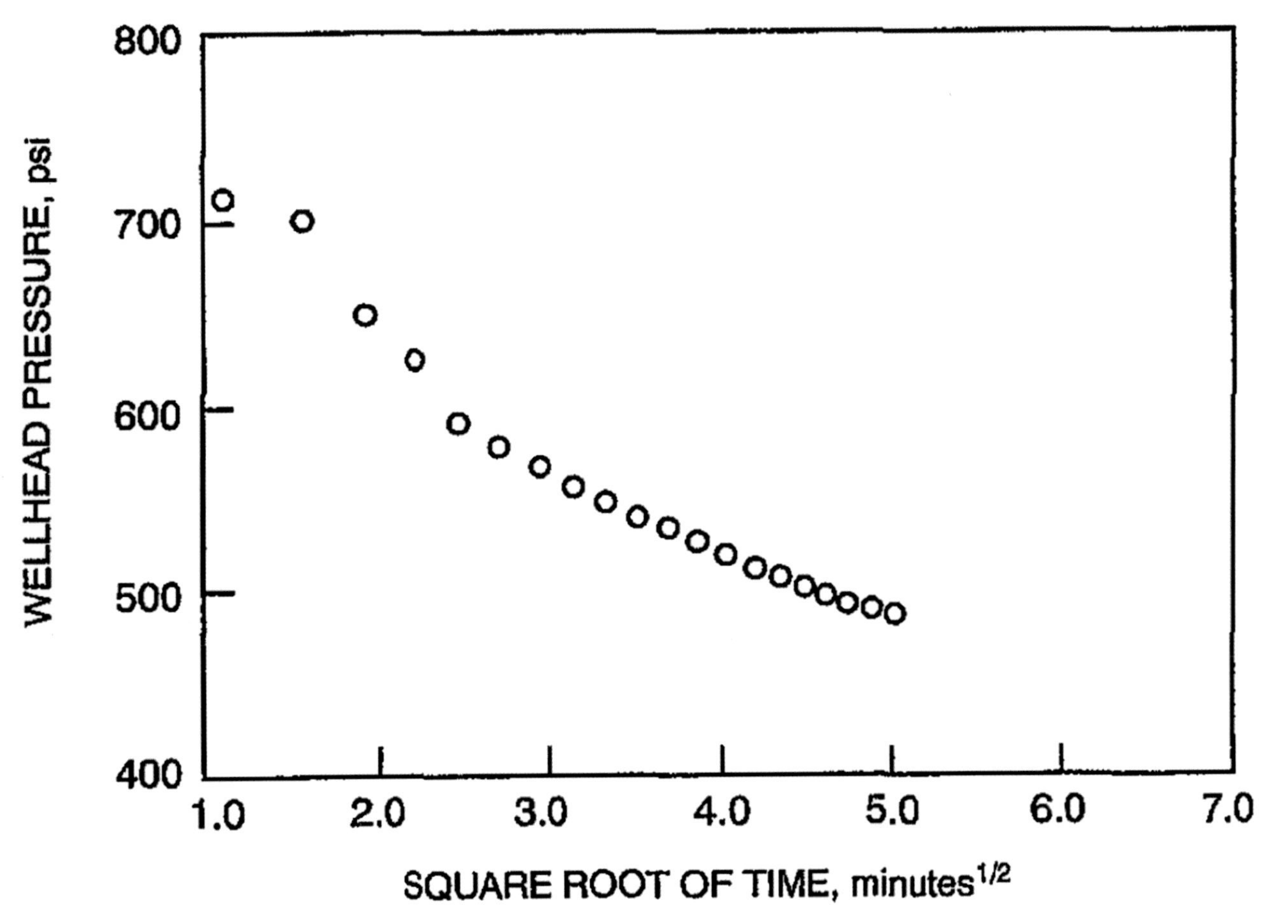

Figure 11-1 Square root of time plot showing the inflection occurring due to packed proppant near the wellbore.

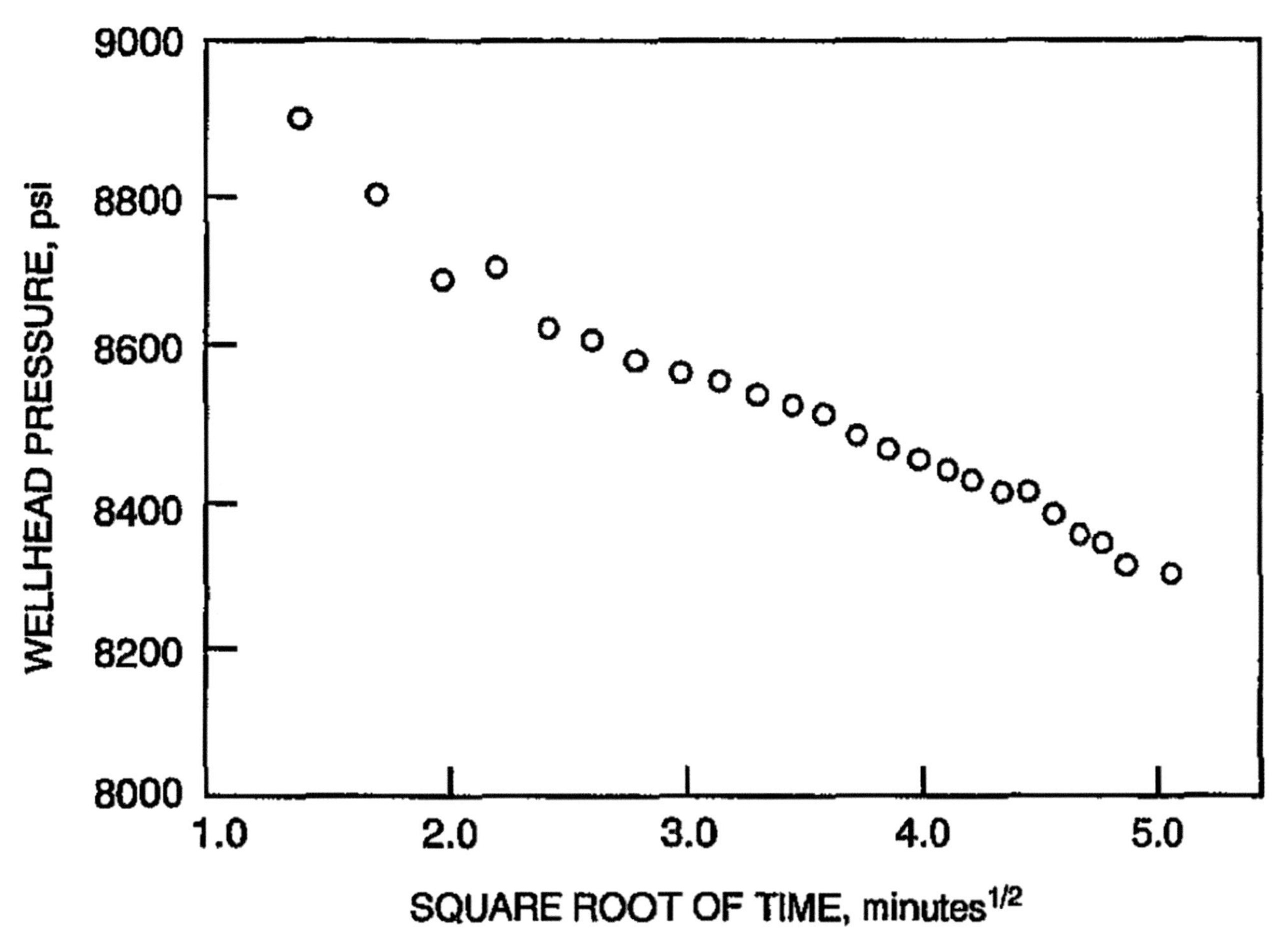

Figure 11-2 Another illustration afforced closure flowback. This well was flowed back at approximately half a barrel a minute and showed a deflection indicating a reverse gravel pack phenomenon occurring at approximately 5 minutes and a fracture closure occurring at approximately 20 minutes.

Further evidence of this phenomena was shown by personnel who did not initiate forced closure immediately but waited 10-45 minutes before opening the well. In these cases, they would never note the early deflection in the square root of time plot. We quickly surmised that the proppant was settling near the wellbore to the extent that some of the perforations would, in fact, remain open, thus disallowing the deflection from the reverse gravel pack or packed proppant pack phenomena. We, therefore, believe that it is absolutely essential that flow back be initiated in less than 30 seconds of shutdown. This has caused some consternation from field foremen and other people who are used to taking ISIPs at 5, 10 and 15 minute shutdowns. With a well being opened up in 20-40 seconds, you can still obtain an ISIP. To cover the field foremen we have given them the 5, 10, and 15 minute pressure data, but we have given it to them with a subscript showing that forced-closure flow back is underway. Another point of interest in Figure 11-2, which illustrates a deflection at a little more than 2 minutes of square root of time, is that you see another deflection at about 4.5 minutes square root of time. This particular job was a fairly small treatment in a fairly permeable reservoir. In this case, you can see the forced-closure inflection and then see the fracture closing later. In tight reservoirs, we have noted periods well in excess of 1-1.5 hours closure time after seeing the initial forced-closure deflection, after only flowing back about 5 bbls. of fluid.

Presently on most treatments, we typically are flowing wells back immediately at 1 bbl./minute. In some areas, people are flowing them more aggressively, some exceeding 2 bbls./minute. In other areas where they are not used to this phenomena, they do actually flow them at a little slower rate in the .5 bbl./ minute range. We have not seen any problems whatsoever with sand production at 1 bbl./minute rate with unbroken gel being produced. In relationship to the pressure vs. square root of time phenomena, I personally have very little interest in the initial inflection from forced closure, since we continue to flow the well, if possible, until oil or gas is starting to be produced. We feel the best procedure to follow is to utilize the supercharge from the fracture treatment and flow the well until it either dies or until oil or gas cut hits the surface. At that time, we go into normal flow procedures to make sure that we don't crush proppant or flow the well in too aggressive a way and start pulling in sand. Although forced-closure has been beneficial in slowing down, or in some cases stopping, sand production, it is not a cure-all; and obviously you can flow the well too hard, particularly when producing gas or high-rate oil and gas, and start bringing in sand. It has been our experience that when you start to make sand, you will continue to make sand for a period of time. Early in the forced-closure procedure, people only flowed the well until they saw the deflection and then shut it in. We think that this is a big mistake and feel that you should continue to flow the well on a 24-hour basis. This allows you to remove gel in an unbroken state, negating any damage from residue formed when gel degrades par. One of the biggest mistakes made when implementing forced closure is too aggressive a breaker schedule trying to break the gel extremely rapidly near wellbore. I do not believe that this is at all necessary. In fact, it can. be very detrimental from the standpoint of rapid proppant settling and negating good proppant coverage in the near wellbore area. What we recommend is to have the fracturing fluids remain stable throughout pump time and then degrade back to water at a time sometime after the shut-in of the well. This might be 30 minutes for a short pump-time treatment of small volume, or it might be 3-5 hours for a high-temperature treatment such as conducted in south Texas. It should be noted that we have not seen sand production even when flowing back wells where the fluid we know is cross-linked in the formation. Additionally, Amoco Research, who was concerned about this project, has flowed cross-link gel back through proppant packs and has found no damage to the proppant pack. The startling thing to most people is that you do not carry sand back out of the perforations but it bridges. In fact, I believe it allows you to have a wide, highly conductive propped fracture near wellbore in communication with the perforations and disallows proppant settling. Additionally, this technique is beneficial in negating ongoing fracture growth that occurs after shutdown of the treatment. If you are above fracture extension pressure, then the fracture will continue to grow outward with some closure at the upper and lower sections of the fracture. This ongoing fracture growth can have very negative effects on propped fracture conductivity due to smearing of proppant and reducing effective conductivity of the proppant pack.

Obviously, monitoring of the pressure vs. square root of time will allow you to monitor fracture closure. If very long fracture closure times occur, you need to downsize future treatments, minimizing pad volume. Obviously, you can conduct mini-fracs and hopefully do a better job of minimizing pad through evaluation of a mini-frac prior to a treatment. But it should be noted that the forced-closure technique, although after the fact, does allow for similar data as is seen in a mini-frac. There has been some criticism of this technique in negating pressure data achieved by watching falloff of the pressures in a shut-in condition. In my opinion, without a downhole gauge, particularly with a constant flow rate, you can achieve the same or perhaps more meaningful data than surface data from a high temperature well trying to monitor falloff of pressure vs. time.

In summary, the forced-closure technique accomplishes the following:

1. Significantly reduces proppant production by reverse gravel packing and disallowing open perforations where high velocity can occur.
2. Allows removal of a great deal of the fracturing fluid in an unbroken state, negating any damage from potential filter cake development or gel residue that would occur upon breakdown of the polymer.
3. It takes advantage of the inherent supercharge from the fracture treatment. We have noted more than 50-60% of the load recovered in older pumping wells by utilizing this technique.
4. Minimization of proppant settling near wellbore.
5. In most cases, continued flow will remove all proppant in the wellbore (i.e. underflushed proppant), eliminating the need for proppant cleanout on many treatments.

Theory

Over the past several years, most fracture design experts and theoreticians believe that the majority of hydraulic fractures are radial and vertical. With the advent of high viscosity, highly efficient fracturing fluids, we believe (particularly with the tight formations) that there is a very large potential for proppant settling prior to fracture closure. For those of us who have attempted to-even with mini fracs-design perfect fractures so that a screen-out occurs on the flush, we realize the great difficulty in accomplishing a perfect design. We feel that in utilizing forced closure, we not only have an additional tool to optimize the fracture treatment, but also to evaluate how well we did, if in fact we had a mini-frac prior to the treatment. In Figure 11-3, there is an artist's illustration of half a propped fracture. This would represent a perfectly designed treatment where proppant is suspended throughout the fluid. Figure 11--4 illustrates proppant settling occur ring with an open fracture. In many cases, this particular figure does not show that extreme proppant settling can actually occur where the proppant is not even across the producing interval. We believe very strongly that excessive pad volumes in tight reservoirs can result in proppant being in (at best) the lower portion of the interval, and in some cases covering none of the interval, certainly near wellbore.

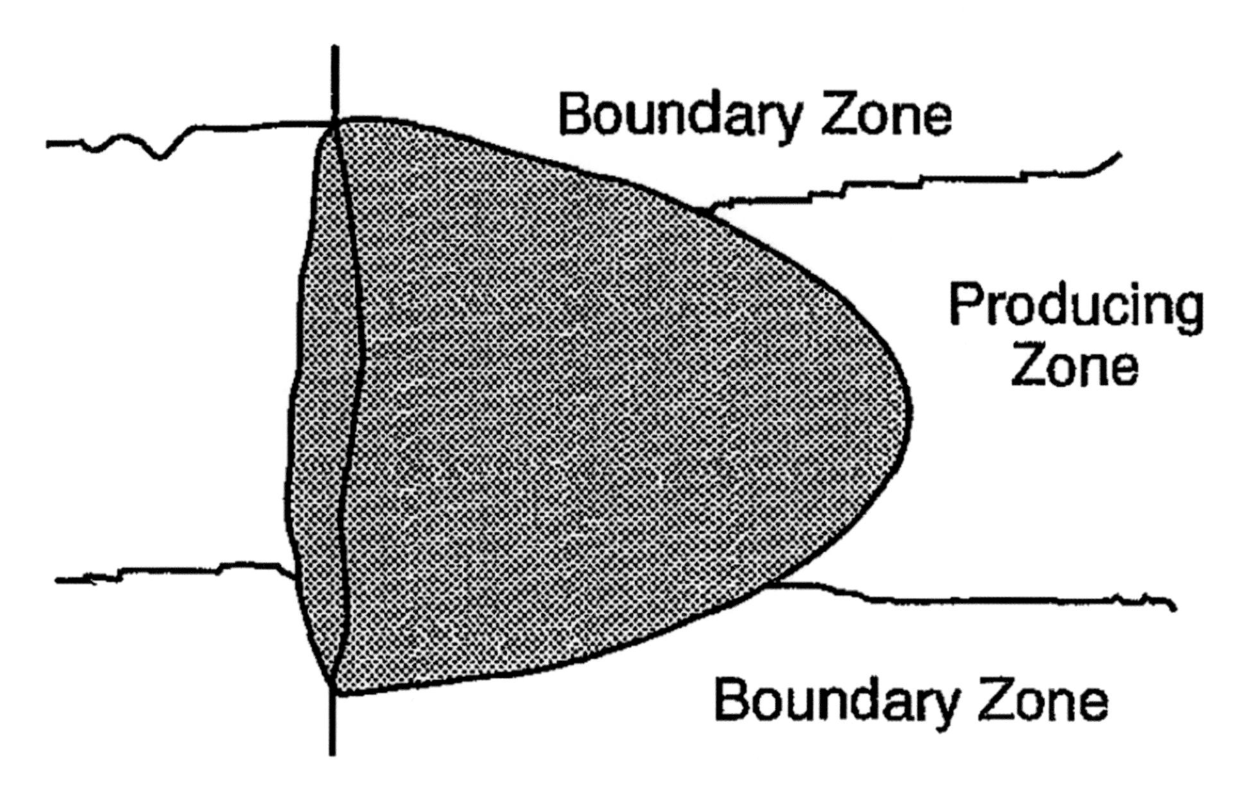

Figure 11-3 Idealistic illustration of half of a packed fracture.

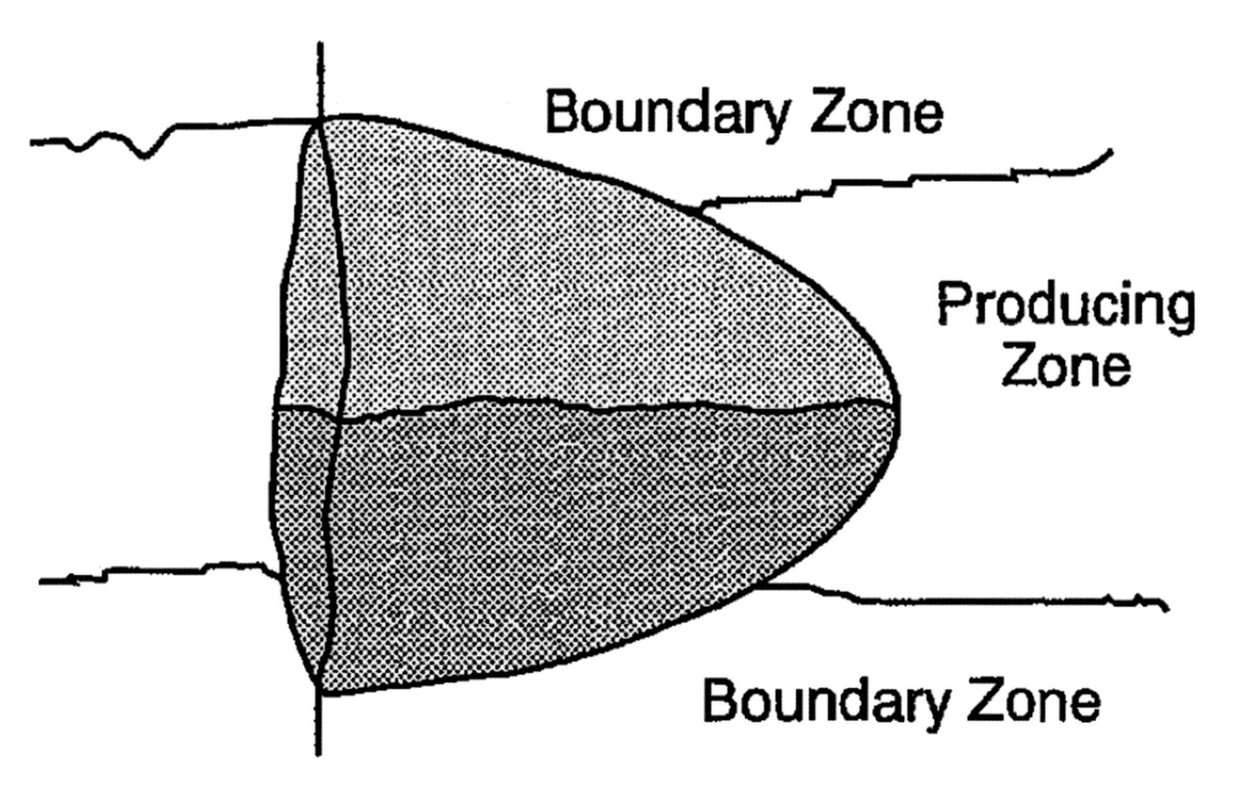

Figure 11-4 Proppant settling in an idealized case. We feel in many cases that the settling may actually occur, leaving the proppant outside of the zone.

From another standpoint in such areas as west Texas where fairly thick oil zones are treated, if you see considerable amounts of proppant settling, then that proppant may in fact settle out into areas of high water saturation or into the water contact. This is the reason I believe people over the years have talked about fraccing into water. What occurred was not that they had fracced into water, but that they only propped the part of the zone where the water contact was and not the oil-producing interval. This has been proven by going back and refraccing high water ratio wells with treatments specifically designed to prop only the upper part of the section and achieving good stimulation of the oil zone. In Figure 11-5, we show the settled proppant fracture with a portion of the fracture where closure does not occur. It is my opinion, if you will note on Figure 11-5, the open area is where a great deal of sand production occurs on fracture treatments. If a portion of the fracture does not close, this causes an area of very high infinite conductivity and a "flow over area" where proppant can flow directly out of the well. The same phenomena occur on treatments where wild fluctuations in proppant concentrations occur at the tail end of the fracture treatment. This allows void areas and results in proppant reduction at the end of the treatment.

It should also be noted that one of the things that we advocate very strongly is enhancement of the treatment with larger concentrations of proppant. In a paper presented in 1990, coauthored by Bill Arnold with Phillips Petroleum, we gave many case histories of a combination of forced closure, high proppant concentrations, and intense quality control. The higher proppant concentrations have the advantage of enhancing the viscosity of the fluid itself, slowing down proppant settling and resulting in more highly conductive fractures. Figure 11-6 illustrates an idealistic version of the ideal proppant pack with forced closure with all the proppant throughout the producing zone. It should be noted that if one utilizes too much pad or effectively too efficient a fracturing fluid that even forced closure is not sufficient to assure you of negating excessive proppant settling. As stated earlier, an excessively aggressive breaker schedule can also result in a rapid settling rate of proppant, perhaps out of the zone of interest even with forced closure. Proper design is critical and forced closure is a means to enhance a fracture treatment design, but it is certainly not the solution to problems therein. As an illustration of the effects of forced closure or lack thereof on fracture geometry, Figures 11-7, 11-8, and 11-9 illustrate what occurs in relationship to fracture conductivity in a moderately tight reservoir with and without forced closure. Figure 11-7 represents the Palmer model in a pseudo-3D mode, which shows proppant pack conductivity at the end of pumping.

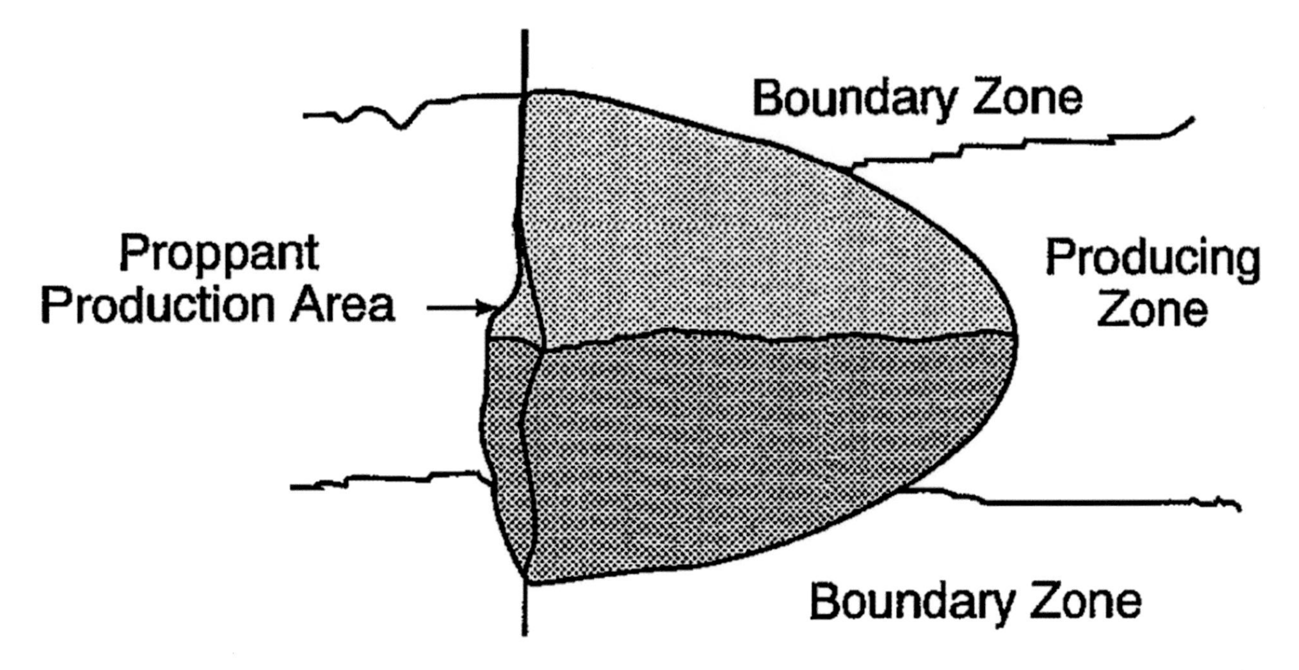

Figure 11-5 Closure of the fracture after settling in an open fracture illustrating where we feel a great deal of proppant production does occur in a high-conductivity flow-area, illustrated by the proppant production area.

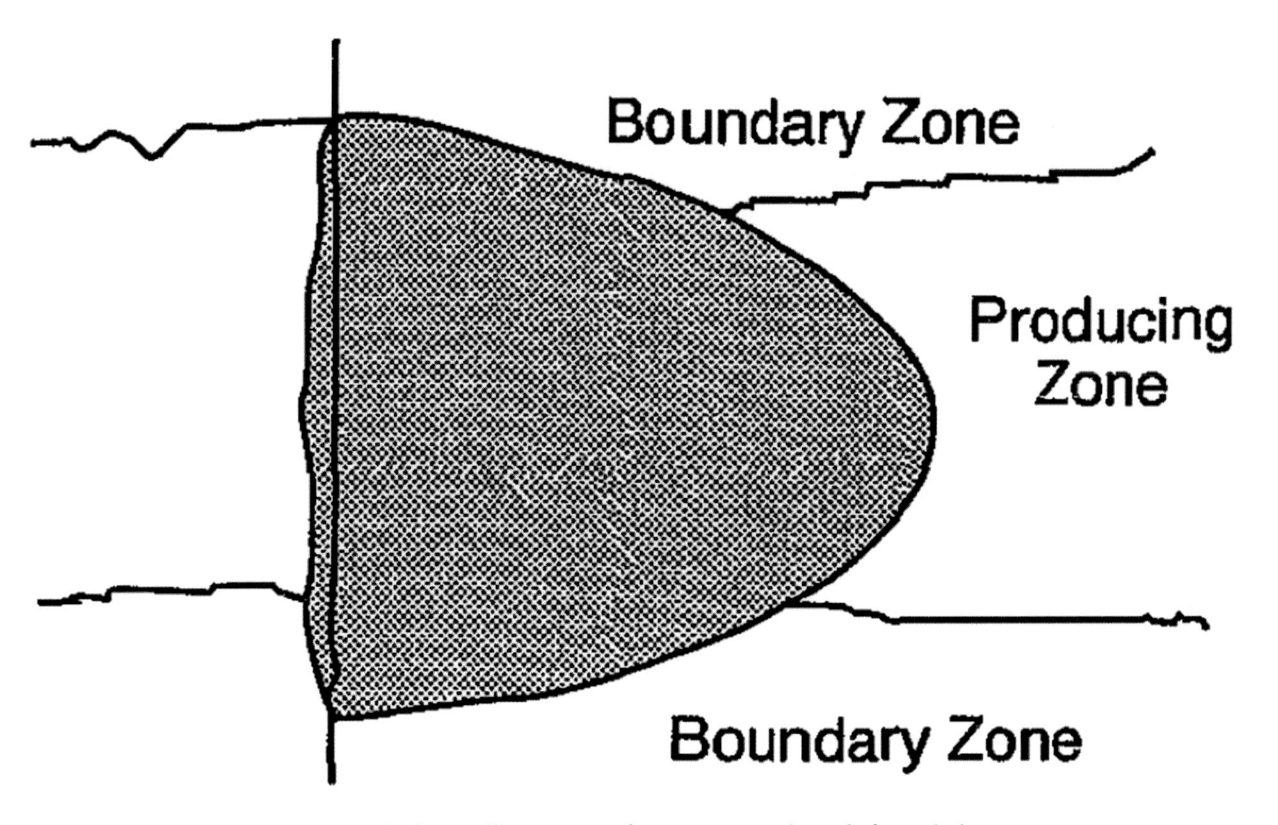

Figure 11-6 Idealistic illustration of proppant pack with forced closure and an optimized pad volume negating long closure times.

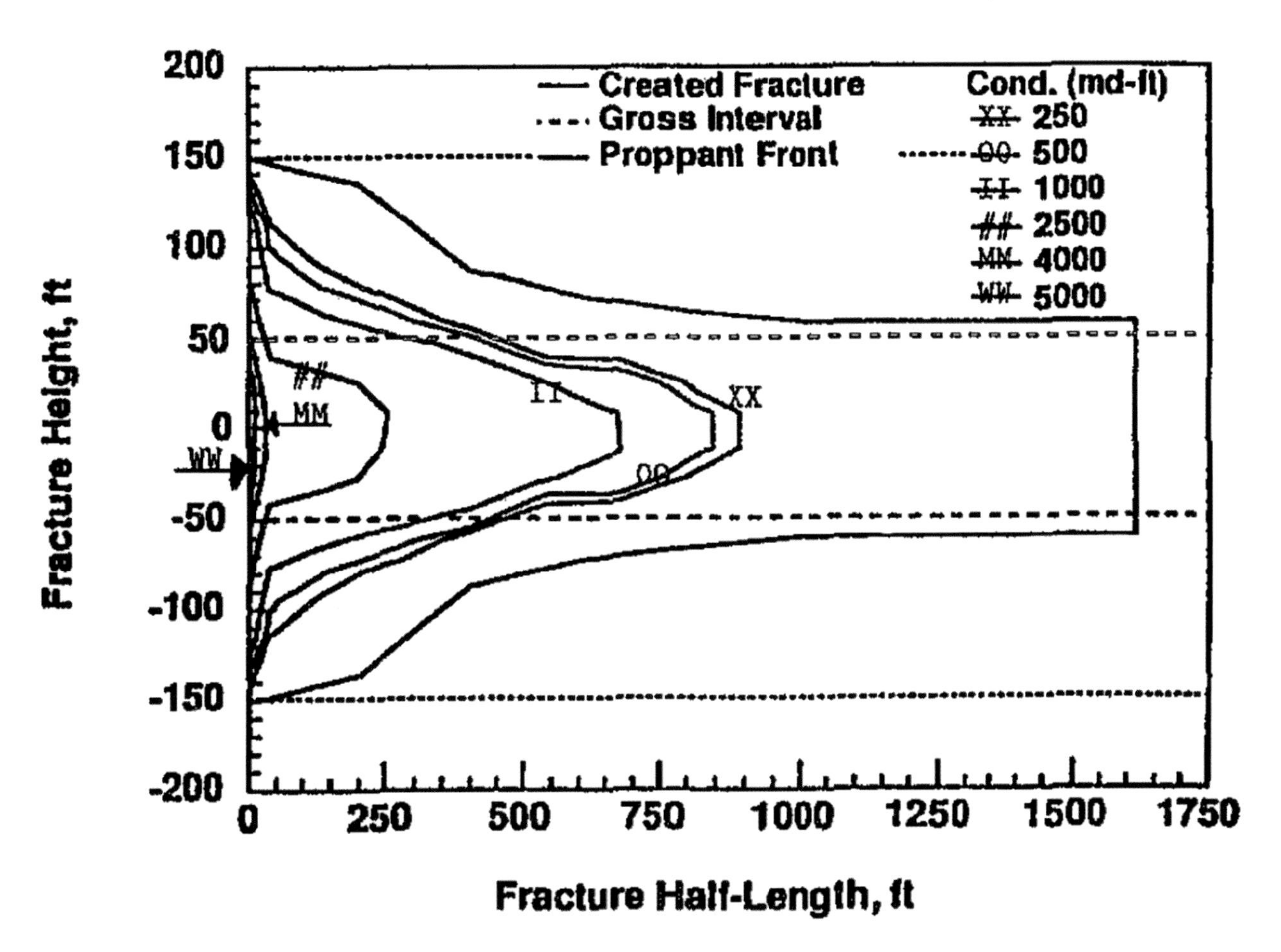

Figure 11-7 Fracture conductivity (MD-FT) end of pumping.

Figure 11-8 illustrates proppant pack conductivity if the well is shut-in to closure. With a long closure time, the proppant smears outward, and you effectively lose a great deal of effective propped fracture. Most fracture design engineers would agree that an effective conductivity of less than 1000 md/ft. will yield little or no stimulation. You effectively lose between Figure 11-7 and Figure 11-8, 100+ feet of effective fracture conductivity due to smearing. Additionally, higher conductivity areas such as 2,500 md/ft are similarly reduced in the near wellbore area. Figure 11-9 illustrates the same fracture treatment flowing the well back to closure at 1 bbl./minute.

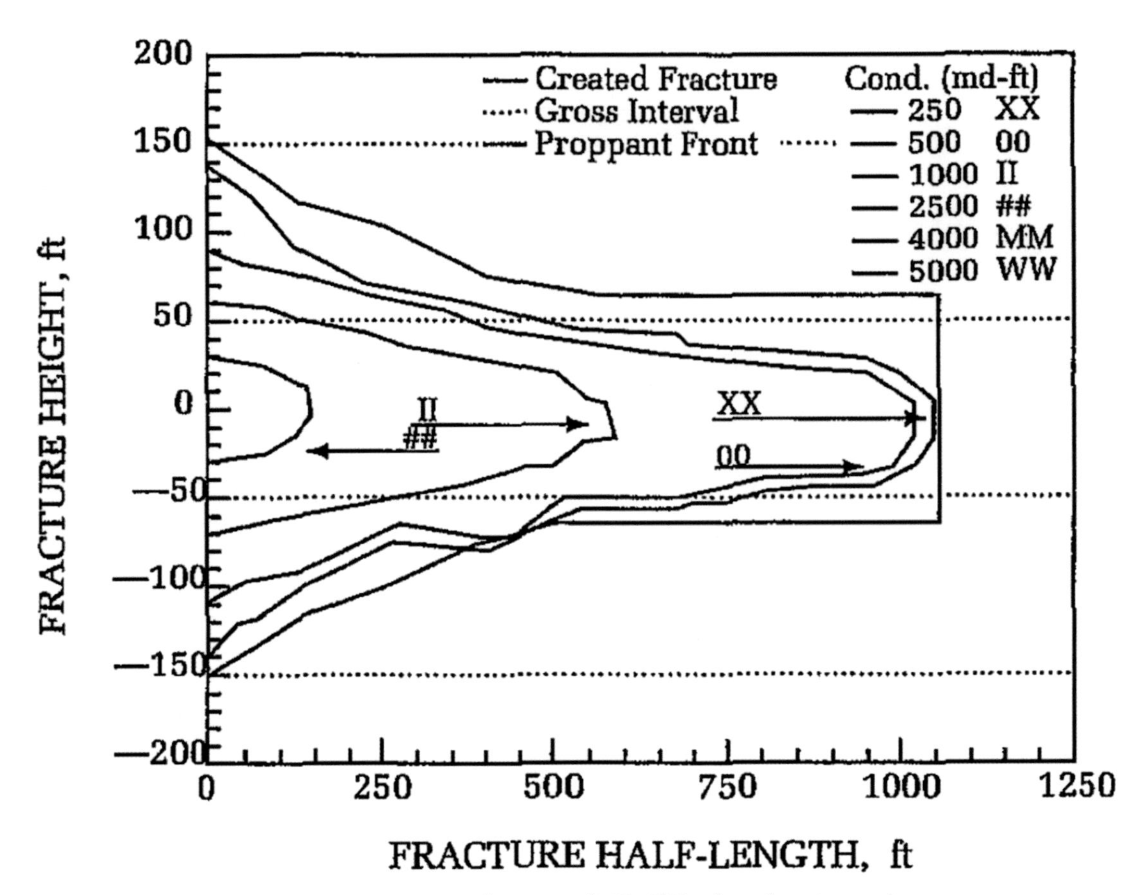

Figure 11-8 Fracture conductivity (MD-FT) after shut-in to closure.

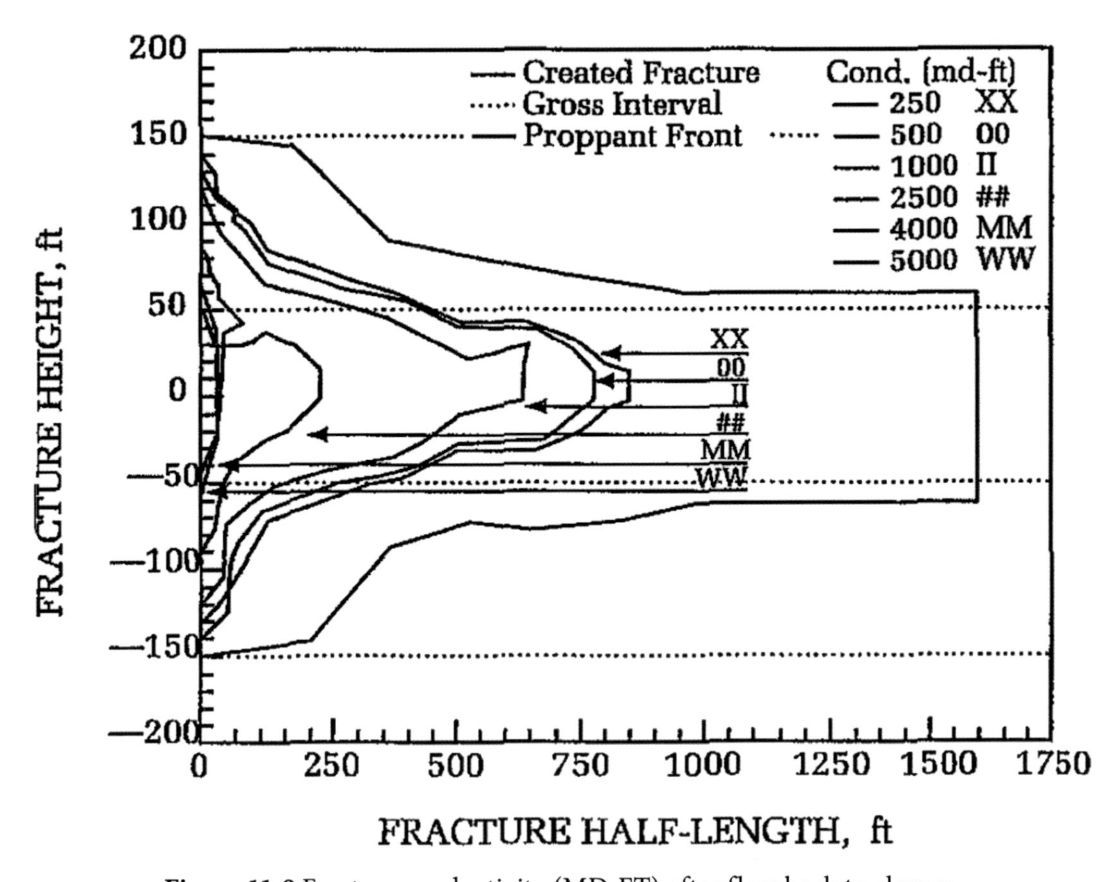

Figure 11-9 Fracture conductivity (MD-FT) after flow back to closure.

As noted earlier, for very tight formations or very large pad volumes, forced closure is not sufficient to overcome poor design criteria. For this particular case, the flow back to closure was extremely beneficial in maintaining effective propped fracture lengths and enhancing near wellbore conductivity.

Forced closure implementation procedure

It is imperative the implementation of forced closure be done in a safe and prudent manner. Most typically, the flow back occurs through the frac line upstream of a blocking valve to isolate equipment. This allows breakdown of frac equipment while the initial stages of forced closure are occurring. This is particularly the case if you are utilizing a tree saver. We do not flow back underflushed proppant through the treesaver, but we typically flow back 1/2 to 2/3 of the tubing volume and shut the well in for the shortest possible amount of time while the treesaver is removed from the tree and then reinitiate the flow. A typical rig-up is shown in Figure 11-10. ·

On occasions where a well-testing company is on-site, they will typically have a dual manifold with a positive choke on one side and a variable choke on the other. The reason for this is that the initial flowback goes through a variable choke; but when it is time for the sand to hit the surface, we will typically flow through a 1/2 in. choke to allow the sand to be produced out without eroding or destroying a variable choke. Once the sand is produced out and clean fluid is visible, we then switch back to a variable choke or, for that matter, a positive choke with the correct size to allow for our expected flow rate. As stated earlier, most people now flow their wells at 1 bbl./minute. We have had little problem with sand production and this allows for fairly rapid removal of fracturing fluids from the well even where the treatments are quite large. In cases where wells are normal or geopressured, you will typically flow the well until such time as oil or gas is produced at the surface and, of course, flow will continue. At that point the operator will take over adjusting gas or oil flow rates depending upon downhole closure pressures or production capacity. From the standpoint of stopping produced sand and maintaining well control, it is advisable to choke back on the well once gas starts being produced, since there is a tendency for sand to be produced if excessive gas expansion is occurring downhole. Additionally, we want to be relatively certain that a minimal closure pressure is placed on proppants in moderate-to-ultra-deep wells. One very important consideration in flowback is that care must be taken that the flowback lines are properly staked down to insure that if gas comes to the surface that the lines do not become mobile and injure personnel. Staking of lines should follow service company guidelines.

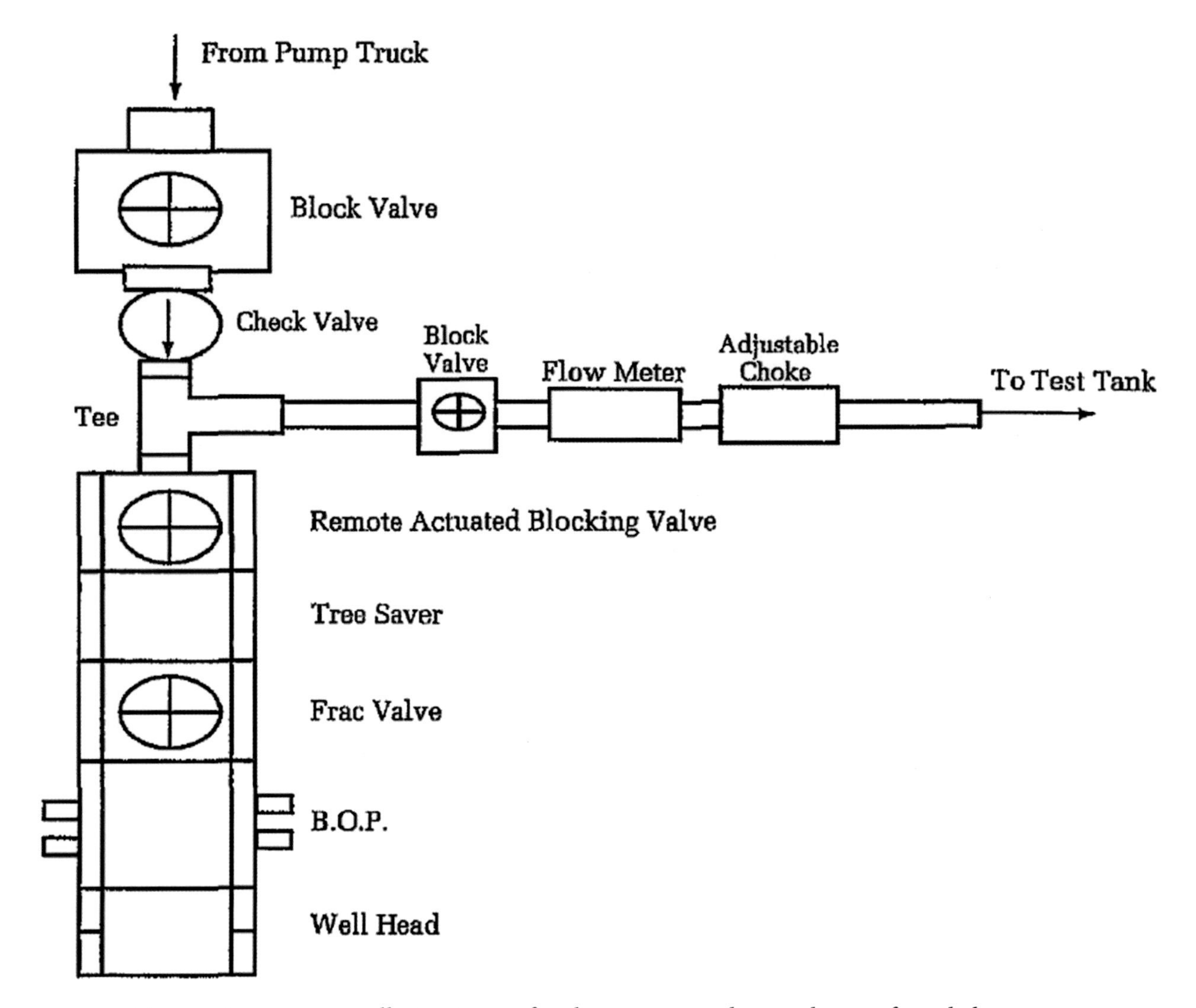

Figure 11-10 Diagram illustrating surface layout required to implement forced closure.

In other cases, where one is working with a well-testing company, they will typically supply manifold with chokes included. In that case, then a simple T off of the line with blocking valves is directed toward the flowback manifold supplied by the testing company. It is always a good idea to have some means of backup for in-line flow meters. In the field, we have utilized 5 gal buckets to meter the flow of the fluid and have also utilized differential pressure charts when we were flowing back foam or gas to measure the flow rate. Unless you are intensely interested in utilizing the pressure vs. rate data for analysis of the treatment, then maintaining an absolute constant rate is not a necessity. Key points of the flow back procedure are as follows:

1. Be sure that the choke downstream of the blocking valve is at least partially open prior to opening the block valve post frac. We have had numerous occasions where if the choke is closed that it cannot be opened and you cannot flow the well.
2. Flow back should be initiated as quickly as possible and certainly within 30 seconds to 1 minute of pump shut-down.

3. If a screen-out should occur, forced-closure procedure is still followed, taking necessary precautions on abrasion effects of the proppant going through chokes. It is not normally a good idea to go through a variable choke with proppant-laden fluid so you will have to use positive chokes in line.

In the case of wells that are underpressured and when an· energizing mode is not used, forced closure is still beneficial. These wells will typically die off in flow rate fairly rapidly, although we have seen the supercharge effects allow production of large percentages of load in such wells. The procedure on these underpressured wells is to flow the well at the same rate until the well dies. At this point, you just simply shut the well in. Another point from the standpoint of practicality and safety is that we have been flushing foam fracturing treatments and energized treatments with straight-gelled water with no energizing medium. This allows for much more controlled flowback procedure and evaluation of flow rate prior to any gas hitting the surface. It also makes most customers feel better about displacement volumes rather than trying to calculate nitrogen or CO_2 volumes, particularly when pressures are changing, perhaps from an eminent screen out. Forced closure is also beneficial, allowing one to use a larger underflush volume, since with forced closure you are bringing the sand out of the well and there should be little problem with fill up of rat hole, particularly if one cross-links the underflushed slurry outside the perforations. We typically underflush wells 2 bbls. or more. It should also be noted that forced closure is a recommended procedure for curable resin-coated sand. We have found that the technique works very well in allowing for better grain-to-grain contact of the curable resin-coated sand. We do not at all feel that forced closure is the solution for sand production problems in all cases such as where a lot of acidizing has been done on carbonate reservoirs and in areas where perforating techniques are such that perforations are stretched over wide intervals. You need to utilize sand that will help control flowback. We feel very strongly that forced closure will aid in the control of flowback with curable resin-coated sand.

Guidelines

1. Flowback upon shutdown should be immediate. Any delay in the flowback can be extremely detrimental. If settling of the proppant below the top perforations occurs, then the effects of forced closure on negating proppant flow are lost.
2. Overly aggressive breaker schedules can be detrimental to the final proppant pack, particularly in tight reservoirs. The service companies' utilization of encapsulated breakers allows the ability to be fairly aggressive on breaker and not be detrimental to' the proppant pack in most cases. You need to utilize intense quality control procedures to be assured that there is stability for pump time and that too aggressive a breaker schedule is not detrimental to the proppant pack near wellbore.
3. Production of unbroken gel at the surface is not something to be alarmed about because most of the fluid that comes out of the well for the first 2 or 3 tubing volumes will have not seen sufficient heat or time for complete degradation of the gel. I do not recommend incorporation of high-concentration catalyzed breakers or enzymes to accelerate breaking because of the aforementioned settling of proppant. We feel that it is indeed beneficial to bring as much gel unbroken out of the well as possible because this negates any damage from that gelling agent whatsoever.
4. It should be emphasized to the customer that you should never leave a flowing well unattended. One of the major reasons that it has taken so long for forced closure to be accepted is the fact that

people have to stay there with the well and flow it back rather than shutting it in and leaving the well overnight.

5. Field results have proven the tremendous benefits achieved by this procedure. We have documented cases where offsetting wells on 40-acre spacings were and were not force closed with tremendous differences in productivity. We also have documented cases where one well was flowed for a short period of time and shut-in and the other was flowed to clean up, again with substantial differences in productivity. Although this procedure somewhat complicates the fracturing operation, there are many advantages to it from the standpoint of enhanced production for the customer. It can only end up being beneficial to service companies in the long run.
6. Forced closure is recommended for all wells that have perforated intervals. It has been used in open hole sections but, in this case, there must be bridging of the proppant in the formation, and we have very little experience at this and do not necessarily feel that this is a good procedure to follow. Forced closure, from the standpoint of reduced proppant flowback, is greatly optimized by short, perforated intervals where constant velocity is seen upon flowback. Sand production has been noted where there has been a tremendous amount of acidizing in carbonate reservoirs behind perforations, and we have also noted problems with proppant flowback where multiple intervals are perforated across long sections. These proppant flowback problems have not been sufficient to not recommend forced closure. The only real situation where forced closure is not particularly recommended is for open hole sections.

CHAPTER 8

FORCED CLOSURE (2022)

Chapter 8 is the updated, original Chapter 11 in the second edition. Again, I will be brief as this very successful technique has been conducted on many thousands of wells since 1986. After reading over the second edition, I did note some things that we have improved with time and experience with the process. This procedure can be followed on single stage low viscosity fracs allowing for the advantage of super charge from the frac treatment, but the process is primarily designed for conventional crosslink gel treatments. In our industry semantics sometimes enters into real understanding of what is being done in a process. When I first suggested the process, I was working at Holditch. We decided to utilize the process on a few jobs and then report back on our monthly, in company seminar. Another One of the other employees went out to test the described process and I did the same. When we presented our results to the other employees, there was a tremendous shouting match relating to whether the process was viable. My presentation was completed using a group of slides illustrating the success I had achieved. I discussed the advantage of using the supercharge from the treatment, rapid cleanup, and the first impression that we were in fact negating sand production. The other gentlemen jumped up and said the process was a joke and he felt the wells made more sand and that we should never do forced closure again. After a significant amount of shouting time, it was decided that we should go through our procedures in detail and determine why one way was working and the other not.

To shorten the story, the problem, which I have found with many other companies and individuals, was he thought immediate flowback was when the service company had rigged down. Following his definition of immediate flowback most of his wells were not opened quicker than 2 hours and sometimes longer. I was flowing back wells in 30 seconds or less.

I will list the new or perhaps clarifying guidelines learned since the previous edition:

1. Immediate flowback is where one opens the well seconds after the ISIP. Longer delay will result in not only loss of supercharge but will many times allow proppant to settle out of zone. Good communication is required between the service company and the flowback crew to assure the procedure is safely conducted.
2. Experience has shown that flowback up tubulars should be in the range of 2-3 barrels per minute. Flowback up casing must be in the range of 4-6 barrels per minute. Equivalent rates should be followed, utilizing proper chokes, for energized treatments.
3. A final recommendation regarding flowback of unbroken gel. We learned early on with forced closure that we could produce viscous fluid without carrying sand from the formation, if a well is

> flowing and making gel, continue to flow and get the gel out of the reservoir. Shutting the well in to allow the gel to break is based on antiquated theories and has no basis in reality.

I would also state that this process caused massive consternation amongst field individuals who had been ingrained with the idea that you must let the gel break prior to flowing the well.

On one occasion, I had the field foreman sharpening his knife as we flowed the well back as he was certain I was flowing all of the proppant out of the well. When I asked him why he was sharpening his knife, he indicated that I would lose my claim to be qualified to be a male if the proppant came out of the reservoir. I still am classified as male, so we did not flow proppant.

As stated in the 2nd edition one of the major problems with getting field personnel to follow the process was it affected their attendance at the local happy hour. On multiple occasions I had to stay and flow the well back. As people began to utilize the process properly, they found it not unusual to have a fractured well going down the production line before sunset, where previously the process would have taken several days.

I want to refer you to bad flowback lines which are extremely dangerous. Illustration 19 show really bad flowback manifolds and stakes which will become aerial javelins if high pressure gas hits the poorly stake manifolds the last slide shows a rig anchor which if the proper way to stake down flow back lines and also to tie down lines when pumping an energized treatment.

Commonly used in the industry, are tie-down restraints. They restrain iron to the extent of the length of the tie-down. It is of my opinion, they are helpful in non-energized fluid pumping operations, but if one were to be conducting energized fluid operations, all treating iron will need to be anchored down with tie-down anchors. Images at the end of the chapter show good and bad anchoring of lines.

Image A

Flowback lines improperly restrained with iron stakes.

Image B

Properly restrained iron used in energized pumping operations and flowback. Proper restraints involve use of rig anchors, normally used in drilling rig operations.

Image C

Homemade flowback operation, waiting for disaster.

Image D

'Rib-And-Spline' line restraint system includes smaller slings and loops that wrap around pressurized iron at critical points such as union connections. In the event of a line failure, the entire dynamic system absorbs the energy release while keeping all of the iron and components tied together. Source: Cuadrilla Resources

CHAPTER 9

JOB IMPLEMENTATION GUIDELINES (1994)

Approximately two years ago, I was asked to make a presentation concerning tools and techniques utilized in hydraulic fracturing. Out of this request, I decided to prepare a paper that was subsequently presented at the Southwestern Petroleum Short Course titled, "A Variety of Tools Assist in Fracture Stimulation of Oil and Gas Wells." For those who are interested, this paper was presented in 1991 at the Southwestern Petroleum Short Course in Lubbock, Texas.

I felt that this particular presentation in a broadened form would become a beneficial part of this book. Many of the things covered about this subject are not discussed or published and may be of some benefit to the reader.

Fracturing fluids available to the industry

Table 12-1 lists commonly available fracturing fluids. I will not discuss fracturing fluids in detail (see Chapters 5 and 6), but I will make a few comments about each one of the fracturing fluids mentioned in Table 12-1.

In Table 12-2 conventional linear gels are described. Linear gel is basically a fracturing fluid that has the viscosity anywhere from a light syrup to a heavy honey-like consistency. This is in direct comparison to the rigid cross-linked gels that will be discussed later. These fracturing fluids have been in use since the early 1960s and today they have found a rebirth in many areas. Their broadened use is due to their simplicity and ease of degradation. In addition, we know a great deal about their friction pressure down various tubulars, viscosity profiles, (etc.). Additionally, they are quite a bit cheaper than cross-linked fracturing fluids and have the ability to carry high sand concentrations in many applications. The major disadvantages of linear gels are fairly rapid proppant settling, particularly at low concentrations. Most operators have opted away from linear gel systems in their fracturing because of their supposed inability to create significant fracture width, allowing for the placement of high proppant concentrations.

Table 12-1 Available Fracturing Fluids
• Conventional linear gels
• Borate cross-linked systems
• Polyemulsion
• Foam fluids and cross-linked foams
• Delayed cross-linked fluids using metal ligands
• Aluminum phosphate ester oil gels
• Energized systems

Table 12-1 Available Fracturing Fluids
• Used since the early 1960s
• Increased usage today due to:
➢ - Simplicity
➢ - Ease of degradation
➢ -Known rheology
➢ -Low cost
➢ -Can transport high sand concentration
• Disadvantages
➢ -Rapid proppant settling
➢ - Lack of created fracture width
• Linear gels in combination with other fluids yield valuable tools in fracture design
• Disadvantages

There are many applications where linear gels are the fluid of choice. In the technique that will be discussed at the end of the chapter termed "pipelining," we have been able to easily place well in excess of 10 ppg proppant in formations utilizing linear gel in combination with cross-linked pads.

The next type of fracturing fluid that is finding not only continued use, but also a much wider use is borate cross-linked fluids. The borate system, as described in Table 12-3, was among the very first cross-linked fracturing fluid ever used. Historically, by a matter of weeks or perhaps one or two months, borate cross-linking fluids followed the very first cross-link fracturing fluid; namely, a cross-linked antimony gel. Very quickly, other service companies immediately began utilizing borate systems. Early on, these fluids suffered greatly because the service companies had no known mechanism to degrade these fracturing fluids at low temperatures. Many plugged wells ·resulted. Additionally, a lot of these treatments, because of our lack of understanding, utilized 60-80 lbs/1000 gal gels, and many treatments were aborted due to the inability to pump down casing or tubulars. Today, with the new technology that has been developed in relationship to borates, this fluid is becoming perhaps the fluid of choice for fracturing most formations.

Table 12-3 Borate Cross-linked Systems

- Among the first cross-linked fluids
- Suffered early on because of lack of degradation mechanisms-many plugged wells
- Early jobs used 60-80# per 1000 gallons resulting in gel-outs and excessive pump pressure

Today borates have perhaps become the frac fluid of choice.

Developments that have occurred to enhance its use are catalyzed oxidizer breakers for low temperatures, high pH enzyme breakers, temperature stability enhancement utilizing high concentrations of boron ion and large quantities of high pH buffers. Additionally, the use of encapsulated breakers and the inherent lack of shear sensitivity of this particular fracturing fluid has made it an excellent choice for placement of high proppant concentrations in all types of formations.

By way of warning, it should be noted that unless you utilize totally delayed cross-linked borate, it is usually unwise to pump a concentration greater than 35 lbs./1,000 gals. A completely cross-linked 40# borate gel, particularly one that has moderate-to-high boron concentration, can easily cause a gel-out from excessive friction pressures. This is particularly the case in large casing in deep reservoirs. We have conducted a large amount of testing on borate systems and have been able to successfully stabilize 35# gels for extended periods of time at temperatures above 260°F. Conversely, the service companies have been working hard and have developed controlled delayed cross-link systems that can give cross-linking well past pipe time and/or controlled systems for cross-link times up to 20 minutes at temperatures of 90°F. Not all service companies have this degree of control, and you should certainly query the service company if you're going to use a polymer concentration exceeding 35 lbs./1,000 gals. Field experience has shown that at 35# or less that the chance of gelling out or excessive friction pressures is quite low.

It should be noted that a 35# gel is many times the minimum concentration to use unless intense quality control is going to be conducted on location. Thirty-pound gel concentrations can be used and create a functional, adequate crosslinked gel system. However, if you are using conventional borate systems and you mistakenly obtain less than 30 lbs./1,000 gals, then it is difficult to have a stable fracturing fluid when you start adding breakers. The reason for instability is due to the very sensitive nature relating to minimum amount of polymer and crosslinker required for stability.

Another fracturing fluid, as shown in Table 12-4, is termed "polyemulsion." This old fracturing fluid, developed by Exxon Production Research in the late 1960s, had a rebirth, particularly in the west Texas area in the mid-to-late-1980s. This fracturing fluid is an excellent proppant transport medium, has excellent fluid loss control, and due to the fact that two-thirds of the fluid is oil and only one-third is low concentration linear polymer, it is a relatively inexpensive fracturing fluid. The major disadvantage is high friction pressures that occur if you attempt to primp down anything other than casing. It is indeed the fluid of choice for oil reservoirs for small fracturing treatments treating down casing. Due to safety reasons, we do not recommend using this fluid where you exceed three frac tanks on location. It would be somewhat unwise to say that we cannot safely handle two or three tanks of oil on location with tank batteries on most producing wells in the United States.

The other category of fracturing fluids listed are foam fluids and cross-linked foam. Foam fracturing got its start in the early 1970s, and it continues to be used widely today. We believe the primary use for foam fracturing or cross-linked foam fracturing fluid is in moderately underpressured reservoirs. Service companies have all utilized either nitrogen or $C0_2$, and some service companies use combinations of nitrogen and $C0_2$. We strongly recommend the use of cross-linked foams for enhanced proppant transport and suspension as well as for better creation of fracture width. We like very much to use constant internal phase as the primary mode to allow us to place high sand concentrations. We never recommend energized foam fracturing fluids for normal or geopressured reservoirs. Over the years many people· have proposed that foams be used in water-sensitive formations. We believe quite strongly that there are almost no water-sensitive formations. Problems noted historically with water-sensitive formations were directly related to not degrading either linear or cross-link gel fracturing fluids and the subsequent damage that occurred due to this problem. With the advent of encapsulated breakers, you are able to utilize water-based systems in virtually any formation successfully. Table 12-5 illustrates some of the aforementioned comments.

The next category of fracturing fluids is delayed cross-linked fracturing fluids using metallic cross-linkers. Table 12-6 is a summary of these delayed metallic cross-link fluids.

Table 12-4 Polyemulsions

- Old frac fluid-developed by Exxon in the late 1960s
- 2/3 hydrocarbon, 1/3 linear gel
- Low cost
- Excellent proppant transport and suspension
- Excellent fluid loss control ·
- Suffers from high friction, applicable in shallow wells down casing

Table 12-5 Foam Fracturing and Cross-linked Foam

- Primary use should be in moderately underpressured reservoirs
- Nitrogen or $C0_2$-some binary
- We recommend cross-linked foams for enhanced proppant transport and suspension.
- We recommend the use of lower quality (constant internal phase) to place higher sand concentration.
- We *do not* recommend energized or foamed fluids for normal or geopressured reservoirs.

Table 12-3 Borate Cross-linked Systems

- First cross-linked fluid was antimony system of cross-linked guar.
- Some chromium and aluminum was used with titanium cross-linked systems dominating in the 1970s.
- In the early 1980s, delayed titanium and zirconium systems were introduced.
- Presently, zirconium cross-linked guars, HP guars, and CMHP guars are the frac fluids of choice where borates are not used.
- Unless severe near wellbore tortuosity exists, we recommend cross-link times in excess of pipe time.

The early part of the 1980s was the time that delayed cross-link fracturing fluids were found to be the fluid of choice for treating down long tubular good where a great deal of shear was placed on the fluid. It was discovered that many of these fracturing fluids used before had a majority of their viscosity destroyed by pumping down large tubulars at high rates. By simply delaying the cross-link mechanism, you could achieve much higher downhole viscosity, better proppant transport, and so forth. Today, we are rapidly moving to a time where delayed cross-link metallic fracturing fluids, particularly CMHPG with zirconium, is becoming the fracturing fluid of choice when you do not use guar borate systems. Obviously, guar titanate and guar zirconium are still being used in many areas. Additionally, HPG, titanate and zirconium fluids are still being used. What we are seeing today is a trend toward delayed cross-link zirconium carboxymethylhydroxypropylguar fracturing fluids. These fluids have the inherent ability to be utilized at low and high pH, depending upon temperature stability needs.

The interesting thing about low pH CMHPG fluids is that they have been improved to the extent that the low pH fluids (i.e., those below 7, sometimes as low as 4 on some of the service companies) do have stability for fracturing formations up to around 260°F. At the temperatures between 200-260°, the hydrolytic mechanism becomes an excellent degrading mode for these fracturing fluids and they are self-degrading. Below 200°, the incorporation of delayed breaker systems make these fluids have excellent clean up, and field results have proven them to be an excellent choice for placement of high concentrations of proppant in a low damaging mode. There have been many occasions where operators have gotten away from borate systems, and they feel that they get better clean up with these fracturing fluids. The borates require internal breakers where many of the low pH zirconium CMHPG systems appear to degrade on their own at temperatures above 200°F with the small amount of hydrolytic degradation occurring due to the low pH of the fracturing fluid.

Tools for successful hydraulic fracture stimulation

The first tool (and one about which I preach long and hard in fracturing schools) utilizes my concept of Pump More Sand or PMS. We feel very strongly that the use of aggressive proppant schedules, and that averaging somewhere between 7-8 ppg on most treatments, has yielded tremendous success in many areas across the country. Many authors have published papers illustrating their success with higher proppant concentrations. There are still a few areas of the country that still purport to get excellent results with low sand concentrations (i.e., 2-4 ppg). We feel that the only thing that is useful in hydraulic fracturing is the proppant that you place in the well. Because we are dealing in a real world where there is always imbedment, there is always fine movement and there is always crushing. Anything that you can do to enhance the conductivity of the proppant pack, particularly in the early stages of the treatment, will assist you not only in clean up of the proppant pack as the fracturing fluid is produced back, but also greatly enhance the final conductivity of the proppant pack for the history of the well. If you do nothing more than increase proppant concentrations, we feel that there is strong evidence to show that great benefits are achieved. Tables 12-7a and b illustrate what some people would term an "aggressive proppant schedule" for a polyemulsion treatment. In fact, we are now starting many treatments at 5 ppg and very quickly getting to 10 ppg. We feel that many times the cheapest thing used on the fracturing treatment is the proppant and that is particularly the case with sand. If you do not place a conductive fracture in the proper place, no benefit can come from the stimulation treatment.

Another interesting thing noted across the country is that many times we have achieved at least equal (if not better) results with smaller proppant than you would see or note from calculations of dimensionless conductivity. As an example, many people in oil reservoirs would look very closely at using 12/20 if not larger proppants for oil reservoirs. We have found in many areas that 20/40 and/or 16/30 has yielded better results than the larger proppant. We felt early that this was due to quality-control problems on the proppant. Since a great deal of effort has gone into QC-ing proppant and the quality of all types of proppant has increased greatly over the last few years, these results are more probably due to better proppant transport than with the smaller sand. We are not advocating the use of 100 mesh or 40/70 for conventional reservoirs, but we have seen better production results in some wells where we have used 20/40 rather than 12/20. In many areas, we have compromised and are using 16/30, where without our field experience we would be looking toward larger proppants.

Table 12-7a Polyemulsion Treatment-Aggressive Proppant Schedule

Stage	Fluid Type	Fluid Volume (gals)	Proppant Concentration (ppg)	Stage Proppant (lbs)	Injection Rate (bpm)
Pad	Polyemulsion	21,000	0.0	0	30
1	Polyemulsion	3,000	2.5	7,500	30
2	Polyemulsion	5,000	5.0	25,000	30
3	Polyemulsion	7,500	7.5	56,250	30
4	Polyemulsion	15,000	10.0	150,000	30
Totals		51,500		238,750	

Table 12-7b Cross-link Treatment-Aggressive Proppant Schedule

Stage	Fluid Type	Fluid Volume (gals)	Proppant Concentration (ppg)	Stage Proppant (lbs)	Injection Rate (bpm)
Pad	35# X-Link	6,500	0.0	0	20
1	35# X-Link	1,000	3.0	3,000	20
2	35# X-Link	1,500	6.0	9,000	20
3	35# X-Link	2,500	9.0	22,500	20
4	35# X-Link	3,500	12.0	42,000	20
Totals		15,000		76,500	

It is always best for us all to look at field experience and tie that together with our knowledge of relative conductivity to come up with the best results possible in a stimulation.

The next area of discussion is one that has been covered in several chapters, both in chapters relating to quality control and particularly in the chapter on enhanced quality control. We believe that by testing the fluids at bottom hole static conditions of temperature and shear and assuring that the fluid is stable throughout the pump time at bottom hole temperature and then subsequently break back to water viscosity (i.e., 2-3 centipoise at 511sec-' at ambient temperature) that we can be assured that the fracturing fluid is functional. We believe, and have shown with a great deal of correlatable data, that more than

70% of the wells that have been pumped in the last few years have resulted in some degree of unbroken gel or poor proppant transport due to the lack of intense quality control. Another technique covered in great detail in a previous chapter is forced closure, which is a technique that continues to find wider and broader use in hydraulic fracturing. At a recent SPE forum meeting held at Snowmass, Colorado it was the conclusion of the entire group that forced closure was a functional procedure. Although many people did not understand how it worked, it was something that appeared to decrease sand production and enhance production of wells. This technique has been in use for over 7 years and primarily refers to immediate flowback of the well at aggressive rates. Presently, most wells are flowed at 1 bpm immediately, not taking into account any enhanced breaking of the fracturing fluids. We feel that producing back unbroken gel is a tremendous benefit and allows us to produce back fines. By getting the fluids out in an unbroken state, you negate problems due to residue and colloidal solids. This procedure is one that requires a little more effort by field personnel and negates everyone leaving the location to go to happy hour. Someone has to stay at the site and monitor the well as you continue flowing it overnight. I feel that the ability to take advantage of super-charge and be assured that all of the perforations in the near wellbore are connected with the proppant pack are further reasons to utilize forced closure on every frac treatment where perforations are utilized.

Another procedure that I feel has become more accepted in hydraulic fracturing is a truly realistic estimate of fracture height growth. For many years, personnel perhaps wanted to believe that their hydraulic fractures were contained within the interval to be treated or believe that fracture height could be controlled by slowing the pump rate or lowering viscosity. In reality, most hydraulic fractures are at best radial in configuration and sadly many of them are even more uncontained than radial mode. By having a true understanding of the lack of containment of hydraulic fractures, you can do a better job of planning for enhancement of well spacing, drainage, and so forth.

The best example of this occurred in east Texas where people ran 2,000,000 pounds sand-fracturing treatments when they were drilling on B40-acre spacing. They found they were not successfully draining at this spacing, even though models indicated more than 3000 ft fracture lengths. They subsequently have drilled on 320's, 160's, 80's and many are looking at 40-acre spacings while still pumping 2,000,000 lb. fracturing treatments. Obviously; they were not seeing contained fractures, and their fracture lengths were much shorter than what the fracture models were indicating. These models were not at fault; it was the data placed in them by people hoping for fracture containment.

Over the last few years, we have done a lot of investigation utilizing stress tests, digital sonic logs, arrayed seismic, and many other tools to evaluate fracture height. The vast majority of wells that we investigated have no barriers allowing control over fracture height. What we must do, then, is to be realistic in our estimation of propped fracture lengths and plan our ability to produce oil or gas built around relatively short propped lengths.

Another technique that has found widespread success has been minimization of the perforated interval. This topic, discussed in Table 12-8, has been widely applied in many areas. The basic premise is that instead of perforating across an entire interval, you pick the lowest stress area in a particular interval and only perforate 10-20 feet. This low-stress area in a particular interval may, in fact, be barren rock and have little or no porosity of permeability.

Table 12-8 Minimization of Perforated Interval

- Any inclination in the wellbore or average inclination of in-situ stresses will result in multiple inclined fractures where long perforated intervals exist.
- Multiple inclined fractures severely dilute propped or etched fracture length and limit drainage of reservoirs
- We recommend 10-20 feet of perforated interval in the center of the interval to be covered by the fracture.
- In tight formations, perforation placement should be based more on a stress profile than permeability, porosity, or water saturation.

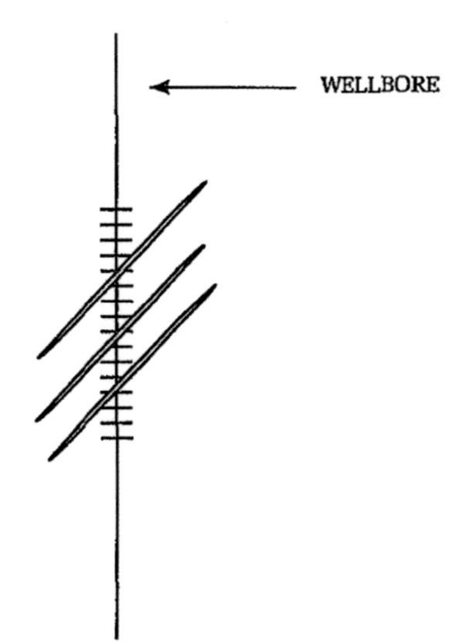

Even with minimal inclination of the wellbore or principal stresses (i.e., 5° or less), most fractures only cross the wellbore. With long perforated intervals, the probability of multiple inclined fractures is very high. Multiple fractures dilute the effectiveness of the propped fracture, creating much less drainage area.

By limiting the amount of perforated interval, you can be better assured of obtaining a single, plainer vertical fracture. We believe that many times when you perforate across an entire long interval (300-500 ft.) that you end up with multiple inclined fractures, as shown in Figure 12-1. We have documented cases where we have gone back into wells and isolated multiple intervals and found separate definitive fractures that have crossed the wellbore in an inclined mode. In reality, very few fractures would exactly follow the wellbore; and with any deviation in the wellbore or deviation in stresses or deviation in the dip of the reservoirs, you can readily see that the possibility of multiple inclined fractures is very high. Additionally, any variation in rock properties, stress, and breakdown mode makes this a much more probable phenomenon. By utilizing a short perforated interval you can optimize fracture length by minimizing the creation of multiple-inechelon hydraulic fractures.

Another technique that is utilized in design and implementation of hydraulic fractures is a procedure termed "surgical treatments." Since most fractures are radial, one control mechanism for propped fractures is selective placement of perforations. By perforating above or below a producing interval and fraccing up or down into the zone, you can eliminate proppant production problems from excessive perforation and acidizing. This is a common occurrence in west Texas, where we know that poor fracturing treatments were conducted on intervals. These wells have been acidized multiple times, and because they are carbonations intervals and large cavities exist, there simply is no way to control proppant flowback. What we have done, as is illustrated in Figure 12-2, is set a bridge plug above the interval and perforated just above the bridge plug. In many cases, the interval in which we are perforating is a barren interval; but by understanding the stress profile existing in the formation above and below, we can create a radial fracture that reaches down to the water contact and propagates upward a similar distance. By doing so, we have been able to very successfully stimulate older wells.

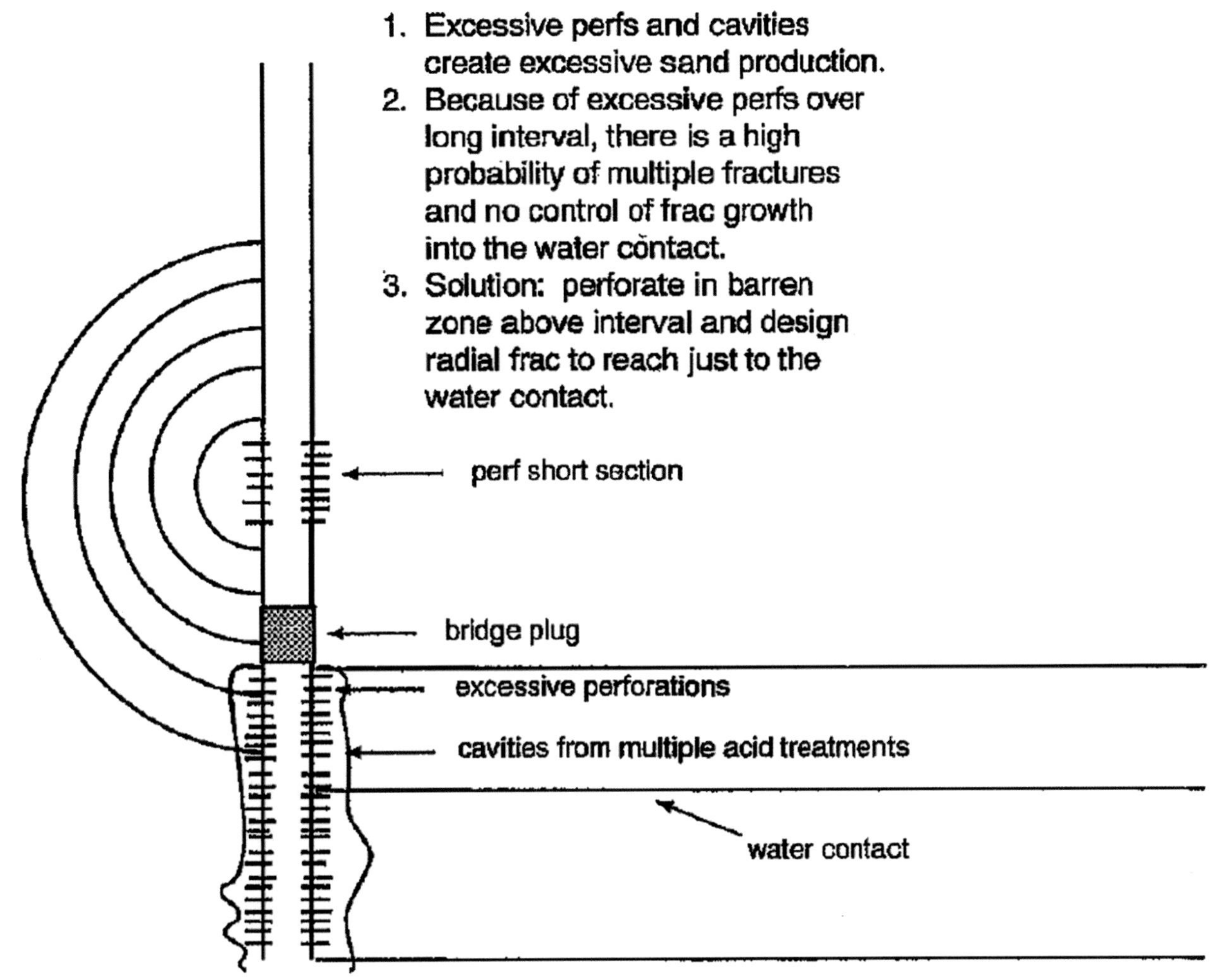

Figure 12-2 Example of a surgical treatment where perforations are shot in a barren zone above the interval allowing radial growth of the fracture down to the water contact.

Obviously, this is a somewhat controversial practice for many people when we do perforate outside of typical producing intervals. This technique has been very successful in west Texas in treating San Andres and Clear Fork formations.

The next technique is a fairly new procedure, termed "pipelining." This procedure has been in use for some 3.5 years, primarily in west Texas and southeastern New Mexico. This procedure was developed to selectively place proppant in intervals utilizing differential viscosity. Selective placement of proppant is a result of a technique similarly used in fracture acidizing where one pumps viscous pads and follows with low-viscosity acid. This viscous fingering technique has proven to be very successful in allowing selective placement of proppant in rather thin intervals where there is nearby water or gas contacts. Field results have shown long, highly conductive fractures created with a fairly small amount of proppant pumped at high concentrations.

With this technique, we utilize a high-viscosity, cross-linked gel to create fracture width and to create, in most cases, a penny-shaped fracture. We design the treatment so that we create whatever length is required with the penny-shaped fracture and then utilize a low-viscosity fluid to place proppant with viscous fingering effects. By perforating only, the selective part of the interval (typically 10-20 ft), and having some differential rock properties that exist in the zone itself compared to bounding layers, we are able to selectively place proppant in a rather thin zone and not prop zones above and below. Figures 12-3 and 12-4 illustrate a typical conventional propped fracture and a pipeline propped fracture.

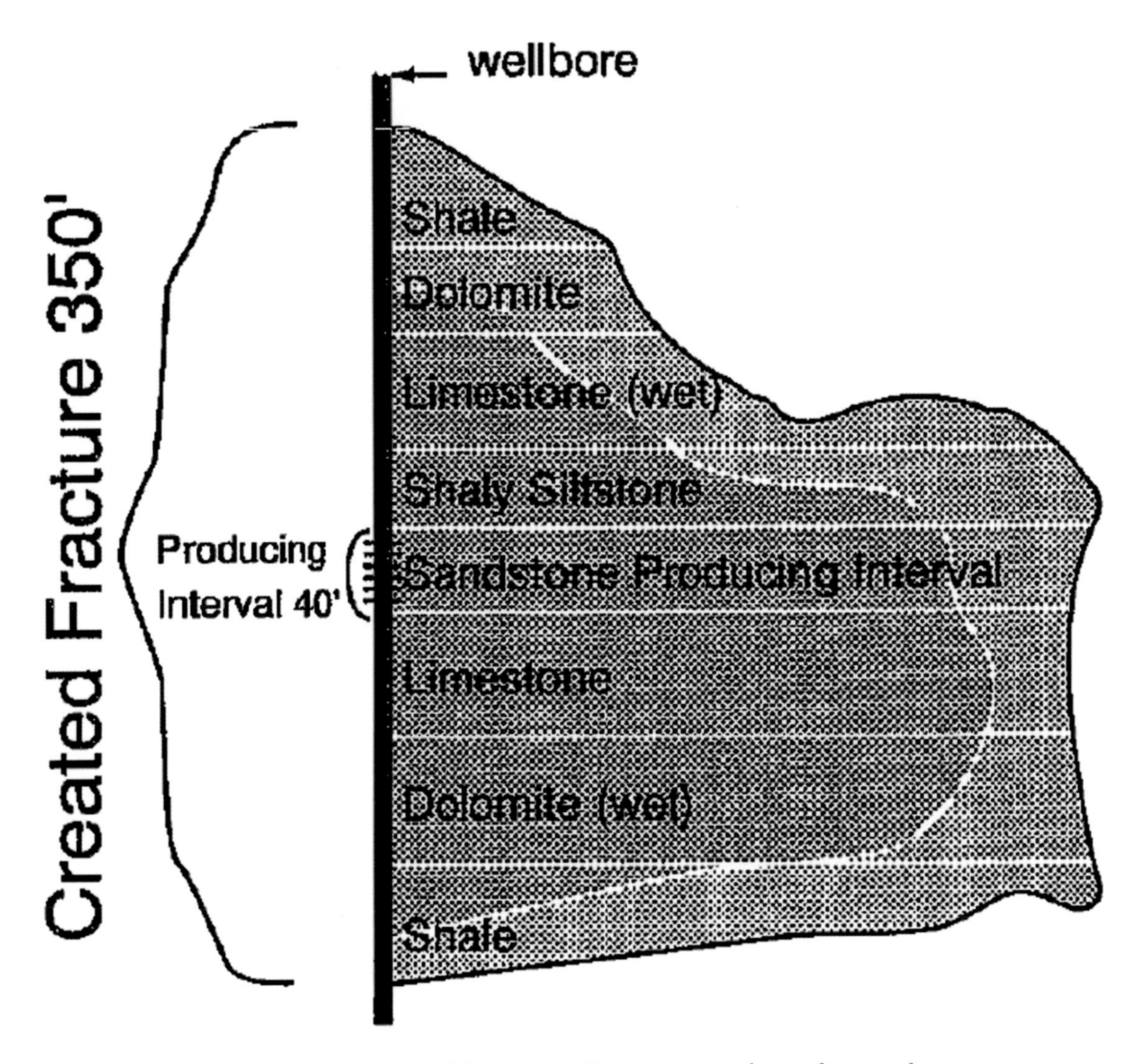

Figure 12-3 Conventional propped fracture without any significant fracture barriers.

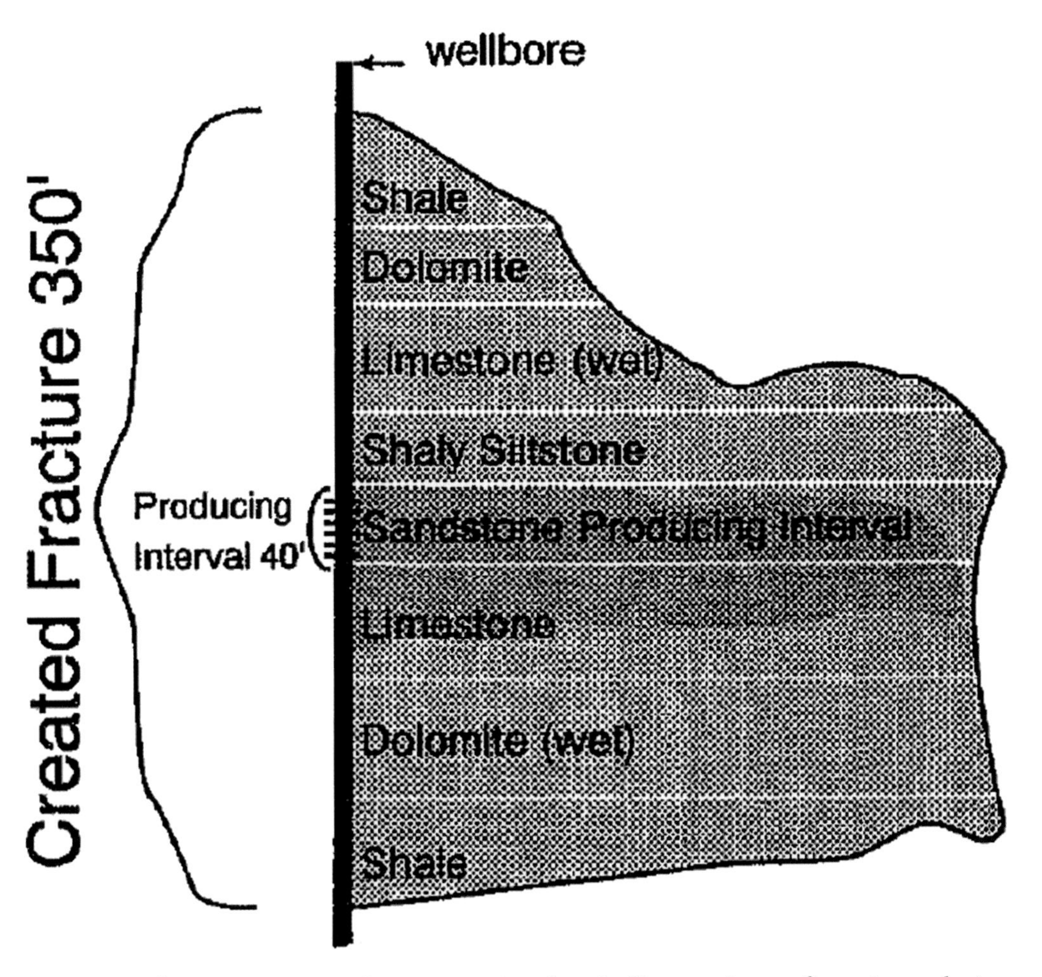

Figure 12-4 Selective proppant placement using the pipeline or viscous fingering technique.

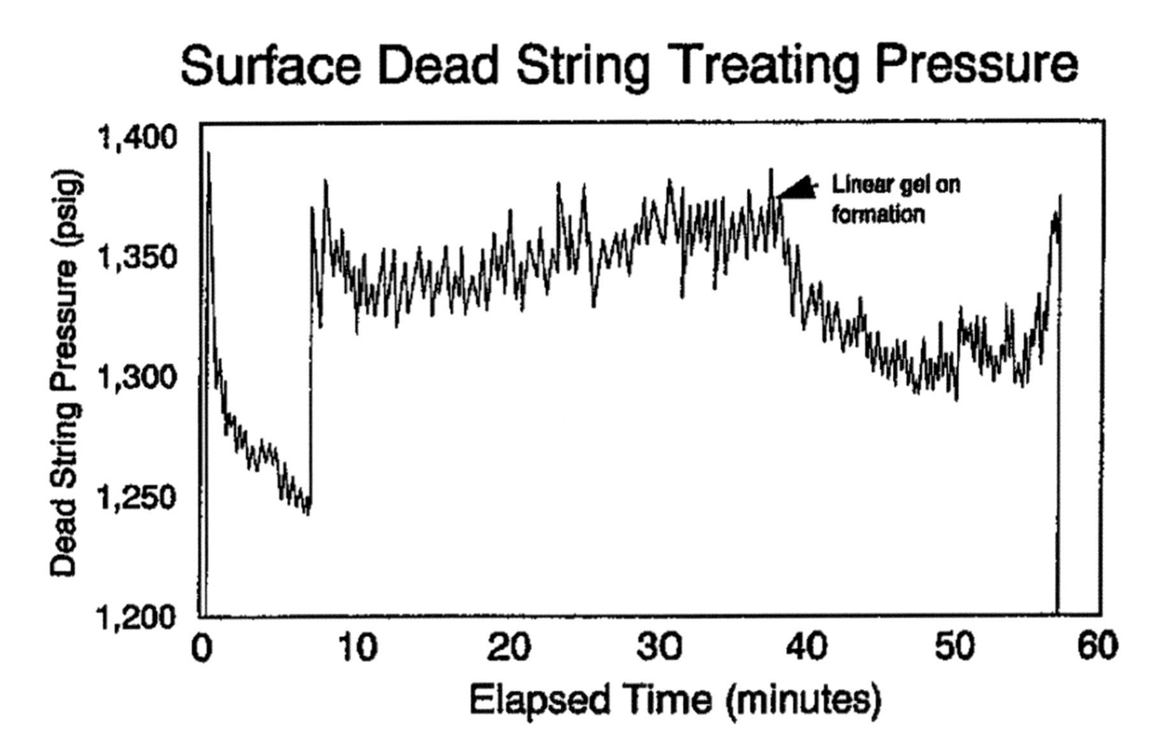

Figure 12-5 Pressure versus time of measured pressure from a dead string on pipeline fracturing treatment.

Figures 12-5 and 12-6 illustrate net pressure plots for pipeline fracture treatments. Note on the pressure plots the decrease in pressure when the linear gel containing sand hits the perforations; and in the case of Figure 12-5, a rapid increase in net pressure toward the tail end of the job. Figure 12-7 illustrates one of the important characteristics that we feel is absolutely essential for successful pipelining fracturing treatments, namely, differential frac width. This width difference is caused by variations in rock properties between the producing interval and the bounding interval.

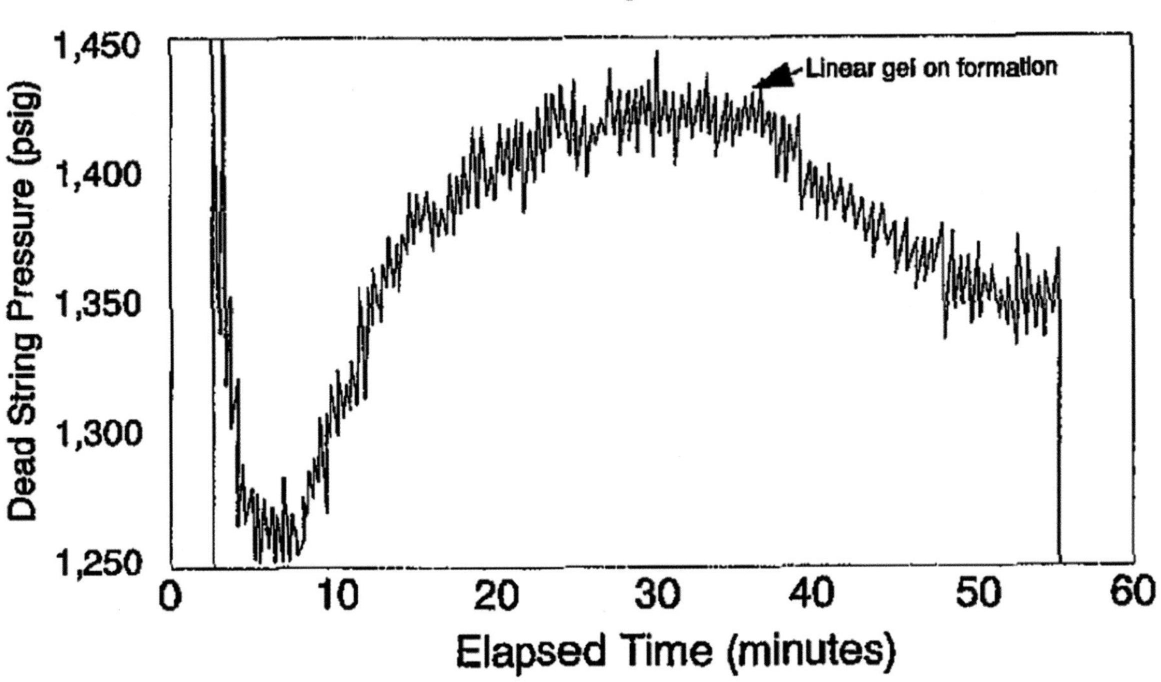

Figure 12-6 Pressure versus time plot of dead string measurements during a pipeline fracturing treatment.

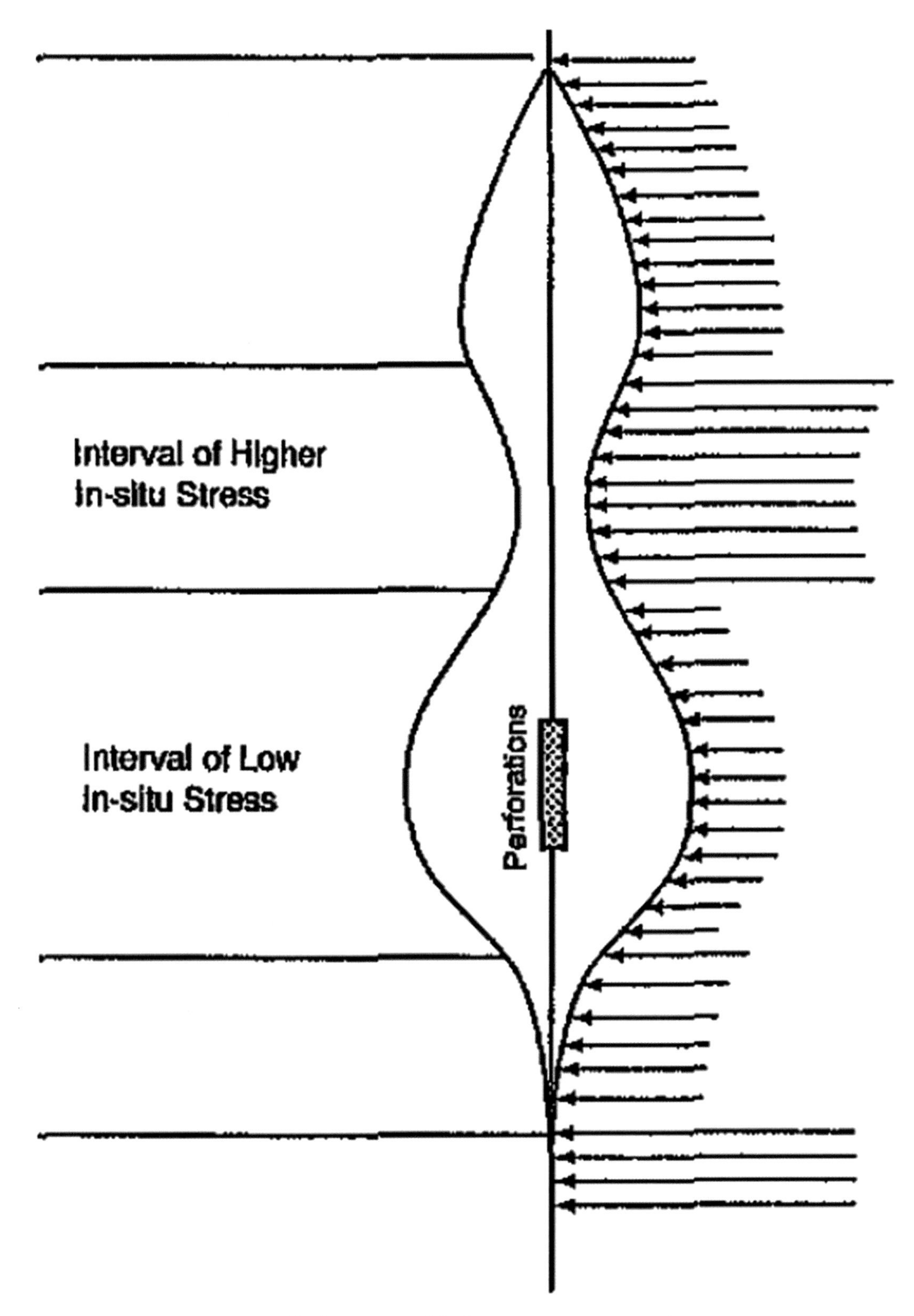

Figure 12-7 Variable fracture width due to variation in rock properties which is beneficial in the pipeline fracturing technique.

Pipelining is a successful means of selective proppant placement. Specific design techniques have been developed whereby you can optimize the amount of proppant placed and minimize proppant placed in bounding intervals. Table 12-9 illustrates a design of a pipeline fracturing treatment.

Table 12-9 Pipeline Treatment-Aggressive Proppant Schedule

Stage	Fluid Type	Fluid Volume (gals)	Proppant Concentration (ppg)	Stage Proppant (lbs)	Injection Rate (bpm)
Pad	35# Linear	30,000	0.0	0	20
1	35# Linear	1,000	2.5	2,500	20
2	35# Linear	1,500	5.0	7,500	20
3	35# Linear	2,500	7.5	18,750	20
4	35# Linear	6,000	10.0	60,000	20
Totals		41,000		88,750	

Figure 12-8 shows the results of a field study done utilizing pipelining fracturing treatments. Note the wells are holding up quite well. We have been able to stimulate oil production without increasing water rates. We have done over 20 buildup analyses with the pipeline fracturing treatments and have shown long, highly conductive propped fractures; many of them have exceeded more than 200ft with less than 50,000 lbs. of sand. The wells appear to hold up very well and they consistently produce at lower water and gas cuts than the offset wells. Offset operators have been using low-viscosity, low-rate treatments and low sand concentrations, trying to stay out of water with little success.

The basic concept that we are using is to accept the fact that there are no barriers to fracture growth in many of these formations, but there is differential frac width. We take advantage of differential viscosity and the fingering that ensues to allow us to selectively place proppant. In most cases, the treatments are relatively small and we utilize a combination of intense quality control to be sure the fluids are within specification and utilize aggressive forced closure to be sure that the treatment is flowed back very quickly to negate any proppant settling that might go into a nearby water-producing area.

Another technique that has been successful when implementing fracturing treatments is to be open to respond to occurrences that go on during the treatment. For many years, personnel have been reluctant to vary pump rate, sand concentrations, and so forth because a computer design has been used. Most of us are aware that the vast majority of input data information in fracturing treatments is estimated. On many occasions, the only known data put into the frac design relates to the name of the well, the date, and perhaps the depth. With this in mind, when I am out on a treatment, and pressures start turning upward in opposition to what I have normally seen, I rapidly increase the pump rate, assuming I can stay within pressure limitations. By doing this, in the vast majority of cases we are able to change the slope of the pressure and get the treatment away.

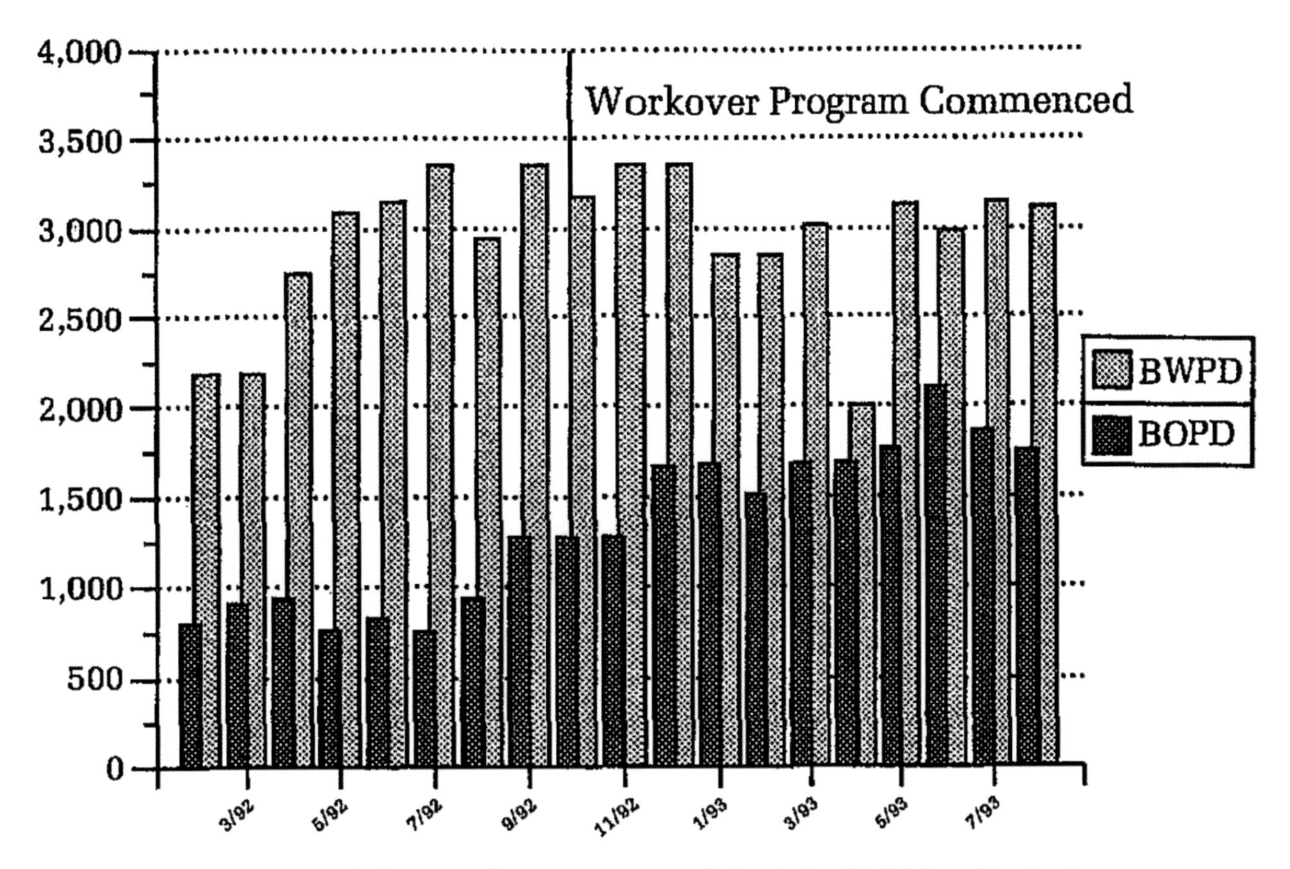

Figure 12-8 Phillips Petroleum Company---Cabin Lake Oil & Water Production

Secondly, when I have designed a treatment for multiple-stage sand concentrations, I am always open to change sand concentrations built around pressure responses we see downhole. If we are trying to create a highly conductive propped fracture near wellbore, many times we will increase sand concentrations at a more rapid mode than what is shown in design, even if we have done the design ourselves. For obtaining long, highly conductive fractures in very tight reservoirs, you must take into account the effects of increasing sand concentrations on negating propped fracture length.

I believe that the intelligent supervisor of a fracturing treatment should always respond to pressures that he sees during the treatment, rather than depend on design information that is many times estimated, based around offset well information.

Another area that I have placed very little credence in is real time Nolte Smith plots using calculated bottom-hole pressure data. I would prefer to utilize surface treating pressures and monitor increases and decreases of that pressure in relationship to what is going on in the tubulars or casing. Most of the information that gives us calculated bottom-hole treating pressures is erroneous, and the variations and changes in pressure that occur in relationship to net pressure can be misinterpreted. In fact, these changes are misinterpreted regularly by personnel who simply don't understand what is going on downhole or who are not aware of changes that may be going on with the fracturing fluid in real time.

On a fracturing treatment I carefully monitor delta pressure versus time of the surface treating pressure. If I note an increase upward when I should be either maintaining a constant pressure or seeing a decrease in pressure, I very quickly respond (if possible) by enhancing pump rate. On rare occasions, I have actually dropped the rate and found that the problem was indeed friction pressure, and I was able to complete the treatment at a lower treating rate. The latter is by far the least common. I do feel one of the biggest mistakes made in hydraulic fracturing is trying to follow designs without utilizing information from the formation during the actual treatment. Our industry desperately needs real-time, bottomhole pressure data from wireless devices, which presently are being tested. Again, I believe that one of the biggest mistakes made is to try to make intelligent decisions based upon calculated bottom-hole pressure data.

CHAPTER 9

AVAILABLE PROCESSES WHICH HAVE ACHIEVED SUCCESS WITH CONVENTIONAL FLUIDS (2022)

The present chapter, with a new name, was in the 1994 edition titled "Job Implementation Guidelines". On some of the previous chapters, I did a little bragging that I had covered the subject well and most of the previous chapter was still applicable. I must admit that, in this chapter, I changed subjects for some reason. Most of the old chapter was about fluids and processes but, for some reason, I got off into single point perforating and fracture height growth. I have covered the single point entry and height growth earlier. In the new chapter, I will attempt to stay with fluid systems and processes. There are basically two types of conventional crosslinked gel systems readily available to the industry, these two types of gels presently available have base polymers of guar, hydroxypropyl guar or carboxymethylhydroxypropyl guar.

The two dominant crosslink systems are guar Borate and Zirconium crosslinked CMHPG. If you are not familiar with crosslinked gels refer to the 2nd addition chapter 6, illustration 6.2. From an information standpoint, crosslinked gels were introduced to the industry in late 1968. The very first product was an 80 #/1000 guar gel crosslinked with Antimony developed by Halliburton. This system termed, *My-T-Gel*, was quickly followed by competitors pumping 80 #/1000-gallon borate guar gels. Halliburton published pictures showing a 3-inch steel ball held by the gel and another with a long gel tied in knots. Dowell's, now Schlumberger, counterpart was named *Wide Frac*. The first 2 pictures at the end of the chapter show the knot and money suspension by early crosslink gels.

As an aside and, I believe, interesting from a historical perspective, there was a dramatic difference in belief between Halliburton and Dowell on optimum pH for frac fluids. Wayne Hower, a researcher with Halliburton who was behind the work relating to KCL's function in clay control, also had done regained permeability testing with low and high pH fluids. His work indicated significant clay swelling with high pH fluids. His work led to Halliburton having a policy negating the use of high pH fluids. There was a significant time, several years in fact, that high pH fluids were forbidden to be formulated or used by Halliburton. Dowell, who was the major competitor for Halliburton, on the other hand, embraced high pH fluids with their use of high pH Borate fluids. Both companies published papers showing their rationale and absolute proof of their contentions. Both publications showed proof related to controlled laboratory tests. With the test of time, it was shown that for the vast majority of cases, the base pH of the frac fluid has little or nothing to do with final productivity. Even for those fluids that contain buffer, which I highly recommend, the ultimate pH of the fluid is dramatically changed once in the formation. Like the case of water sensitivity, there are unique exceptions where massive amounts of swelling clays might be affected by high pH or conversely a carbonate rich reservoir might be softened by low pH fluids. I try very hard to let the test of time make the final decision for what is viable for the industry.

For those of us who now regularly use borates with 20-45 #/1000-gallon system, we would anticipate many plugged wells from the 80 # borate system particularly when there were no internal functional breakers for the high pH borate system below 130 F. This period in fracturing was a wild and crazy time with everyone caught up with the fascination of fluids with jello like consistency. Indeed, many wells ended up plugged with unbroken base gel or crosslinked gel. These early fluids created tremendous production problems particularly in areas where most of the reservoirs were low temperature i.e. 70-130 degrees F. Halliburton's *My-T-Gel* was low pH and enzyme breakers, properly added, would degrade the gel back to water. On the other hand, Dowell, BJ, and the Western companies' systems did not have any known breaker mechanisms for temperatures below 130F in the late 60's.

There was a story circulating in the early days of crosslink gels that a super salesman with Dowell had sold a very large *Wide Frac* job and then talked the customer into pumping a very large acid job to clean up the plugged fracture. I need that salesman in my company. One of the troubling assumptions by many, well into the early 90's, was the belief that the gels would eventually break on their own, regardless of temperature. Also driving this problem was the notion that breakers could create screenouts, which was the ultimate nightmare for both the service companies and the operators. Both of these created a large business for well cleanups where it was obvious that gel had not broken.

From a historical standpoint, there were multiple metals used for crosslinking. A very popular one was titanate crosslinkers used with both guar and Hydroxypropylguar. For a short period of time aluminum and chromium were also popular.

A challenge was put forward by the legal department for patents at Halliburton. Someone in management stated that only borate, titanate and antimony were the only metals that were functional. By using redox chemistry, the research group demonstrated that all metals, including gold, silver and platinum could be used to crosslink guar, guar derivatives as well as cellulose and polyacrylamide polymers. To not lose your readership through extreme boredom, I will state that with rare exception that various forms of boron and ligands of zirconium dominate our industry for crosslink fluids. Since there has been little or no research done in the last 20 years on these fluids, I expect many of the previous systems will be reinvented. As is the case with competing with local sand in the propping agent industry, it is very hard to compete with guar gum which is very inexpensive product and with competent chemists it has replaced derivatives at bottom hole static temperatures over 300F. In the previous edition, I listed the service company product lines. At the present rate of total shutdowns and mergers this would be a work in progress for years. Below I will list the common polymers and their uses:

1. <u>Guar</u>

   Guar gum, which is simply a ground up bean which thickens water, has been around in the industry for more than 55 years. Although it is basically the same base product, it has been improved by cleaning up the evil residue and an additional grinding which yielded more viscosity. In the second edition of my book, I have charts relating to 511 reciprocal second dial readings. These types of charts are hard to come by and the dial readings vary widely by supplier. If your guar viscosity is as good or better that the 2nd edition charts you can proceed with the job.

   As stated earlier, unless you are dealing with bottom hole temperatures exceeding 300F, your crosslink gel of choice should be Guar Borate. Some service companies switch over to CMHPG Zirconates at temperatures above 270F.

2. Hydroxypropylguar

HPG, due to derivatization of the bean guar, is more expensive than the base product. For a significant period, the late 70's until the mid to late 80's HPG was touted as the savior of the oilfield yielding much less residue and higher temperature stability. As is the case today for various political beliefs, much harassment and judgement were laid upon those who ran guar rather than the clean HPG. Methanol diagnostic tests were conducted on location to be sure the high residue guar was never pumped. The methanol test was done using approximately 30 % alcohol and if the polymer precipitated it was guar and not HPG which was soluble above 50 %. The last image shows suspensions of Guar, HPG, and HEC. A substantial number of papers were presented extoling the virtues of HPG and service companies were threatened if straight guar was in their systems.

During the very hard times of the early 80's, one major supplier was caught supplying a mixture of 75 % HPG and 25 % guar. Because of this nefarious behavior, the supplier was shut out for six months which was a major loss of sales.

Hopefully, those reading will know I am being somewhat facetious. To give some credit to those who followed the pied piper over the cliff because of the virtues of HPG (myself included), There were benefits other than the lesser residue that were beneficial to the service companies and the operators. These benefits included higher viscosity and less complications in crosslinking. It also moved our industry away from a very generic product to a man-made material that allowed for profitability, due to the use of technology in our industry. Because of the profitability of the HPG, significant improvements were made to the guar industry. These included developments of dispersible products and several chemicals were developed that improved its stability to temperature and many other non-related chemicals which proved very beneficial to other guar products as we moved forward.

As discussed in the chapter on quality control, the work of the Gas Research Institute showing almost impossible sensitivity to chemical loadings with HPG proved to be its death nell. With research still booming, the development of CMHPG gave the industry a product that was deliverable not only in the laboratory but in field operations.

Additionally, over time and evaluation of field results, we came to understand that gel residue was not a real factor in damage and hence the return of Guar borate.

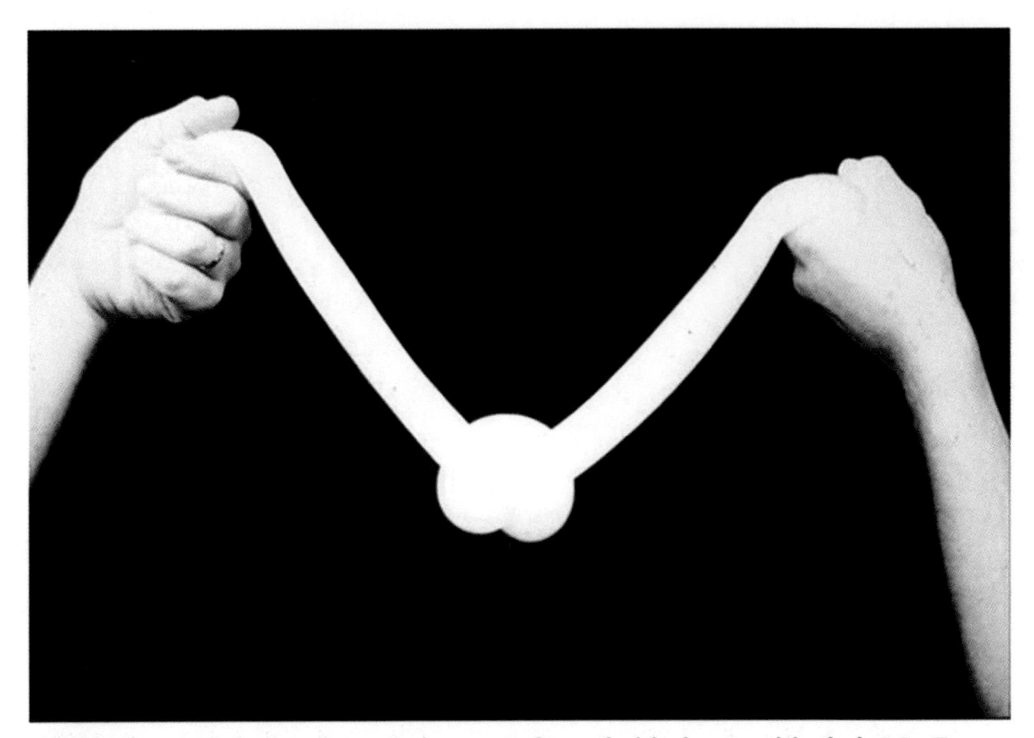

Haliburton insists they have their competitors tied in knots with their My-T-Gel fluid.

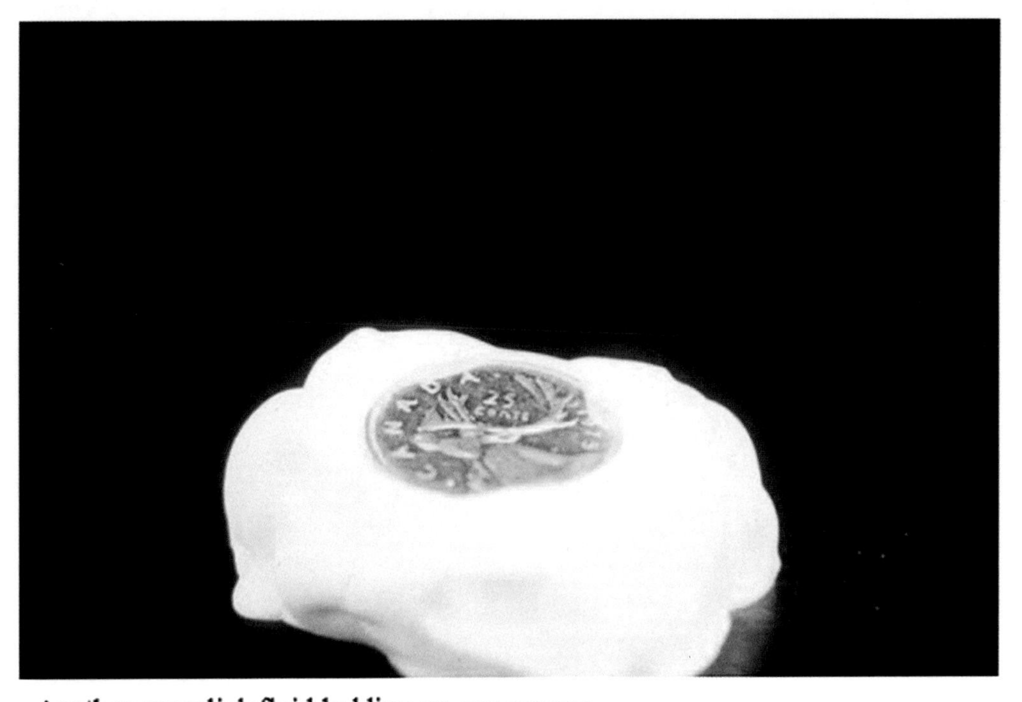

Another crosslink fluid holding up our money.

GUAR

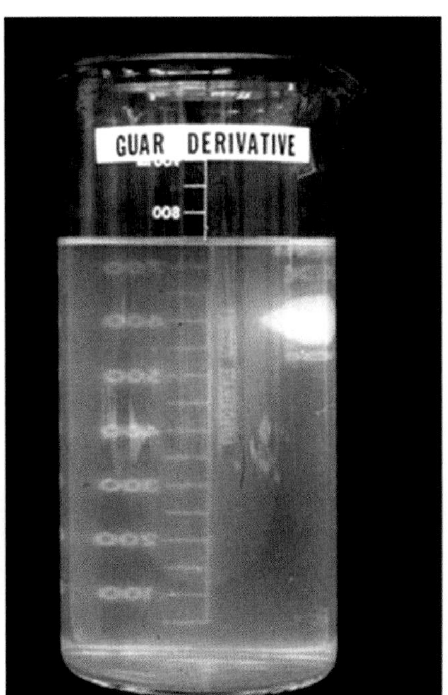
GUAR DERIVATIVE

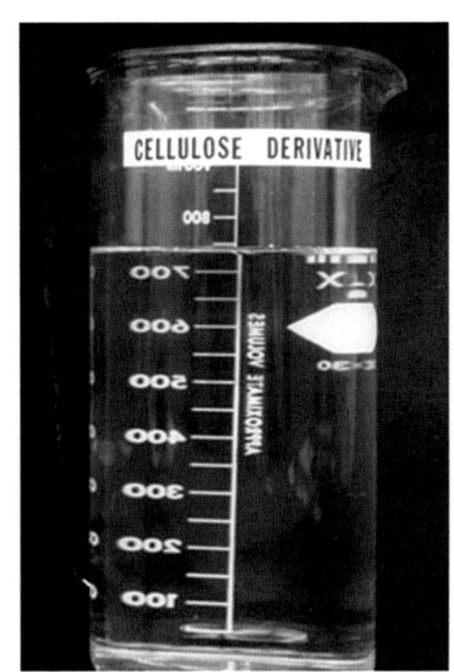
CELLULOSE DERIVATIVE

CHAPTER 10

PROPPANT SELECTION (2022)

I am going to reference a few articles, books, etc which give guidelines on picking proppant for moderately permeable reservoirs. The first is a book by Steve Holditch, Ralph Veatch and George King with a chapter 7 titled *"Proppants and Fracture Conductivity"*. The book published in 2017 titled Essentials of hydraulic fracturing gives in great detail over 68 pages with tremendous detail on proppant selection. The next reference is the latest SPE Monograph published in 2019 with chapter 5 titled Proppants and fracture conductivity. This chapter is less detailed with only 19 pages.

You might want to continue reading chapter 6 immediately following relating to frac fluids by a great author. The third book published in 2007 by BJ titled Modern fracturing has chapter 8 titled Proppants and fracture conductivity. Although in my humble objective opinion has too much emphasis on proppant pack damage does yield proppant selection criteria for conventional moderately permeability reservoirs. Also, for your reading pleasure, the following chapter on Execution of Hydraulic Fracturing Treatments has a great co-author.

Ladies and gentlemen from the middle east would gasp when I say a moderately permeable reservoir is .05 millidarcies and above. I was fortunate enough to live and work in the middle east traveling and working in more than 11 countries where a tight reservoir was less than 100 millidarcies. When I was working in Iraq in the 70's an Indian company had a contract to explore and develop wells for their own profit and for the Iraqi's. One part of their contract stated that wells that would not produce 10,000 barrels per day with a small acid job, were considered dry holes. One of the major reasons I was in Iran was to stimulate some poor wells in the Bangistan reservoir which would only produce around 4,000 barrels per day. We were able to stimulate these wells, but they were considered marginal when we stimulated the reservoir and achieved 16,000 barrels per day. See pictures of the Ahwaz 54 frac at the end of the chapter. Safety personnel would be aghast at the iron and rig up but God was with us and the 8,000 psi frac was completed without 2 safety barriers and hand welded high pressure iron. I treated the job using a martin decker chart and a little box with a needle showing the rate.

The Asmari formation in Iran and the Arab D formation in Saudi Arabia around 9,000 feet would easily produce 20,000 barrels per day with relatively high skin factors.

You are probably saying I have digressed away from the subject. Permeability is not a minor thing in hydraulic fracturing. It is EVERYTHING. As stated earlier, hydraulic fracturing started in 1947 but, no real stimulation of reservoirs occurred until the late 60's. Fracture treatments were conducted to break past reservoir damage.

I would also state to the chagrin of frac pack advocates (and I am one) that the upper limit for well stimulation for hydraulic fracturing is around 50 millidarcies. We do need to state that the numbers mentioned for lower end and upper end relate to non-damaged reservoirs.

Dr. John Lee will be pleased when I do not go into great detail about skin damage. A factor of zero for skin basically means the reservoir can produce to its maximum capacity based on pressure and permeability. A negative skin indicates that the reservoir is able to produce better than in a virgin state by enhanced surface area achieved by stimulation. A positive skin indicates the well has been damaged by such things as drilling fluids, incompatible completion fluids, or even unbroken frac gel. The skin factor value is measured by executing a pressure build up on the reservoir. What is occurring with a pressure buildup is shutting the well in and allowing the reservoir to build pressure with time until it comes to an equilibrium. A lot of complex things occur which allow for experienced petroleum reservoir engineers to evaluate such things as permeability, bottom hole pressure and extent of the reservoir. It should be noted that executing a pressure build up on microdarcy reservoirs not to mention nanodarcy formations requires an enormous amount of time and simply are not practical.

Fearing an attack from experts such as Dr. Lee, I will not go into more detail. I will relate that due to a Japanese customer not wanting to pay for two separate lecturers, I had to teach pressure build up analysis in addition to hydraulic fracturing to 20 Japanese students in Tokyo. I was frightened beyond comprehension. The time I went to Tokyo coincided with the movie Black Rain, sometime in the late 1980's. All I really knew about pressure build up was taught to me by one of my fellow cohorts at Holditch and I expected to be skewered by the very intelligent Japanese. I was very relieved when of the 20 students, only 7 could communicate in English and not one question was asked. All of the gentlemen followed the slides and passed the tests given on pressure build up and fracturing with flying colors.

The reason for the story is I do not want to go into further detail on pressure build up and skin factors as my understanding is far less than my knowledge of fracturing.

All that said, having skin factors on conventional reservoirs prior to fracturing is sometimes like having the score before the game starts. From one side a well with a -3 skin and decent permeability, 10 millidarcies will be difficult to stimulate without a very long propped length. On the other hand, the same well with 10 millidarcies with a +7 skin is a tremendous frac candidate requiring only a small treatment to achieve a production increase of close to 7-fold. There has been and continues to be confusion in how fracturing works. The most popular advertisements in the 70's and 80's on oilfield publications were articles stating case histories of wells that were producing 10 barrels per day and they called their local service provider, and they pumped a huge frac with several hundred thousand pounds of proppant and the well after flowback yielded more than 400 barrels per day. The 40-fold production increase cannot come from fracturing unless there is a tremendous amount of damage to the reservoir. With a 0 skin, with either proppant or acid, the maximum folds of increase would only approach 5. On the 10 barrel per day well, it most probably been equally stimulated with a small acid job or relatively small proppant treatment.

Back to the proppant selection of which I gave you references to utilize; I will give my personal selection criteria for moderate permeability reservoirs is as follows:

I will attempt, if at all possible, for conventional reservoirs treated with viscous fluid to give the following guidelines,

1. I want to see a minimum of 1000 millidarcy feet in the proppant pack. I do not utilize a dimensionless conductivity of 10 referenced in some of the publications because with the microdarcy permeability, it is very easy to obtain the dimensionless conductivity value of 10 in very tight reservoirs, and we have found superior results with minimum conductivity as a cutoff.

2. Drawdown or closure pressure is defined as the frac gradient times depth less the bottom hole flowing pressure.
3. For treatments where we are using viscous fluids and attempting to achieve a dominate fracture, I use the following criteria:
    A. For formations with a closure less than 5,000 psi most of the brady sands or some of the higher quality local sands almost regardless of size will suffice.
    B. For formations with a closure between 5,000 and 8,000 psi, white round Ottawa quality sand should be used. Typical sizing is 30/50, 20/40 or 16/30.
    C. For closures of between 8,000 and 10,000 psi, the use of high-quality resin coat or lighter ceramics are selected. Typical sizes are 30/50, 20/40 or 16/30.
    D. For closures of greater than 10,000 psi then bauxite or even heavier and stronger proppant is used. Typical sizing is 20/40 or 16/30.

The above guidelines are simplistic but have fared well over more than 40 years. As you will note everything changes when we move to the ultra-tight shales and very low permeability conventional rocks. Although the moderate permeability reservoirs mentioned above are not a major portion of the work going on today, there are hundreds of thousands of wells throughout the world that would greatly benefit from viscous fluid stimulation. The malice toward hydraulic fracturing by the media and Hollywood personalities has sadly limited its use throughout Europe and elsewhere. One of the major problems with the types of treatments shown above is there is little expertise within the service companies on preparation and execution. Our company has a very difficult time getting companies to perform jobs such as this and even more difficulty in achieving quality execution. We see the majors concentrating on pad slick water jobs but have had better success finding smaller service companies who specialize on this type of work.

For Nanodarcy shales and ultralow permeability conventional reservoirs, closure does not appear to be a major factor. 4,000-pound crush angular local sands do work well even in conventional closure ranges above 12,000 psi. This is a controversial statement particularly when discussing with high cost proppant companies. I do not believe we are seeing crushing in these very low permeabilities.

The obvious question is what is the permeability range when closure starts? I do know that local sand is widely used with success in microdarcy rock, but I do not believe anyone has a real definitive answer to an exact number. I must say again that I do not believe we are seeing proppant packs in shales and brittle rock. We are getting massive conductivity from our dream of a partial monolayer. Some of the recent papers about this compare 4 #s per square foot data which is never achieved in slick water fracs.

Most everyone is aware that with slick water the proppant size is either 100 mesh or 40/70, with some people running 30/50. Use of larger proppant in slick water has not led to great success for obvious reasons.

Image A

"Ahwaz Special" - Homemade blender used for acid fracs; built in Ahwaz, Iran from an old industrial cleaning unit. The skid to the left of the blender was used as the suction pump, and two downhole pumps that allowed for 40 bpm treatments (not visible) made the blender operable.

Image B

Complete rig up with combo cementers and frac units.

Image C

Alternate vantage point of the frac job; the largest frac operation in Iran at the time.

Image D

This photo shows skid units delivered from the United States.

Image E

Haliburton-Ahwaz district employees on location.

Image F

Illegal surface rig up on the Ahwaz-54 well.

CHAPTER 11

THE SAND/WATER REVOLUTION FOR EXTREMELY LOW PERMEABILITY RESERVOIRS (2022)

There has been no change in our industry that has affected more people than what I term the "Sand/water revolution". There have been multiple books written, most of them false, about how the wild changes in our belief and use of much smaller and lower quality sand at great depths has occurred. In some of the books there were Eureka moments where great technological developments were brought forward. Although there were similar happenings relating to slick water in other areas, I believe it was George Mitchell's fascination with the Barnett that eventually led to the first successful stimulation with water and small sand. Mitchell had pumped dozens of every type of frac imaginable in the Barnett with no real economic success. Eventually someone, not identified, said let us pump the cheapest thing possible which was water and sand. Most people knew that it would be difficult to pump 20/40 so 40/70 was utilized. The results were startling but even with 40/70, screenouts occurred and so 100 mesh (70/140) was used to lead in. Thankfully, there were no frac Gurus around or the move to small sand and water would have not taken off.

The success of waterfracs has flown in the face of conventional thinking. The manufacturers of high strength proppant were uniquely excited when such deep plays as the Haynesville came online thinking that effective closures would sometimes approach 11-12,000 psi. Very quickly the price of bauxite and ceramics as well as resin coated sand pushed operators to run white Ottawa sand and eventually have moved to much cheaper local sand. What was startling and not widely publicized was the small sand at a much lower price was outperforming the more expensive bauxite and ceramics. Multiple papers were written telling us in a foreboding manner that eventually all these wells would die off as the sand crushed. It seemed as if everything that we had been taught about closure and crushing was not true. It was early in the slick water movement where I found myself examining results and attempting to understand why it was working. I cannot pinpoint the actual start for the radical change in proppants and design, but it was at least 18 years and probably closer to 20 that I began to understand at least in my own mind what was occurring.

In my opinion, what is occurring is the slick water because of its low viscosity and the incorporation of small proppant creates a very complex fracture system approximating something like non-reinforced glass breaking. It is my belief that the small proppants bridge and divert further enhancing the complexity. Ultimately what is achieved is the frac persons dream of a partial monolayer of proppant holding fractures open. We always have known that our models, although very helpful in design, were not correct particularly in relation to true geometry. Models used for nano Darcy rock would yield tremendous lengths with significant propped length, using 10,000-barrel stages, approaching more than 1000 feet. By monitoring drilled offsets, 300 feet away, we saw no depletion other than where natural fractures existed. When going back to basics and looking at proppant settling and bridging, there was simply no way to transport even the smallest proppant, 100 mesh, farther than 300 feet. The answer also was backed up with older horizontal near vertical fracs where drill throughs showed many fractures were created in the near wellbore area.

One must have extreme complexity to create the massive amount of surface area in the rock to produce 4,000 + barrels per day from nanodarcy rock. I had one customer tell me that they had done intense studies and found that it was impossible to produce oil and gas from nanodarcy rock. They indeed are correct when massive surface area is not available. I was involved with a study using microdarcy rock and the study showed that when one was below .006 millidarcies in a conventional flow path, i.e., single planar fracture, no production was feasible. This study was conducted with non-fractured sandstone from the Piceance basin. By identifying the 6 microdarcy cutoff, one could predict long term production values on step out wells. This study was based on conventional viscous fluid stimulation.

In relation to picking proppant for ultralow permeability reservoirs, I really believe that with worldwide experience the only presently feasible proppants are 100 mesh technically known as 70/140 and 40/70 mesh sand. Also based on experience, we have noted that sands that have a crush capability of 4,000+ psi are applicable even at great depths. This is based on my belief that with essentially no drawdown no crushing occurs. My estimate is this phenomena is viable in permeabilities less than 6 microdarcies.

My scientific side believes that if one can quantify the rubbelization occurring in the rock based on input energy and the explosive bridging and diverting that is occurring one might be able to scientifically set a perfect sieve size distribution to use. I believe that there should never be a proppant pack in slick water fracturing. I believe we have finally achieved our design dream of partial monolayer with the complex assortment of rocks held open with single dispersed proppant. If one investigates the effective permeability of a 100-mesh proppant pack, you understand why I do not believe in tailing in with high proppant concentrations. Today, in the real world, all we have to use is reported data and so-called optimized results. I will try to list below what we have seen in more than 30 reservoirs around the world. I will state that the guidelines below are subjective and indeed are controversial to many. We have worked for hundreds of companies and gain knowledge from all of them. We have to keep confidential what our customers use but still based on dramatic similarities can make some pretty well documented guidelines. These guidelines are for long laterals. I will make some statements at the end for massive stages we have pumped in verticals.

Guidelines for Success in Slick water long lateral Completions:

1. The first and most important thing one must recognize about slick water fracturing is that answering the question about what is different between slick water and conventional stimulation is simply EVERYTHING. In my opinion when one starts to mix the two technologies the results are jeopardized. The first example that comes to mind is hybrids. Hybrid treatments are still around but are fading into the sunset as production results come forward.
2. Understand that shales like conventional reservoirs are not pervasive. A few hundred feet can create massive changes in rock properties, permeability, presence of natural fractures, pressure etc. "Me-too" designs without common diagnostics may result in economic failure.

3. Volume of fluid is dominant over volume of sand. NPV analysis is essential to achieving profitability. Huge jobs have produced enormous amounts of oil and gas but without profit they must be classified as failures.
4. Small sand has been shown to be the best material. To the best of my knowledge, little success has been achieved with 200 mesh suggesting that a minimal size is necessary to hold the complex fractures open. I believe it is imperative that we know what size distribution is being pumped to have any chance in optimizing future treatments. We have seen little or no benefits to pumping larger than 40 mesh. Many operators are pumping all 100 mesh and others are pumping 100 mesh followed by 40/70. Sadly, we have seen numerous cases where what was pumped was not as specified. A large part of the sand called 100 mesh in West Texas and New Mexico is 50/100 rather than 70/140. We have also seen screenouts occur because the sand was larger than 20/40 when 100 mesh was specified. We strongly suggest that you monitor flowing sand and have sand sieved on location. At least once per stage. As stated before, it does little good to have massive capabilities to calculate what is causing optimal production when we do not know what was pumped.
5. We have achieved success with a combination of what we call "Waterfrac Sweep" and "Counterprop". The Waterfrac sweep technique was borrowed from UPRC before they were purchased by Anadarko. I attended and monitored jobs in central Texas where they were alternating a sand concentration with no sand. Since I was not involved in the design, I asked for the rationale and the field rep told me they had seen increasing pressures and ran sweeps which seemed to alleviate their problem. I also noticed that they ran large volumes of pure pad ahead of the alternating water and sand. They were also using 40/70 sand which was very unusual for that time (around 2001). The answer was they had screen outs with larger sand and had elected to use only the smaller sand. I checked production on this well and offsets and UPRC had the highest EUR in the area. I from that moment on used Waterfrac sweep in all my slick water designs. I also included large early pads to achieve widths on all subsequent designs. With the passing of time we know to use diagnostics in slick water completions to optimize fracs in real time.

The "Counterprop" use came from a conversation with a customer discussing the use of small sands and how they were being applied. We all knew that water did not carry sand any appreciable distance and also knew that whatever was pumped first would be at the wellbore. We therefore for more than 3 years have been designing and running slick water jobs with 40/70 lead ins followed by tailing with 100 mesh. This type of design is not universally used but in my opinion it should be. We have not had a failure compared to offset production in any of the Counterprop jobs.

I have always had a problem with the name "shale" that was placed on early source rock plays such as the Barnett and Woodford. For as long as I have been in the industry, until the early 2000's, the word shale described a soft formation such as the shales above and below the Wilcox in south Texas, none of the successful shale plays in today's world are soft, typically with a Youngs Modulus of greater than 2 million. The much touted and very unsuccessful Floyd Shale is an example of huge expenditures without relying on science. The play, primarily in Alabama was started with headlines in the Tuscaloosa paper showing a core brought to surface bleeding oil. Literally 100's of thousands of acres was leased, and several wells were drilled, and fracture treated without fundamental understanding of critical parameters for success. I had a customer who had 20,000+ acres leased in Alabama. I was called in when core data from the Floyd was presented. The vice president asked me to look over the data and give him a quick recommendation. I scanned the modulus data and told him to sell to a competitor. He was initially upset because of his acreage position but later did sell to one of the largest shale players at the time. That recommendation was perhaps my very best. It should be noted that several operators tested the play and there was a statewide conference to compare results. For most of you who have not heard of the Floyd, there is a basic reason that we as an industry do not publish our failures. The low modulus numbers, just greater than 1,000,000 indicated ductility or basically softness approaching what we normally called shale above and below our conventional plays.

There are relatively soft conventional formations such as the Miocene sands and even parts of the Vicksburg in South Texas, but these conventional reservoirs have inherently high matrix permeability. A combination of soft rock and microdarcy permeability, or lower, is an economic disaster waiting to happen.

The success of shale plays today is a result of brittle rock with interconnecting natural fractures or interconnected fractures created by the slick water treatments with small sand to prop the fractures open. It is common knowledge that many of our so-called shale plays are in fact not source rock plays but are hard rock carbonates and sandstone that are brittle and have formed a trap for hydrocarbons. For those of us in the early stages of shale fracturing, it was forbidden to get into the condensate or oil window because everyone knew you could not produce liquid hydrocarbons from ultra-tight reservoirs. Again, early on we outsmarted ourselves not understanding that massive surface area through complex fracturing would allow for some of the highest rate condensate and oil wells in multiple areas of the world. Perhaps another case where the evil intuitive reasoning problem kept us initially from attacking the most prolific areas of hydrocarbon production. To my knowledge, the first Eagle Ford well was almost plugged because of massive core data indicating a 2-300 nano Darcy perm. This well mistakenly drilled in the condensate window and, to supposedly kill the deal, the pressure analysis engineer announced in the meeting that the condensate was retrograde. Because we had already had success in the Northern Barnett Shale play, they decided to go forward, and the rest is history.

There have been no changes greater than what is happening in the use and handling of sand proppants. First of all, as mentioned earlier in the chapter, the vast majority of sand now used relates to proximity of the sand to the reservoirs being stimulated. Historically we have specified the highest quality sand, mostly from the great lakes area of the US, for deeper wells and Brady angular sand from south central Texas for shallower lower closure wells. These guidelines had survived for many years and if your well was far from the great lakes or Brady Texas you had to pay enormous costs for transportation of sand. Today for our ultralow permeability reservoirs, the sand you are using is typically only miles away and some of it is coming from the surface. Rail transportation is still going on but nowhere near the volume percentage as in the past. Other than man made proppants there is nothing on earth to compare to the round Ottawa sand. The bad news is that for nanodarcy reservoirs it is not required. Many of the suppliers of this very high-quality sand have suffered greatly but many have opened sand mines in areas close to producing areas.

In addition to the success of the lower quality sand and much lower prices there was still an enormous problem in being able to supply a minimum of 10-20 million pounds of sand for each lateral. On the early slick water treatment there was mile or longer lines of sand delivery trucks creating massive demurrage costs. With American ingenuity one approach was to develop transportable boxes of sand which could be stacked on location prior to the job. The boxes were then moved to a belt by a giant forklift feeding directly into the blender. There are 2 dominant companies supplying these boxes. One supplier box holds close to 50,000 pounds' net and the other company is limited to 25,000 pounds. It is quite a sight to see 4-800 boxes stacked on a location awaiting a frac. We have been moving boxes of sand since near the beginning of fracturing, but this idea was an enormous improvement on the classical Mountain Movers, Sand Hogs etc. etc.

The other means of handling sand was more of a "back to the future" thing. This technique involves huge upright gravity storage bins on location prior to the pumping equipment arriving on location. These types of storage units were used in the 80's but obviously fewer quantity and lower volume. There have been improvements in onsite storage using other techniques than air loading.

All of the new techniques, to my knowledge, are incorporating means to drastically reduce the amount of silica dust on location.

The logistics and volumes of both fluid and sand are enormous. For non-fracturing folks the typical frac pumps downhole a large milk trailer every minute. The average amount of sand in a 2-mile lateral is between 15 and 20 million pounds. This would be the equivalent of 375 standard dump trucks of sand pumped beneath the surface. We are indeed moving mountains. There is plenty of sand but of concern of many is the volume of water. There is and has been, from the beginning, water cleanup and recycling moving forward at a fast pace. The industry is moving away from fresh water as chemicals have been developed that are compatible with even with high brine content.

We are showing pictures of various sand storage at the chapters end. If you are interested in this sort of thing, go to Sand Box, Solaris, HICrush, Stack or Propx websites to get more information. There are more suppliers of both upright storage and onsite box storage and I apologize to them for not mentioning their names. Even some of the service companies have their own storage equipment.

CHAPTER 12

QUALITY CONTROLLING ACID TREATMENTS (1994)

Acidizing treatments have been conducted in the oil industry since the early 1920s. Most of these have been simple, small hydrochloric acid treatments in carbonate or sandstone reservoirs to remove damage that was probably deposited during drilling operations. Many hydrofluoric acid treatments have been conducted in sandstone reservoirs at matrix rates to react with clays in the formation or with muds lost in the formation. Additionally, treatments have been carried out using either pure acetic or formic acid. These weak organic acids have been used as perforating mediums or as acidizing systems in high-temperature wells. Commonly, blends of these acids with hydrochloric acids have been used in deep, high-temperature operations. The combination of hydrochloric and acetic or hydrochloric-formic yields a retarded or delayed reaction acid system.

When I teach at hydraulic fracturing schools, or schools just involved with quality control or stimulation, I have a trick question that I always try to ask to emphasize something I consider very important in stimulation. The question I ask is, "What do you want to do to the formation with a hydraulic fracturing treatment?" It is certainly not uncommon to have young individuals raise their hands and tell me that what we want to do is increase the permeability. I always enjoy that because I get to tell them that only God can increase the permeability of a formation. We have little to do with that. Other typical answers are creating a conductive fracture, making a bigger wellbore, creating a proppant pack with high relative conductivity, and so forth. All of these answers, other than the one about enhanced permeability, are good answers, but not the one for which I was searching. The answer to the question, "What do you want to do to the formation with a hydraulic fracturing treatment?" is "nothing." This always brings startled looks to the students' faces but I would prefer to conduct a hydraulic fracture treatment on a reservoir without acid being pumped ahead. This is particularly the case for carbonate reservoirs. The explosive reactivity of acid, particularly in high-temperature carbonate reservoirs, will normally create cavities and create large amounts of fines near wellbore. It would be much more beneficial to have a competent fracture near the perforations than to have craters, fines, or softening of the rock in the near wellbore area.

In the ideal world, you would like to create a fracture much in the mode of a person going downhole, and with a chisel, create a smooth wall parallel plate-type fracture where we could place propping agents and produce all of the fluid back without filter cake development, without fines being generated, and so forth. Perhaps one of the biggest mistakes made in the hydraulic fracturing process is to pump large quantities of hydrochloric acid ahead of fracture treatments.

It has always interested me to see people pump thousands of gallons of hydrochloric acid ahead of a fracture treatment in a pure sandstone reservoir. When I ask people what the acid is reacting with, commonly they state that it is reacting with cement. Again these people are startled when I tell them that acid (other than an instantaneous reaction that can occur with detritus material in the perforation tunnel) will not react with cement. Major service companies have conducted extended pump tests with thousands of gallons of acid through cemented perforation tunnels and have seen no dissolution of cement. This is because cement is self-passivating. An oxide layer is created fairly quickly and no more cement is removed. The next question is, "How much acid should be pumped in relationship to breakdown of perforations?" Like most great scientific discoveries (such as the peanut butter and chocolate discovery that developed into Reeses Cups® by two people running together), many great discoveries in the oilfield have occurred because of mistakes or because of necessity. On one particular job in Alabama, where we couldn't break down the perforations, we then discovered that the service companies were all extremely busy and that it would be 2 to 3 days before we could get acid to location. A wise old individual was on site and suggested that we utilize a dump bailer to place the acid over the perforations. The dump bailer in question could only hold 8 gallons of acid. We proceeded to a local swimming pool outlet, bought 4 gallons of 32% acid and then proceeded to dilute it to 15% with water, ran the dump bailer into the well across the perforated interval, and opened the dump bailer. The 8 gallons of acid were sufficient to allow us to pump at 40 bpm at 2,200 psi. Prior to that, we could get no entry whatsoever at 5,000 psi.

My basic recommendation is, if you are going to fracture treat a well, utilize big enough guns and get enough penetration so that acid breakdowns are not required. I certainly am not opposed to ball outs but recommend ball outs with KcL water. In sandstone reservoirs, you can break any perforations and make sure all the holes are opened every bit as well with KcL as with hydrochloric acid. There are many documented cases where HcL has been damaging to many sandstone reservoirs because of clay shrinkage and fines movement. The best fluid that you can put on a formation-with the exception of fracture acidizing-is a fluid that is not reactive.

When this subject comes up in fracturing schools, many individuals tell me about the tremendous cost if the well will not break down and they have to go get acid to spot. If you can't break it down, you can't get acid to bottom. The calculations for determining how many holes are open when using Newtonian fluids and pump rates and breakdowns are very simple and straightforward. I am in favor of using hydrochloric and other acids where required, but to use them just as a matter of course ahead of fracturing treatments can be very damaging to the potential success of the fracturing treatment.

Basically, quality control of acid treatments follows the same lines taken in quality-controlling fracturing of cementing treatments. The same high pressures typically exist; the same rig-up considerations occur. There are, of course, some exceptions for quality control on acid treatments. When using any of the mentioned acids, safety needs to be the foremost consideration.

Acids are dangerous

All of these acid systems are potentially dangerous because personnel are around high-pressure iron, which can spray or leak. The major danger from hydrochloric acid is getting it in your eyes. Hydrochloric acid can be rinsed off easily with little permanent damage as long as it does not enter your eyes. Rapid loss of eyesight can occur if hydrochloric acid is not very quickly diluted and removed.

Formic acid is extremely dangerous and vigorously attacks the skin and extremities. The same goes for strong acetic anhydride used as the concentrate for preparing weaker solutions of acetic acid. A common name for weak acetic acid is table vinegar. The vinegar that you put on your salads is somewhere around 4% acetic acid. Strong acetic acid is 90% or more.

Formic acid, on the other hand, is virulent. It is a major component of bee or ant stings; therefore, it is not something you want on your skin or sprayed around a location.

Hydrofluoric acid is also very dangerous. It attacks the skin and penetrates the bone marrow. In low concentrations, it can be diluted and washed off. The primary danger lies in utilizing any of these materials in a concentrated state.

Concentrations

The oil company engineer should be aware of the acid concentrations utilized and the amounts required to prepare specific acid solutions on the location. See Appendix Ill. Different concentrations of these strong acids are available. Many service companies use 20 or 22 Beaume' hydrochloric. Some use glacial acetic, while others use acetic anhydride as the concentrate for acetic acid. Generally, service companies use 90% formic acid concentrate for preparing the weaker acid. Though some service companies use concentrated hydrofluoric acid, we strongly discourage this practice because of the danger to personnel. Granular ammonium bifluoride is readily available and combined with HcL forms HF acid with much less danger to the personnel involved.

Precautions

Follow the basic rig-up precautions mentioned earlier for a fracturing treatment, including placing check valves, using non-rigid iron, and checking the iron for maximum pump rate. When using multiple trucks, the suctions hoses should be covered to protect personnel from spraying fluids if one of these hoses ruptures. People in the immediate vicinity of the location, whether they are with the oil company or the service company, should be required to wear goggles to protect their eyes in case a rupture or spray occurs.

There are other precautions for acid treatments compared to gelled-water or gelled-oil treatments. Do not overlook volumes on the site. Most acid treatments are small ones run out of 4,000- or 5,000-gal transports. Climb onto these units and gauge the tanks yourself for their volumes before and after the treatment.

Second, obtain samples to take back to the laboratory. Take these samples only after the tank has been thoroughly stirred to ensure homogeneity. These samples can be used as a cross-check if problems occur later. Get these samples in either plastic or glass containers, of course, and not in reactive metal vessels. Use only plastic vessels if HF acid is utilized.

Evaluate acid strength

If you are limited to a single controlling tool, it should be a hydrometer (see Appendix IV) for evaluating acid strength. Table 13-1 illustrates percent hydrochloric acid versus specific gravity. If you have the capability, titrate the acid against a known constant base to evaluate the percentage of hydrogen ions present. You can have a high specific gravity solution, indicating a high-strength acid, and indeed have little or no acid because of the presence of weighting materials.

This does not mean the service company is trying to cheat you. A weaker acid may have been supplied by their commodity source, or people may have diluted the product incorrectly or sent it to the wrong location. Another important consideration - not just for taking samples but also for doing the job well-is that the acid in the tanks, whether small transports or large frac tanks, must be adequately stirred just before pumping. The major reason for agitation is that acid inhibitors tend to float out.

Acid inhibitors

Acid inhibitors function primarily because they plate-out on the metal's surface. If these products are plated out on the metal's surface, obviously they cannot be totally soluble in the acid solution. Many acid jobs pump little or no inhibitor in the initial portion of the acid and run high concentrations of potentially damaging inhibitors late in the treatment. At this late stage, any corrosion that might have occurred during the treatment has already taken place, and the acid inhibitor can act only as a potentially damaging media near the wellbore.

Kinds of treatments

There are many different types of acidizing treatments. We list six with some specific recommendations or questions that should be asked about these treatments. Some of the questions relate to quality control, and others relate to specific design to make sure the proper treatment is conducted.

Matrix acidization

Probably the most important criterion for a matrix-acid treatment is that it be conducted at less than frac pressure. Acid should penetrate the matrix radially and not through a hydraulic fracture. A matrix-acid treatment is normally utilized just to remove damage near the wellbore. Determine the frac gradient exactly and make sure the fluid's hydrostatic head plus any pressure exerted will not exceed the frac pressure. Say the fracturing gradient turns out to be 0. 7 psi/ft. for instance, in a 10,000 ft. well. You might put in a safety factor and not exceed a pressure of 6,500 lb. This pressure would be the fluid's hydrostatic head plus the pressure exerted upon the fluid to pump it into the formation.

Typically, with a 15% hydrochloric acid solution and the previous criteria, you cannot exceed 2,000 lb. pressure at the surface without fracturing the formation. There are, of course, many situations in a carbonate reservoir when you might be required to exceed fracturing pressure slightly to break down perforations or to initiate the fluid's entrance into the formation. Once the formation starts taking fluid, immediately drop the pressure and the pumping rate below fracturing pressure to complete the treatment.

Matrix-acid treatments are designed to remove damage near the wellbore. If you fracture the formation, very possibly you will fail to remove any damage at or near the wellbore or to clean out the perforations. Many matrix-acid treatments fail because they are pumped at pressures higher than that required to break down the formation.

Because it usually is run at very low rates, a matrix-acid treatment requires special care in the concentration and blending of inhibitors. Fairly low inhibitor loadings can be utilized in high-rate treatments where a great deal of cool-down occurs. For matrix-acid treatments, especially if there is a large volume of acid, take particular care that the inhibitor loadings are designed to protect the pipe for minimum corrosion at bottom-hole temperature conditions.

Work with the service company representative in his design for ball sealer or diverting agents to ascertain if these are useful at the low-pumping rates in matrix treatments. In reference to ball sealers, over the years I have developed a rule-of-thumb that has worked very well for me in reference to ball sealer functionality. I have found that on the low end 1/10 of a bbl./min./perforation up to ¼ bbl./min./ perforation with 1/4 bpm/perf or greater being a good value and 1/10 being the low end that I have been able to achieve ball outs utilizing these rate criteria 95-98% of the time. I have had heated arguments with individuals concerning this, but quite frankly I have had great success with even up to 1.6 gravity weighted ball sealers utilizing these criteria.

One of my pet peeves is the definition of a ball out. The typical comment from a treatment is, "We had good ball action." Frankly, a ball out is achieving maximum pressure at zero rate (i.e., you have shut off all of the perforations). Typically, I run 25-50% excess balls and utilize the previously mentioned rate criteria. With this in mind, if I have a decent cement job, it is very seldom that we ever fail to achieve a ball out.

In most cases, problems with achieving a ball out are due to not dropping the balls or to personnel not continuing to pump properly and maintaining rate while the balls are seating. Gradually, service companies are starting to get either kick outs for their pumps or hydraulic actuated pop-offs to allow us to conduct ball out treatments safely and to continue to pump once we achieve maximum pressure without exceeding pressure limitations of the pipe. Good pump operators can accomplish excellent ball out by running their business properly, and I have conducted literally hundreds of these types of treatments.

I am not a particular fan of lightweight ball sealers, although in situations where we have literally hundreds of perforations that may be the only option. The problem with lightweight ball sealers is that many times if personnel don't really watch exactly what they are doing, they may end up where they can't pump in the well, nor can they flow the well. I had an interesting situation occur in Iran. As is typically the case, we had 200-300 ft. of perforated interval shot at 4 shots/ft. I was working for a service company, and they were having a large argument about whether to use 25 or 50% excess balls. At that point, I had been in the country long enough to know that I wanted to leave, and I wasn't particularly tactful. I raised my hand and indicated that I knew exactly how many balls would be required to get ball action on the treatment. The individuals were arguing about whether they should run 1,500 or 1,800 ball sealers on the treatment. Our pump rate would be limited to approximately 20 bpm, so I saw no opportunity whatsoever for achieving any ball action unless a great majority of the holes were not opened early in the treatment. I raised my hand and told these individuals that what was required was 4.5 bbls of balls. They were very startled and looked at me strangely and asked how in the world I could come up with that. Of course, the way I came up with the volume was calculating the rat hole volume plus the volume to the top perforation. For some reason, these individuals were not very amused. Achieving ball action or getting any functionality out of balls at such a low rate with so many perforations was impossible. Filling the casing was really the only way possible to achieve true diversion.

Diverting agents used at low rates often give unacceptable results. One of the more interesting things that I have observed over the years in stimulation treatments has been the running of diverting agent to stage hydraulic fracturing treatments as well as slugs of diverting agents in acid treatments, etc. Now that

I have a better understanding of hydraulic fracturing, I know that there is little or no success with any of the fracturing treatments utilizing diverting agents.

If we had used common sense, we would have known that running a pound or two per gal of a bridging diverting agent in a cross-linked gel would have little or no effect at the perforations or certainly not in the formation. The cross-linked gel, linear gel, and so forth is designed to carry materials through the perforations. I have found the only way to assure good diversion is to run bridging materials, such as rock salt or benzoic acid, in the nonviscosified acid or water solutions. Most field personnel do not want to run benzoic acid or rock salt or other flake material in straight water or acid because of a tremendous amount of line vibration occurring. This is the way I ascertain if indeed diverting agents are functioning; namely, by whether the lines are jumping 1-2ft off the ground into fine material. f no line vibration is occurring; you can be relatively assured that no diversion will occur downhole.

I had a station manager call me and inform. me that he had solved the problem of line vibration on his diverting agent treatments. He informed me that he had taken his benzoic acid flakes to a local feed store and had them ground. Since he had started utilizing ground benzoic acid, his treatments had gone much smoother, and they were getting identical results as they had seen before. The bottom line to that discussion is he was not getting any diverting before but at least his treatments were going smooth.

In my experience, most successful applications of diverting agents have been using them at approximately .2 lb./gal continuously throughout acid treatments conducted at matrix rates. In one very large study, we were able to pump acid at matrix rates of approximately 5 bbls/min. across long, perforated intervals and utilized .2 lb./gal of coarse benzoic acid flakes. Although there was a lot of surface vibration, downhole gauges recorded excellent diversion as well as spinner surveys, post-acid treatment indicated we had accomplished excellent matrix acidization of the completed interval with this technique.

I am very much opposed to slugs of diverting agents in gel. If you want to bridge or divert, I believe you are better off to run it continuously or run fairly large slugs in low-viscosity solution. Anytime you viscosify, you are negating the effect of the bridging agent by making it easier to carry through the perforations.

One of the things that plagues our industry is the fact that people shoot far too many perforations. Excessive sand production, poor control of entrance of the fracture treatment, and many other problems are caused by personnel who feel that they must perforate at very high density to be assured that they are in communication with the formation. It has been my experience in treating wells that much more successful stimulation, particularly from the standpoint of hydraulic fracturing, occurs if you do what I term "single-point entry fracturing" (i.e., shooting a minimum number of holes across a short interval and then hydraulically fracturing into the zone). As discussed in Chapter 12, this allows for the creation of a single, plainer vertical fracture rather than multiple ineschelon fractures. If an acid fracturing treatment is to be conducted or if a hydraulic fracturing treatment is conducted containing proppant, then you should minimize the feet of perforations open to minimize the possibility of creating more than one hydraulic fracture.

Fracture acidizing

Fracture acidizing treatments require much larger quantities of hydrochloric or organic acids. These treatments are commonly conducted out of standard frac tanks. A major hazard is excess corrosion in the frac tanks. This can lead to leaking tanks and loss of the acid. Corrosion can result in high iron concentrations in the acid, which can be detrimental to your well. Excessive corrosion rates sometimes occur because many of the commercially available acid inhibitors are incompatible with high iron concentrations in the acid solution.

Ask your service company to utilize lined frac tanks, if possible, or be certain the proper amount of corrosion inhibitor is added before putting the acid in the tank. Take precautions to prevent leakage at the manifold from the pressurizing pumps or blenders. There are many acid frac treatments that run several additives continuously; fracturing blenders are needed on the location. These blenders present a safety risk, since an operator stands over an open tub into which acid is pumped at high rates. Acid can splash or splatter personnel. Again, be sure the people in the vicinity wear goggles and protective gear so no one is injured.

One of the commonly overlooked problems in acid fracturing treatments is the incompatibility of other systems utilized before or behind the acid. Quite often on acid frac treatments, viscous and/or cross-linked gels are run ahead of and used as spacers between the acid stages. Viscous gels are followed by acid in these treatments. This format allows for viscous fingering ofthe acid through the gel systems yielding a heterogeneous flow pattern on the fracture face for better productivity. This technique has been very successful.

Time and time again on viscous fingering acid treatments, I have seen personnel amazed because they couldn't get cross-linking on the gel systems that were pumped as spacers or as pads. This is because the frac gels were contaminated with acid. Many of these systems work at neutral or high pH; if they have any acid contamination from leaking tanks, the pH is very low. Therefore, take great care that there is no contamination between the various fluid tanks on the location. Do not have a common manifold between the gel tanks and the acid tanks.

Most service company blenders are rigged up so the acid by-passes the tub if additives are not being added continually. If this is the situation, do not go through the blender tub with the acid.

Acid fracture treatments normally are conducted at higher pressures and higher rates than matrix treatments. Consequently, more pumping equipment and greater care are necessary to be absolutely sure all the pumping equipment is pressure-tested prior to the treatment. In one instance, an oil company required the service company to test all lines with hydrochloric acid. This was a mistake because a pressure test typically lasts about 15 minutes. Virtually none of the current treating iron in the industry can withstand such a test. The longer you test at high pressures with acid, the greater the chance of complete failure. One company spent an entire day trying to test lines and get a straight line chart with 28% acid. They finally gave up and tested the lines with water.

I unequivocally do not recommend testing lines with hydrochloric acid. Almost invariably, you will have some minimal leakage during an acid treatment. The fluid loss additives used to control fluid loss downhole are often used sucessfully to control small leaks as you pump acid. Severe leaks should not be allowed during acid pumping because they generally get worse during the treatment. Be open-minded and use common sense when watching dripping leaks during a high-pressure acid frac treatment.

A frequent error in acid frac treatments is the failure to use proper fluid loss control additives. This is because few people screen-out on an acid treatment. In fact, many times the treatment is pumped at a lower pressure and a higher rate than it would have been if fluid loss additives are used. Unless it penetrates far into the formation and remains active, acid is useless as a stimulation medium. Work closely with your service company to be sure the proper fluid loss additives are used on your acid frac treatment. ·

The most common thing that occurs on an acid fracturing treatment is that once the cool-down pre-pad or viscous pad has been pumped and the acid hits the formation, a precipitous pressure drop occurs.

Commonly, everyone on site is elated and they are saying that the acid is doing its job. Quite frankly, this is most probably indicative of the fact that very high leak-off is occurring, and instead of achieving deep penetration within its fracture systems, you are simply leaking off the acid in the near wellbore area. A properly designed acid fracture treatment should treat exactly in the same manner as a properly designed proppant fracturing treatment.

Recently, some very successful acid fracturing treatments have been conducted in the United States with oil external emulsified acid. In these treatments, the oil company and service company report no decreases in pressure as the acid hits the formation. In fact, in many of the treatments they have seen an incline in pressure indicating some fracture containment. As most of you know, a precipitous decline in pressure along the lines of classical Nolte Smith theory indicates either fracture height growth or a precipitous amount of leak-off. In the case of acid, where we see large drops in pressure, in most cases it is high leak-off. Obviously, it also could be opening perforations as is seen in matrix treatments or break down treatments.

As discussed in Chapter 12, I have found very few areas where there are distinct fracture barriers. Since the majority of hydraulic fractures occur in a radial mode, and because of the high reaction rate of hydrochloric acid and the very small amount of cool-down that occurs in the fracture face, the highest pump rate possible is probably the best choice. This will allow penetration of the acid at fairly large distances, although this can be somewhat negated by accelerated reaction rate due to turbulence in the fracture.

Although I am a soft-spoken, non-opinionated. individual, I will express a "somewhat" controversial opinion. I have found very few occasions where I could not achieve more economical fracturing situations utilizing proppant than I could with acid. The problem with hydraulic fracturing with acid (other than the aforementioned leak-off problems) is that it is extremely difficult to achieve heterogeneous etched fractures that will remain open under any significant closure stress. Obviously, there has been a lot of success over the years with fracture acidizing. The typical response, in many cases, is high early production and very rapid fall off due to fracture closure. A tremendous amount of success has been achieved in carbonate formations all over the world with hydraulic fracturing with proppant. Some personnel have stated that hydraulic fracturing with proppant will not work in naturally fractured formations. If the reason for failure is the fact that the fracture parallels the natural fracture systems, then most probably fracturing acidizing will not succeed and the only chance for success will be with horizontal drilling. Drilling a horizontal wellbore perpendicular to the fracture systems has worked well in the Austin Chalk.

Once again, on acid frac treatments, be sure the tanks are adequately stirred. Any additives run on the fly must be monitored closely. On this type of treatment, you are usually running acid retarders and acid thickeners. The quantities and addition rates need to be worked out very closely with the service company to accomplish your objective with the treatment.

For an acid frac treatment to work well, pump concentrated or live acid far into t4e formation so it can react, etching the face of the fracture. Obviously, you cannot achieve etching if the acid leaks off near the wellbore or into fractures or rugs. The object is to create a massive fracture system with large surface areas open for production.

For deep acid penetration, retard the acid. Many additives are available from the service company; namely, sulphonates or phosphate esters that work chemically and viscosifiers that work physically. These products are used successfully. They will not function if they are applied incorrectly.

Since the writing of my first book, I have become more convinced of the fact that the very best retarded acid available is an oil external emulsified acid. This acid system combines viscosity, which comes from the emulsion, but also contains an oil external phase. This product has been used widely in the Middle East as a very common acidizing fluid for high-permeability reservoirs. It also was used widely as an acid fracturing fluid. Quite recently, Bass Enterprises and Dowell Schlumberger published a paper, SPE 26581, verifying the applicability of this particular acid in many domestic reservoirs. If I am going to utilize a fracture acidizing system, my first choice (if at all possible) is to utilize an oil external emulsion system. This system suffers from relatively high friction pressure, but some recent work within the last 10 years has created some pumpable oil external acids. These systems can also contain organic acids as well as the aforementioned sulfonate and phosphonate retarders.

Acid treatments require rigorous quality control. This is because you never know if anything is going wrong on a treatment. The only possible problem that can be observed is a pressure increase if the service company is not running friction reducer.

You can see things going wrong on a proppant treatment because a screen-out may occur. A drastic decrease or increase in friction pressure means little or no cross-linker is used or too much cross-linker is used on a treatment. You can detect wide variations in proppant concentrations on the pressure chart, either from the hydrostatic changes or from the densimeter. On an acid treatment, there are few surface monitoring indications. The service company may or may not be running the proper additives, with little indication either way. Therefore, the oil company representatives must work closely with the service company to ensure the proper additives are utilized.

Gelled or cross-linked gelled acids

Over the years, service companies have supplied a multitude of gelled acid systems. Prior to the 1980s, service companies utilized the commonly available viscosifiers for waters such as guar and cellulose to viscosify acids. In fact, these gelled acids were used as fracturing fluids, particularly in the Texas Panhandle for many years. They were proclaimed as being a very successful system. In reality, they were successful in not doing a great deal of damage because they, in most cases, did not have any viscosity whatsoever once they got to the bottom of the hole. The stability of cellulose or guar or even hydroxypropylguar in any acid concentration above 3% with any temperature at all is very minimal. In most cases, most of the gelled 3% acid jobs had no viscosity whatsoever prior to entering the perforations. The fracture treatments with proppant conducted in those areas were conducted with very low sand concentrations, and they were in fact utilizing water with a minimum amount of friction reducer from the rapidly breaking thickeners. If you should choose to run such a system, the best approach is to gel up the water and dilute the acid with the water as you go downhole. In these cases, you will at least get some friction reduction as you pump down casing or tubulars. As we got into the late 1970s and early 1980s, the development of carboxymethylhydroxyethylcellulose gave us a slightly more stable gelling agent, and one of the service companies introduced xanthan gum as a standard gelling agent for acid. CMHEC has very little stability above 10% acid. Xanthan, on the other hand, is an excellent gelling agent for acid for up to 15% hydrochloric. Above 15%, it degrades very rapidly. The problem with xanthan gum .is that no degradation occurs unless the temperature approaches 200°F. CMHEC degrades quite rapidly at minimal temperatures even in the 3-10% acid range. Weak acid fracturing fluids used with CMHEC work quite well as long as you don't exceed 90°F.

Xanthan gum was widely used for a long period of time as a gelling agent for acid and quite obviously was stable. Plugged formations resulted from xanthan being pumped into many wells where there was insufficient bottom hole pressure to flow the thickened acid back out of the formation. The thickened acid gel, which would indeed not degrade, when combined with fines created by the acid could create a very strong viscous plug that could damage the formation. The major problem with any viscosified acid system that does not contain internal breaker, which is by definition stable gelled acids, is that you must have sufficient bottom hole pressure to remove it. Otherwise, it shouldn't be pumped into the well.

There is indeed very complex chemistry involved in coming up with a degrading mechanism for polymers that are stable in acid. Presently, most of the service companies have available acrylic-type polymers for gelling acids. These polymers give you a very strong, stable gelled acid. The problem with them at temperatures of less than 300° is that there is no degradation of the polymer other than some thermal thinning. You must be absolutely sure that there is sufficient energy in the formation to remove these materials; otherwise, they will plug up pore spaces and cause more damage than good.

Cross-linked acids

Over the years, many service companies have come up with cross-linked acids. Early on these consisted of CMHEC cross-linked with high concentrations of zirconium. Other systems have been developed that were acrylics cross-linked with metallic cross-linkers. The CMHEC systems suffered from the fact that they were not particularly temperature stable and broke down rapidly in the presence of minimal bottom hole temperature. The acrylics suffer from a more serious problem because there isn't a mechanism for breakdown of the actual polymer. Service companies have technology to controllably break down the cross-link, but the acrylic polymers do not break down until very high bottom hole temperatures are reached. If you choose to utilize a cross-linked acid as a fracturing fluid, be sure that there is a mechanism for degradation of not only the cross-link, but also of the polymer. Or, be absolutely certain that you have sufficient bottom hole pressure to flow out of the formation and the pore spaces very high-viscosity polymeric solutions. The biggest mistake people make in relationship to pumping polymers into the formation is the lack of understanding that non-Newtonian fluids have very high viscosities at very low shear rates. The shear rate within the pore space of the formation may be 1/500 of the shear rate as seen in a flowing pipe. This very low shear rate can translate into a syrup-type viscosity being well in excess of 1,000 cps. in the formation. When you plug this kind of viscosity into Darcies radial flow equation, it becomes very obvious that very high pressures are required to remove it from the formation. To quality control gelled or cross-linked acids, you need to ascertain acid strength and do the similar things that you do with gelled fracturing fluids; namely, compare the viscosity of these base gels with the service company charts. But more importantly, as stated earlier, you need to do intense quality control with these acid systems and be absolutely sure that they will degrade and upon degrading do not leave insoluble materials in the formation.

With the earlier edition of this book, people were still trying to batch-mix conventional polymer gelled acids. With the passing of time and as our industry has gotten more intelligent, hopefully this has passed. Most of the gelled acid systems utilized today are primarily acrylic and are properly applied in relatively high-pressure, high-temperature formations. We are seeing some new technology that relates to special acid systems that trigger yielding high viscosity as the acid is thinned on the formation, controlling fluid loss and yielding diversion. You should never pump an acid system into a formation without doing testing at insitu conditions. This involves having cores from your formation and having it react with the acid at high temperatures. Service companies are happy to do this kind of work in conjunction with large acid fracturing treatments.

Hydrofluoric acid

A hydrofluoric acid treatment in the oilfield is a combination of hydrochloric acid and ammonium bifluoride, creating a 12% hydrochloric-3% hydrofluoric acid, or variations such as 6% HcL-1.5% HF. This system is used for matrix acidizing wells that normally have been damaged by large amounts of water-based drilling muds. Some wells may also have a lot of naturally occurring clays in the formation that react with hydrofluoric acid to improve conductivity. A hydrofluoric acid treatment should never be used on an undamaged sandstone well or on a limestone reservoir. The reaction products of hydrofluoric acid with limestone yields detrimental reaction products like calcium fluoride. The results are detrimental to the productive life of the well.

Hydrofluoric acid is typically used in the moderate-to-high-permeability sandstone reservoirs, such as those on the Louisiana gulf coast. Y au must have enough permeability in the formation to pump into it at matrix rates. The most obvious use for hydrofluoric acid is when a high volume of drilling mud has been lost into the formation. This drilling mud can be removed by pumping hydrofluoric acid. Always pump a hydrochloric acid spacer ahead of the hydrofluoric acid treatment to react with and neutralize any carbonates in the reservoir. This spacer also sweeps away any salt water that is incompatible with hydrofluoric acid.

Note that potassium chloride water is also incompatible. In flushing or overflushing the fluid, use ammonium chloride water, which is compatible with hydrofluoric acid. Unfortunately, many wells have been mistreated with hydrofluoric acid, sometimes causing irreparable damage. The only real candidate for a hydrofluoric acid treatment is a moderate-to-high-permeability well with inherent or externally created damage. Hydrofluoric acid cannot improve the productivity of a tight well or an undamaged reservoir. The only obvious choice in a very tight sandstone reservoir is a proppant fracturing treatment.

Retarded acids

There are many retarded acids marketed by service companies. Various retardation means have been mentioned earlier, such as viscosifying the acid, which slows down the migration of the hydrogen ion and thereby slows down the reaction rate at the fracture face.

Retarded HcL acids are functional only in fracture acidizing; no retardation function is really applicable in matrix HcL acidizing. Once the HcL acid is in the matrix of a formation, the reaction rate is so fast that the most severe retardation mechanism would only retard the acid for a microsecond. Therefore, when we speak of retarded acids, we are just speaking of fracture acidizing. When you go beyond viscosifying the acid, consider chemical retardation.

There are two types of chemical retarders. One is a sulphonate used in low temperature wells (up to 200°F), and another is a phosphate ester retarder used for higher temperatures. The lower temperature acid inhibitor always requires the presence of oil. The mechanism of retardation is oil-wetting. Without oil, the sulphonate retarder cannot function. Typically, oil or another hydrocarbon is run ahead of or added simultaneously to the acid treatment to yield the retardation needed for deeper penetration of the acid. The high-temperature retarder also supposedly works as an oil-wetter. This type of retarder functions well in dry gas wells in the presence of organic acids. Perhaps the organic acids themselves change the wetability, but the actual mechanism for the high-temperature retarder must also include some shielding of the formation face in addition to oil-wetting. The basic retarder mechanism is to plate-out on the face of the rock, whether low temperature or the high temperature, and, either through oil-wetting or just physical shielding, protect the rock surface from acid reactivity.

When applied properly, both. of these chemicals are functional retardation mechanisms. If the low-temperature retarder is utilized without any hydrocarbon present, it cannot function. The high-temperature retarder gives some retardation, even without the physical presence of hydrocarbons.

Other means of retarding acid include using organic acids combined with hydrochloric, such as mixtures of acetic and hydrochloric. Combinations generally include 15% HcL with 10% acetic, 25o/o HcL with 10% acetic, and 7.5% HcL with 10o/o formic. Various combinations work synergistically, yielding a much longer reaction time. This type of system is used in deep high-temperature wells where long reaction times on the face of the rock are required.

Quality controlling these mixtures of acids is more difficult than just measuring the hydrogen ion content of a straight hydrochloric acid solution. Gather a sample of these and take them back to a laboratory to see that the proper concentrations are used. Once again, adequate mixing is a necessity. (See Appendix V, Acidizing Quality Control at the Wellsite.)

There are means to retard the reaction rate of hydrofluoric acid in matrix treatments. These involve in situ formation of the hydrofluoric acid by pumping chemicals (like fluoroboric acid and aluminum salts) that react to form hydrofluoric acid deep in the formation. You also can pump alternating phases of ammonium bifluoride in water and follow it with HcL. This is known as "sequential acidizing."

These techniques are functional in removing deep damage, but they are relatively expensive. Quality controlling such treatments is similar to standard hydrofluoric acid treatments with emphasis on assuring all chemical components are present and added properly.

Rules
General

1. Follow basic guidelines for any pumping treatment.
2. Recognize the danger of acid.
3. Install hose covers and have personnel wear eye protection.
4. Take special care when using formic, acetic, or hydrofluoric acids.
5. Take samples and evaluate acid concentration.
6. Inventory all fluids before and after the job.
7. Have the service company agitate the acid to disperse the inhibitor prior to pumping.
8. Use lined storage if possible.

Acid Treatments

1. Do not exceed frac pressure on matrix acid treatments.
2. Use proper amount of inhibitor, based on bottom hole temperature conditions.
3. Make sure all additives are compatible.
4. Follow design guidelines.
5. Control of fluid loss is of major importance in frac acidizing.

Types of Acids

1. Emulsified
2. Gelled or cross-linked
3. Retarded
4. Hydrofluoric
5. Organic

CHAPTER 12

ACID STIMULATION (2022)

I spent some time reading the 1994 version of acidizing and found it pretty complete with lots of valuable information. I will therefore keep this short and to the point.

We have recently seen multiple cases of operators unaware of the problem with solubility of acid inhibitors. Acid inhibitors work by plating out on the tubing or casing to be treated and are only dispersible in the acid. It is quite common for inhibitors to float out in transports and tanks on location. This means that the first acid pumped may not have any inhibitor to protect the casing or tubulars. At our company we insist the any acid be properly stirred before pumping or that the inhibitors be added continuously as the acid is pumped. Also, there is still present in our industry individuals who think acid should be left to react across the perforations. Unless you are treating a dolomite and your bottom hole temperature is less than 120 degrees Fahrenheit, the acid should be pumped past the perforations to protect the casing. Most of the slick water laterals treated today have bottom hole temperatures well above 150 degrees Fahrenheit. As I believe was discussed in the 1994 version, acid should only be used as the last resort ahead of a frac treatment.

In our present age of figuring out how to spend less and achieve more, I would strongly suggest that massive amounts of monies are wasted pumping acid ahead of all frac stages regardless of the pressures or lithology. It is not unusual to see 100,000 gallons of 15% acid pumped on a 50-stage lateral. The typical answer I get is it is a safety factor to be assured all perforations are open. First, if your perforations are not open, you are going to have to find a coil tubing operator or spend money on a tractor to carry more perforating guns into the well. If you have low injectivity 500 gallons will cover a 500-foot interval. In addition to pumping too much it has always impressed me with the diverse additives put in the acid. The costly additives such as inhibitor (needed) and non-emulsifier, (not needed), iron control (not needed) and many times other additives that are small volumes but add up in 50 stages. It is my recommendation that a lined transport of acid with no additive be on location if required. Inhibitor can be added on the fly and other additive are not required. My rationale for no additives is that the 500 gallons of acid and reaction products will be vastly diluted by the 10,000 + barrels of frac water pumped behind it.

I almost was attacked by a gentleman (stretching the definition of the word gentleman) who ran at me on a location asking who had removed the acid before a ball out in the Piceance basin. He told me in a vulgar manner that our breakdown pressures were sometimes 5 or 600 psi higher than with acid. He became more furious when I told him I did not care. I asked him if our treating pressures were high or that our ISIP had changed. He replied that they were the same but was still amazed at my not caring that the breakdown pressure was higher.

I put this sort of thing in the category of our typical 5-, 10- and 15-minute shut-in pressures taken after most frac treatments. I have asked the question hundreds of times what these values are used for and have never gotten an answer. These stupid measurements are recorded because our fathers, grandfathers and great grandfathers did so. There is no other reason.

I have drastically changed my mind on the versatility of buoyant ball sealers. Properly designed to fall out of the way after a ball out or designed to permanently plug perforations in poorly treated prior stimulations are a great tool in the industry. I have authored 3 patents, either pending or released, that offer great ways to re-frac older wells.

I still believe that there are thousands of carbonate wells that can economically be stimulated with acidizing.

Matrix and fracture acidizing are enormous tools which have been wildly underutilized in the past 20 years. There is a plethora of papers which illustrate the massive amount of research completed which has almost been forgotten in our quest for stimulation of nanodarcy reservoirs. Images are shown at the end of the chapter.

Image A

Western Company of North America acid transport.

Image B

Cudd – chemical totes on body-load for acid treatments.

Image C

Common acid transport.

Image D

Nowsco combination unit, acid transport and pumper.

Image E

Nowsco combination acid blender and pumping unit.

Image F

Continuous acid mixing unit with CAT C15 engine and max discharge rate of 84 bpm; 3 liquid chemical additive pumps and 660-gallon capacity on-board mixing tank.
Source: Jereh Pumping Equipment

Image G

Trailer-mounted acid fracturing pumper; 1,350 to 1,500 HHP Rolligon acid fracturing unit; can also be used as additional horsepower for traditional hydraulic fracturing application. .
Source: NOV

CHAPTER 13

QUALITY CONTROLLING CEMENT JOBS (1994)

One of the most surprising statements I heard upon arriving in the Middle East was the comment from a young Iranian engineer to the effect that a good cement job is better than sex. This engineer had his priorities out of order. What I have found from attending cement jobs around the world, offshore and onshore, is that a smoothly run, properly designed cement job is difficult to perform. Quality controlling cementing is not a major topic in this book. The equipment, chemicals, and actual running of the treatments would fill a book the size of this one.

We do, however, want to give a few guidelines and some general recommendations on quality controlling cement treatments. Cementing treatments are much like having babies-they usually occur between 1:00 and 4:00 a.m. Many industry people say that cement does not set at other hours. On very rare occasions, I have cemented in daylight.

More lab work

Much more laboratory work is done on the average cement job than for quality controlling stimulation treatments, whether fracturing or acidizing. The reason for this is obvious. Most people would be unhappy if they flashed set cement in the pipe while pumping and had to either abandon the well or drill out a tremendous amount of solid cement. Therefore, it is not unusual to have cement-thickening times run on cement utilized for everything other than surface pipe. This is an excellent idea, and this type of work should be done on stimulation treatments as well. Although you cannot cement up a well on a stimulation treatment, you can cause irrevocable damage by not doing the Intense Quality Control discussed in Chapter 10.

Quite often field samples of the cement blends are carried back to field laboratories. These are put on thickening-time testers to see if the cement (1) will set up too quickly or (2) will set up at all. This test also evaluates compressive strength development, free water, and viscosity.

Determine BHT and circulating BHT

Before sampling the cement, the real starting point for a successful cementing treatment is to determine accurately the well's bottom-hole temperature. Then take that temperature and calculate the bottom-hole circulating temperature. Accurate measurement of bottom-hole temperature and bottom-hole circulating temperature can be the difference in a highly successful cementing treatment and a complete failure. A matter of 20-30°F can make a tremendous difference in thickening time, compressive strength development, and/or the possibility of gas breakthrough of the cement, negating the cement barrier.

Choose cement slurry weight

Cement is run to isolate each of the various formations behind the pipe. If this isolation is not accomplished, then the cementing job is a failure. After the bottomhole temperature and bottom-hole circulating temperature have been ascertained, the next step is to determine the weight of cement slurry required. The cement must be equal to or above the mud weight to control the producing formations and thus prevent the flow of oil, gas, and water.

Calculate volume, evaluate annulus

Once the slurry weight has been determined, calculate the volume and evaluate the physical status of the annular space. By "physical status," I mean whether the hole is washed out, whether there is a large filter cake, and so on. If the hole is washed out, it requires a great deal more cement and possibly some special designing for spacers ahead of the cement to separate the mud and the cement.

Space fluids

Over the past few years, there have been great advancements in the development of spacer fluids to run ahead of the cement. These fluids are designed to clean up the hole, remove the mud filter cake, and displace mud out of the hole. Obviously, if a large portion of the mud is left in the hole, isolating the interval will be very difficult. Work closely with your service company in designing the spacer fluid. Examples of the spacer fluid include thin, watery fluids with surfactant mixed with sodium acid pyrophosphate, and various gelled weighted fluids. Some of these are pumped at laminar flow conditions, and others are pumped at turbulent flow. Consult the service company to find the proper spacer for your well's conditions.

Cement weight and additives

After choosing the spacer, then decide on the weight of the cement and on the various cement additives. The first consideration is a cement retarder or accelerator. Common retarders are ligno-sulfonates, but other retarding agents include high concentrations of salt and various sugar molecules. A common accelerator is calcium chloride. There are grades of cement designed so you can use a particular cement grade or grind and eliminate the need for retarders or accelerators.

American Petroleum Institute's grades of cement vary from A to J and are based on grind and on thickening time versus temperature and chemical additives. (See Table 14-1.) In the United States, class H cement covers most temperature ranges. Overseas, class G is a common, broad spectrum cement. Many people use construction cement or class A or B cement for surface pipe or other low temperature conditions and simply add an accelerator.

Now that the brand and type of cement have been selected, evaluate the effects of any additives, such as retarders or accelerators or fluid loss compounds, on free water or thickening time of the cement. Most cements are designed to be run with a specific amount of water. Do not exceed the ranges of the water ratio. If you run too little water, then the cement slurry will be extremely viscous, and incomplete reaction will occur. If the water ratio is too high, then the cement ends up very thin, and there will be free water and settling of the cement. Therefore, stay within a recommended weight range during cementing operations.

If you have to run a lighter cement, additives can lessen the cement's weight. These additives vary from simple bentonite for some weight reduction up to hollow glass beads and foam slurries, which allow quality cement without free water down to 9 lbs./gal. or lower.

Table 14-1 Applications of API Classes of Cement

API Classification	Mixing Water (gal/sack)	Slurry Weight (lbm/gal)	Well Depth (ft.)	Static Temperature (°F)
A (portland)	5.2	15.6	0 to 6,000	80 to 170
B (portland)	5.2	15.6	0 to 6,000	80 to 170
C (high early)	6.3	14.8	0 to 6,000	80 to 170
D (retarded)	4.3	16.4	5,000 to 12,000	170 to 260
E (retarded)	4.3	16.4	6,000 to 14,000	170 to 290
F (retarded)	4.3	16.2	10,000 to 16,000	230 to 320
G (basic)[8]	5.0	15.8	0 to 8,000	80 to 200
H (basic)[9]	4.3	16.4	0 to 8,000	80 to 200

Due to their nature and their ability to take up water, there are polymer additives that allow you to mix water-free cement over a broad range. Other additives are added to the cement to control leak-off or fluid loss. The basic theory here is quite similar to that used in fracturing. If you are leaking off or filtering off water, then the cement becomes much denser, more viscous, and quite possibly the thickening time will decrease radically. Many cementing jobs have been terminated prematurely because the cement exhibited poor fluid loss control. The cement had dehydrated, resulting in premature shutdown of the treatment. Many of the fluid loss compounds added to cement act as retarders themselves. This is because of their chemical natures and of their similarity to the ligno-sulphonate sugars used to retard cement. Most polymer additives increase the viscosity of the cement. Some of them give you a wider working range before free water appears.

Oil company and service company representatives usually confer and select a particular blend. This blend may be selected based on previous experience, or it may be something new. Once the blend has been selected, take samples, mix them, and conduct thickening-time tests on the cement and additives. After the lab samples have been evaluated and approved, sample and evaluate field blending.

One of my pet peeves in the industry is that people continually seem to use one particular cementing blend, regardless of failures that have occurred in the area. Over the years, oil field service companies have developed very good lightweight cement that will negate fracturing of the formation or loss circulation problems that can occur. Because of some short-sighted individuals believing in cost-cutting procedures

---

8 Can be accelerated or retarded for most well conditions.

9 SPE Monograph and Cementing by Dwight K. Smith

in cementing of a well, they would rather use a heavyweight cement than to use a new generation lightweight cement to solve this problem. The difference in the costs may be less than $1,000. What I am recommending is that you should not necessarily just accept the recommendation of the local salesman on a cement in a particular area. You should investigate such lightweight cements as those created by the addition of lightweight spheres and additionally, combined with these, new generation thixotropic additives. There is available from service companies today good, high-compressive cements with weights as low as 9 lbs./gal. There is probably nothing more critical to the completion of a well, or in fact, stimulation of a well, than having good zonal isolation. To pinch pennies on primary cementing can cost huge sums of money in later stages of the well.

Similarly, I have been involved in situations where the opposite situation has occurred in relationship to using lightweight cements where they are not required. Most oilfield personnel feel quite strongly that you should utilize as high a compressive strength cement as possible opposite the producing zone. This allows you to approach the rock properties of the formation with the cement itself. If you are going to conduct hydraulic fracturing operations, acidizing or other potentially destructive practices, then it's best, if at all possible, to use very heavy, and conversely high-compressive, cement opposite the producing zone.

For a good customer, I recently discovered that they were utilizing a lightweight cement across a long (approximately 2,000 ft) interval where multiple pay zones existed. It seems that a service company representative had sold them a relatively lightweight, moderate compressive-strength cement where they should have been using heavyweight, high-compressive strength material. A good rule-of-thumb is-to utilize across the producing interval as good a quality cement as possible. From the standpoint of the previous paragraphs, you can then use new technology, lightweight cements above the zone and achieve good zonal isolation. Where you are going to conduct acidizing or fracturing operations and the possibility of channeling, fracturing, or etching is going to occur, then you should (if at all possible) utilize cements that will approach the strength of the formations that they are being used to isolate.

Not unlike unbroken gels in the formations or emulsions that are caused by poor quality control, a poor primary cement job is also almost impossible to cure. Squeeze cementing is a black art, and the ability to place cement and to isolate zones once some amount of cement is behind pipe is a task that requires heavenly intervention. The best thing that can happen in relationship to poor cementing is that there is no cement behind pipe; you can then utilize techniques such as suicide squeezes or the identical process renamed "wrap around squeezes" where one can circulate cement behind pipe with a retainer or a packer between perfs below and above the packer. Some people do not recognize that in perforating an interval and pumping cement that you are simply fraccing the well and have little chance whatsoever of achieving any zonal isolation. Insufficient zonal isolation can lead to many problems in stimulation, including multiple ineschelon fractures and severe tortuous paths that can lead to very early screen-out in a fracturing treatment.

Take samples and test them

Once that the field blend is in place in bulk storage, obtain representative samples for laboratory testing prior to pumping the cement. Not all wells have laboratory testing on the blends prior to pumping downhole. If you consistently treat the same depth range with the same additives, then lab testing on every well may not always be necessary. However, I recommend saving samples and spot checking to ensure that quality control on both the cementing supplier and the additive suppliers have been maintained.

All of the quality control work we are talking about often takes place in less than 24 hours. The items that we have discussed take place during the time that the operating company decides where to set the pipe, conditions the hole, pulls out of the hole, and starts to run casing. A great deal needs to happen in a very short time. Frequently, the oil company man wants 24 hours compressive strengths on a blend and allows the service company 8 hours to run them. Service companies are competent, but they cannot accelerate the passage of time.

Run a conditioning trip

Cement jobs often fail because mud could not be removed from the hole behind the pipe. Many times this is attributed to the drilling mud's poor condition. A conditioning trip before running casing is always advisable. The trip is conducted to be sure that the mud has been moved, agitated, and thinned down so it can be displaced from the hole by the spacer and/or cement.

Pipe movement helps

During the actual cementing treatment, some pipe movement (particularly up and down motion) is beneficial for effective mud removal. Reciprocation of the pipe helps remove filter cake and tends to thin mud in static areas, especially in a crooked hole. This movement results in a better cement job. This just is not feasible in many cases, such as tight hole situations or circumstances where pressure phenomena hinder pipe movement.

Loss of circulation

Make sure that you do not lose circulation prior to cementing. It is a common occurrence to go out to a well and mix a cement of considerably heavier weight than that of the mud in the hole if the oil company has been losing mud. If the mud losses occurred because the mud fractured the formation, then the cement will be an even better frac fluid, following at a faster pace. It is quite easy to fracture with cementing fluid.

Any loss of circulation needs to be controlled. If the lost circulation is due to fracturing, then redesign the treatment so the hydrostatic head of cement is less than the fracturing pressure. Solutions to loss circulation problems involve the use of lightweight additives, lightening the mud column ahead of the cement or to foam the cement.

Although I spent five years of my life working for a nitrogen oil well service company, I was never a fan and probably will never be a fan of foam cement. When mixed properly, foam cements can form a vital function of isolating zones where lightweight cements are required. However, if mixed improperly, with too little nitrogen, you will break down the zone. If you add too much nitrogen, you actually end up with a permeable cement that allows gas to be produced to the surface negating any zonal isolation. As I have stated earlier, service companies have come up with excellent lightweight additives for preparation of good, compressive, straight cements down to 9 ppg. In every way possible, I always attempt to accomplish zonal isolation without the use of foam cement.

I realize that service companies have come a long way (e.g., continuous-mix treatments) in developing better metering techniques for foam cement. However, the tremendous problems that exist on the surface, as well as metering the additives during the job, make this a procedure that I try to stay away from as much as possible.

Use centralizers

Problems with crooked holes and the lack of centralization of pipe can cause cementing failure. It is almost impossible to get a good cement job if the pipe is against the side of the hole. This topic is a controversial one. There is little chance of zone isolation if you don't get cement between the pipe and the wall. I always recommend that centralizers be used. The centralizers need to be strong enough to withstand the forces of placing the pipe in the ground and to withstand any movement, such as reciprocation of the pipe, during cementing operations.

Implementation of a cement job

There is nearly an unlimited number of types of cementing mixing equipment available. Many companies use cement hoppers feeding into jet mixers that go into a tub. The cement is sucked out of the tub and pumped downhole by triplex pumps. Other companies have dry feeding systems that use either pressure or mechanical means to feed the cement into turbulent mixers that are named Tornado Mixers or Recirculating Mixers, etc. These mixers take this slurry, which may or may not be pressurized, to a triplex pump to be fed downhole.

Since my previous book, service companies have developed, with the use of microprocessor control, better cement additive control, particularly on units that have been used off-shore. Although there is a great number of jet mixer operations still going on in the field, most of the service companies have gone to recirculation systems where better mixing of the cement is achieved. As in continuous-mix fracturing operations, I feel a great deal of pressure is placed on the oil company representative to ascertain that the additives are present, that pilot-testing be done before the job, that pre- and post-job inventories are taken on cement jobs. With the new microprocessor control systems available, it is absolutely essential that postjob inventories be done to be sure that all additives have been pumped in the well.

Density is of primary importance

Whatever the mixing system, the most important property of the cement is its density going downhole. If too little water is added, the cement slurry will be viscous. Then there are the possibilities of high pumping pressures and of flash setting of the cement. On the other hand, if the water-to-cement ratio is too high, a very thin slurry with settling out of the cement and a great deal of free water can result. Nothing on the treatment is more critical than the cement density. The density must be within a few tenths of a pound per gallon of that designed in the laboratory.

Most service companies supply densimeters, which constantly check the weight of the cement. In addition to the densimeter, all operating companies should have pressurized mud cups to monitor the cement. This does not mean the service company would try to cheat you, but it is a good practice to double check what is going on. Too light a cement, along with the problems of free water and poor compressive strength, can cause the well to be out of control if you are trying to cement in a balanced state. Too heavy a cement can cause highviscosity problems, poor quality cementing, and could cause the formation to be fractured.

Pumping rate

Another controversial (but realistic) mode of quality control or quality assurance is that the service company maintain the designed cement-pumping rate. Many studies show that the rate of pumping a cement, in turbulent or in laminar flow, can affect mud removal. Be certain that the service company conducts the cement job at the designated rates. Question the service company to see that it has flowmeters and adequate monitoring devices so you can tell if the flow rates are within the designed limits.

Check quantities

As in a fracturing treatment, check all bulk storage equipment and chemicals before the job begins. Check after the treatment to be sure that all the additives that were to be pumped were indeed used. Many jobs have been conducted where less than one-half of the cement or other additives were pumped. In most cases, this was due to human or mechanical error by the service company, but it certainly makes sense to see that you get what you pay for and that the job is conducted as designed.

Cementing problems vs. stimulation problems

Quality control prior to cementing is far ahead of that art on stimulation jobs. This is because you do not cement up the pipe if you do not quality control an acid job or a frac job. Many failures on cementing operations can be near catastrophic. Although a cemented up long string due to a catastrophic event is indeed disastrous, it is hard for me to believe that this is worse than a totally gel-plugged formation that will not produce oil or gas.

Recently, we have been involved with several of our customers in evaluating stimulation failures. We have discovered that some of the problems relate to lack of any quality control or supervision of cementing treatments. Although a great deal of care was taken in measuring thickening time, compressive strength, and so forth, of cements prior to the treatment, little or no supervision was done on the job site. Although many cement jobs occur in the wee hours of the morning, it is certainly critical that the execution of the cement job be observed and monitored every bit as closely as stimulation treatments. As previously mentioned, this primarily involves assuring that the cement is the correct weight, all additives are added correctly, and the pump rates follow the predesigned schedule.

Use both plugs

Even an excellent cement job with the correct additives mixed at the proper rate, with a good spacer, and with sufficient mud removal from the annular space can fail. This failure can be for reasons other than loss circulation. Cement jobs frequently fail simply because both top and bottom plugs were not used.

As most of you are aware, a cementing head is put on top of the casing. In this head is a wiper plug to be placed ahead of the cement. This plug ruptures when it reaches the bottom of the pipe, allowing the cement to be pumped through it. At the job's completion, a solid plug is set between the cement and the displacing fluid. Often failures occur because a top plug was not run or because the service company representative was not brave enough to bump the bottom plug. Failures to run the top plug have resulted in poor cement jobs on the lower part of the casing. This wiper plug displaces mud ahead of the cement as well as removes mud deposited on the inside of the casing. A thin layer of mud inside the casing can fill enough volume to displace from one to three joints of cement on the backside of the annulus ahead of the bottom plug.

In many parts of the world, the doorstop in the company man's office is the top plug! Most people do not like to run it because they are afraid it might plug off and cause the cement job to fail. Another common practice is to cut a hole in the bladder of the top plug. This is detrimental to displacing the mud ahead of the cement.

In one area of the Mideast, I found almost all of the shoe joints had no cement behind them. After some field evaluation, I discovered top plugs were not being run. The mud used left at least a one-quarter inch layer of cake on the inside of the casing. When the cake is removed by the bottom plug, a large volume of mud is pumped ahead of the bottom plug. This mud is then circulated around behind the casing, leaving no cement at all on the shoe joint and possibly none on one or two joints above it. When the oil company started running top plugs (or wiper plugs, as they are called), this mud was removed before the cement was pumped and, of course, before the bottom plug was pumped.

The oil company representative should be present to make sure both plugs are run on a cementing treatment. There are more complicated plug situations, for instance, when you are dealing with sub-sea applications. Utilize radioactive tracers or, at the minimum, a physical signaling device to make certain the plugs have fallen. Another common occurrence is insufficient bravery to see that the bottom plug is bumped. Most people have great trepidation of over displacing cement and often do not totally displace the bottom plug. This is detrimental to a cementing job because the cement in the last few joints must be drilled out. The casing could be damaged during these drill-out operations, particularly if a lot of cement is left in the casing. Proper quality control and attention to the service company conducted cement job can eliminate a lot of the fear of complete displacement on the cement job.

Cement quality control guidelines

1. Determine accurate bottom-hole temperature and bottom-hole circulating temperature.
2. Determine the required slurry weight.
3. Ascertain the volume of the annular space and the condition of the open hole.
4. Design spacer fluids, if required, based on the status of the open hole.
5. Design the cement slurry, optimizing thickening time, compressive strength, and free-water content.
6. Conduct lab testing of the blend, using lab samples at bottom-hole temperature and pressure (BHTP) conditions.
7. Obtain samples of the field-blend cement and evaluate them at BHTP conditions.
8. Utilize reciprocal pipe movement during the cementing operations, if possible.
9. Catch multiple samples during the treatment for later troubleshooting.
10. Have the service company constantly evaluate the cement with a densimeter and a pressurized mud balance.
11. Always use both top and bottom plugs.

CHAPTER 13

QUALITY CONTROL OF CEMENT JOBS (2022)

The guidelines given in the 1994 chapter 14 are well done and it would be valuable to read and follow these guidelines because there is nothing more important than zonal isolation to not only protect the formation and the frac as well as the environment.

It has never been a problem getting the service companies to do testing of the cement prior to pumping the job. The consequences of a failed cement job are very straight forward and can be catastrophic. Perhaps the greatest challenge our industry has is achieving good zonal isolation in 2-mile-long laterals with the hole never remaining straight and porpoising always occurring. It is my opinion that many of the sophisticated analysis done involving fiber optics and other tools are flawed because of poor cement jobs.

The use of solid centralizers is a good option but does create the possibility of stuck pipe and not achieving placement of the casing to the final measured depth of the well. We had customers who, in light of the problems with good cement isolation, tried to utilize inflatable packers to isolate the wellbore. The inflatable packers at that time exacerbated the problem, because of their sizable I.D., the customer on two occasions was unable to get the lateral to bottom after getting stuck in the curve close to TVD.

I would like to discuss some of my pet peeves about cementing relating to failure to achieve zonal isolation. Over the years there has been hundreds of millions, probably billions, of dollars wasted on so called "squeeze jobs".

Most prudent operators will check out their cement job with a bond log. If there is indications of a poor bond and lack of zonal isolation, the standard move has been to conduct a cement squeeze. Countless times I have reviewed bond logs and if there is indications of any cement at all in the area of the planned squeeze, I recommend against perforating and squeezing. With all my dissertations on the need for zonal isolation why would I recommend not squeezing?

If you are pumping into an area where you cannot circulate cement above the perforations, you are simply doing a cement frac. I have been present where customers have tried to get a squeeze as many as a dozen times only to finally give up with no zone isolation and holes in the casing.

Someone reading this will tell everyone, I did a squeeze and it worked. They should put their money in the lottery. In my early days of working in the field, we had major customers who would squeeze the productive interval even when the bond log either was not available or they did not run a CBL. I do not believe it takes a genius to understand that you are not pumping into a vacuum behind uncemented pipe and that all that will be achieved is creating a cemented fracture.

I would also state that probably the worst thing to do in a lost circulation situation would be to pump cement. Brave individuals from the past have pumped a cement additive that causes the cement to set up between a few seconds to a few minutes. Some of those jobs (Calseal was the generic name) were successful and others had the operators chiseling cement out of lines and pumps. By its very nature cement is heavy and has little diverting capacity. Loss circulation, while drilling, can be a very frightening problem with well control potentially lost and, in my opinion, the products that are readily available on drilling rigs are many times not capable of handling severe losses.

I will digress to one of my more memorable loss circulation stories. As Halliburton was transferring me to oversee technical problems in international, I was sent to North Africa to tour Halliburton's facilities in Egypt and Libya. I was carried all over the Libyan desert in a Pilatus Porter airplane, which is basically an engine with massive wings. This was needed as there were not many runways in the desert. As sometimes occurs, one of the rigs was in deep trouble as they had complete losses and had no more LCM on the rig. The tool pusher was wringing his hands and said he basically was going to close the shear rams on the drill pipe to negate a blowout. I interrupted the conversation and told the tool pusher that he had an infinite amount of fluid loss material available. He almost attacked me but finally asked what I was talking about. I told him to use his front-end loader and dump desert sand in his mixing tank, carefully so as not to stick pipe and run slugs of desert sand to control losses. I did not get the Nobel prize, not even a thank you, but the rig was drilling ahead 2 days later. Sometimes common sense can come in handy.

While I am in the mood for stories, I will tell about my favorite cement job. In Iran, in the early 70's, the Iranian Oil Company supplied the cement and did their own thickening time testing. We had a 9,000-foot-long string of 9 5/8" casing and did these types of jobs almost on a daily basis. In fact, we had preblended the additives and the majority of cement was on location when we found out, for whatever reason, that the pumping time was very short. We figured out the answer and were going to ship the cement back to the station to add additional retarder. I got a call from the Iranian lab telling me that we did not need to re-blend, he would just add extra water to the cement to extend pumping time. I objected very strongly telling the lab manager that we would have a very poor cement and a lot of excess free water. Free water is a big no-no in cementing, particularly in long string pipe, where isolation is critical. It must have been a weekend and they did not want to do more testing, so I, like a good service company engineer, did what I was told and went to location to do the job.

The company man on location had heard about the short pumping time and met me as I got out of my Iranian made Rambler. He immediately did my most unfavorite thing, he put his finger in the middle of my chest and screamed that he wanted us to pump at the highest rate possible or he would see that I was castrated before leaving location.

Like politicians of today I did not want to waste a crisis, so I gathered all of the Halliburton employees and told them to hook up all of the available chicksan bales and stack them on the floor and tie them into the cement head on the 9 5/8" casing. A bale is a full chicksan which will stretch out to more than 12 feet. It allows for pipe movement to occur as pressure from below moves the casing upward. I also told the cementers we would be mixing using our low-pressure mixer and go down hole with 2 pumps on each of the 3 trucks on location. What this translates to is that we were able to pump the crappy cement downhole at 42 barrels per minute. For my frac people reading this, there are virtually no cement jobs done anywhere close to this rate as we would be pumping the pipe out of the hole.

I was at my usual position by the strip chart densitometer when I gave the signal to bring on all the pumps. As the rate accelerated the 9 5/8 casing started upward toward the top of the crown of the drilling rig. I will enjoy until I die, and maybe afterward, the site of the obese company man running down the stairs to the walkway waving his hands and pleading for us to slow down. We did slow down, and we got cement returns, before dropping the wiper plug. The holding pits overflowed, and we had crappy cement all over the location. My next greatest moment was when I took the ticket to the company man and he kept his head down and signed it without saying a word.

Advancements have been made in cementing research allowing for a much higher use of residual fly ash from coal plants and mining, the area of research that is most needed in our industry today is to come up with unique ideas how to better isolate our laterals to achieve stimulation where it is intended. For photographs I took of various jobs in the past as well as of some older and modern cementing equipment from various sources. Jet mixing equipment was used through the 80's, eventually replaced with recirculating cement mixers by all the service companies.

Image A

Actual jet mixer in operation.

Image B

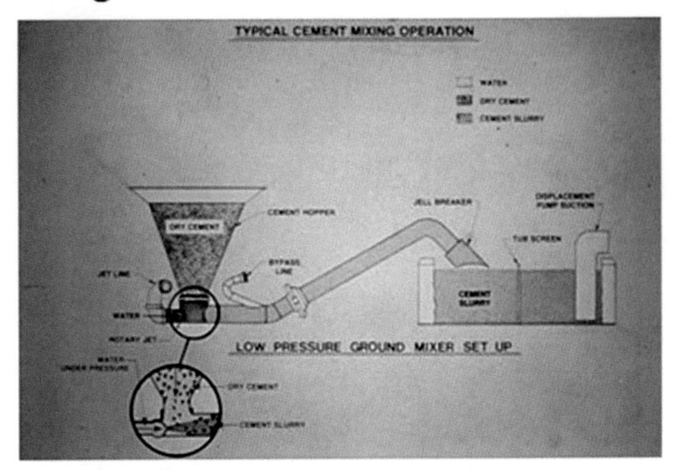

Jet mixer schematic for typical cement mixing.

Image C

Haliburton early cement body-load pump truck.

Image D

Haliburton cement bulk transport.

Image E

Haliburton CPT-ZS4 Cementing Unit; body-load, single pump cementing unit with HT-400 pump, 150-ft of 2-inch 5,000 psi cementing hose with reel system, 2 compartment mixing tank with agitators. Engine is driven by an Allison HT-755 automatic transmission with 350-hhp. Max rate range between 10.8 and 13.4 bpm and 9,000 to 11,200 psi max pressure.
Source: Haliburton

Image F

Haliburton Elite Cementing Unit; automatic density-controlled mixing system with twin HT-400 pumps for over 1,025 trailer horsepower. Built-in dual deck provides area for safe observation, zero discharge manifold and displacement tanks, and Tier II emission certified engines. Cement density optimization software details pre-mix and downhole density for monitoring.
Source: Haliburton

Image G

Schlumberger CBS-965 Electric batch mixer skid; two 50-bbl blending tanks with two 150-hp AC motors.
Source: Schlumberger

CHAPTER 14

MONITORING AND READOUT HARDWARE (1994)

The oil service industry has come a long way in developing electronic equipment to monitor rates and pressures for fluids pumped down oil and gas wells. In the early stimulation treatments, pressures alone were monitored and were only occasionally put on a chart to give a permanent record. Many times the pumping rate, at best, was measured by pump strokes on the high-pressure pumps or was gathered by simply strapping – physically measuring tank volume with a tape – the tanks as the fluid was pumped. As the sizes of acidizing, fracturing, and cementing treatments increased, it became apparent that it would be beneficial not only to monitor pressure and pump rate, but also to strip-chart these pressures and rates for evaluation after the treatment. In addition to pressure and rate, many other factors needed to be recorded. The most important of these factors are density of the fluid, particularly in cementing or proppant fracturing treatments, monitoring the backside pressure, and strip-charting backside pressure during a treatment. Additionally, we should monitor critical additives while the treatment progresses. Examples of these additives are cross-linkers and breakers in complex fracturing fluids. Always record and monitor the energizing gases that are added to the treatment fluid.

Pressure and flow monitors

Figure 15-1 is a circular Martin Decker chart with which you monitor pressure versus time. This type of device is still available on many pumping units. This is a reliable system that requires no electronics. It is simply a mechanical timer that is wound manually and that records with ink on a circular chart. Most commonly, pressure is monitored and strip-charted utilizing linear strip charts and electrical transducers.

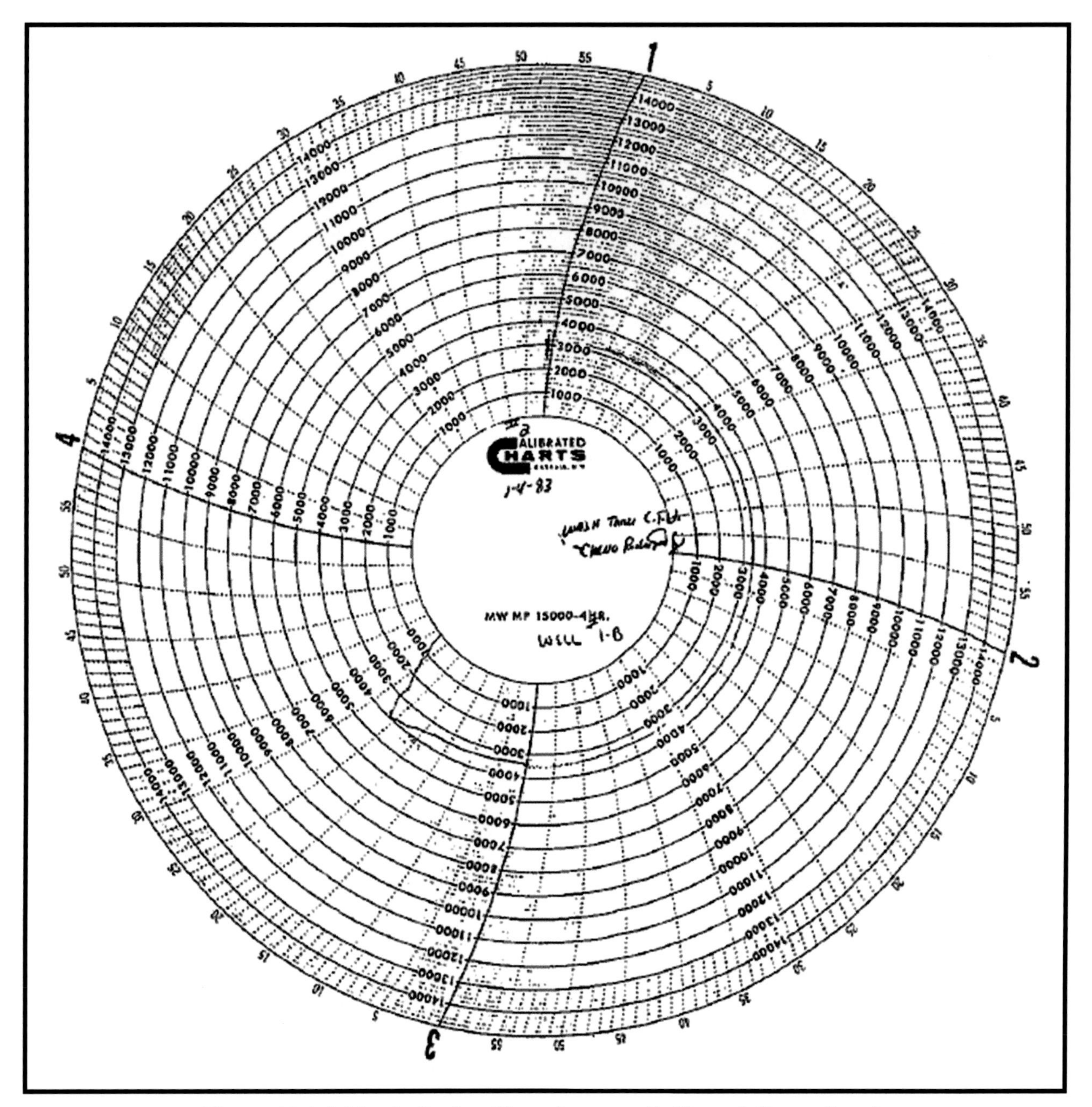

Figure 15-1 A Martin Decker Chart *(courtesy Calibrated Charts Corp.).*

On most jobs, a visual digital monitor is available that yields the same information being plotted on a chart. The simplest devices for monitoring flow rate are systems that count the strokes of the plungers of the high-pressure pump. Dowell typically utilizes this technique on most of their treatments. They have the capability, however, to place in line magnetic flow meters for what I believe is a more accurate determination of flow rate. By assuming a given efficiency and knowing the plunger size, the pump rate can be accurately monitored. The distinct disadvantage of this method is that readout data do not tell you if fluids aren't present. In other words, the device may indicate a rate, even though no fluid is moving through the pump.

Often service companies either supplement these devices or use only flowmeters. These flowmeters measure the fluid flow as it passes through high- or low-pressure treating iron. There are many types of flowmeters available, and formerly the most common was the turbine-type flowmeter (Figure 15-2).

Although quite reliable, this device suffers some deficiencies, especially when you utilize variable viscosity fluids and sand-laden fluids.

It is my observation that the industry is gradually moving toward full-opening mass flow meters. These devices, which early on suffered some electronic problems, now are the flow meter of choice for all water-based fracturing fluids. At the present time, you must utilize turbine meters or other devices for hydrocarbon fluids; but even as this book is being written, full-opening flow meter devices are being tested and utilized for hydrocarbons. One that is available and applicable for low rate and is highly accurate is a correalis meter, which uses a force-mass relationship to allow you to measure rate and density simultaneously.

Density measuring devices

All service companies have radioactive densimeters for measuring the density of cementing fluids and for measuring the proppant concentration of fracturing fluids. If your service company is not doing this, change companies.

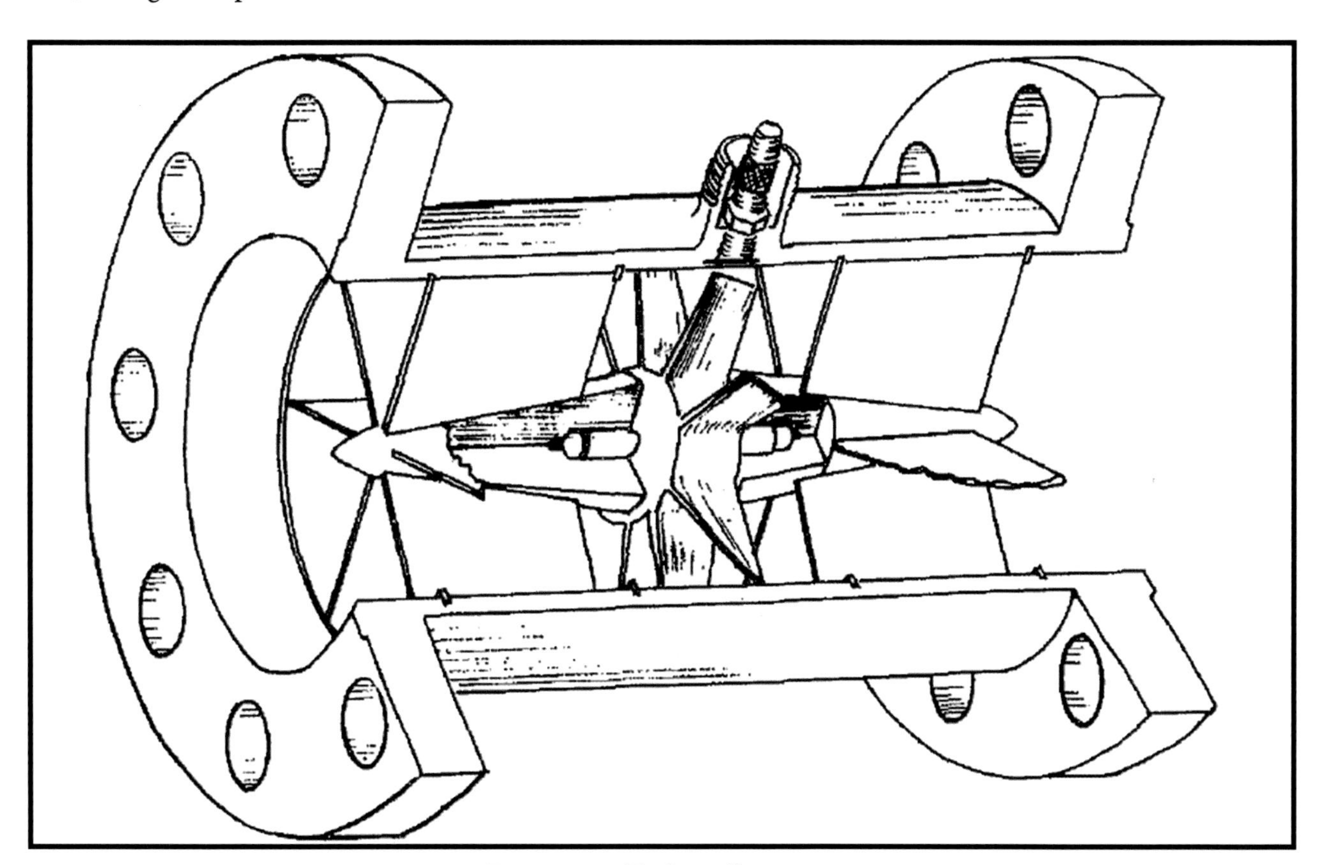

Figure 15-2 Turbine flowmeter.

Knowing the density of the fluids going downhole is a very critical part of the treatment. Density is as important as pump rate and pressure. When these devices malfunction, be prepared to measure manually, using pressurized mud cups, the density of the cementing fluids. For fracturing treatments, most service companies as a backup have the sand screws calibrated so they have a good idea of how much sand they are pumping. With wear, the efficiency of the screws is lessened.

The accuracy or revolutions per minute (rpm) of a screw versus the amount of sand lifted with time is questionable and should be used as a last ditch alternative.

A company such as Dowell that does not use sand screws does have calibrated charts for the slides used which allow a given amount of sand to be dumped into the pod blender. It should be noted that most service companies utilize more than one densimeter. Most of them have densimeters on their blenders as well as high-pressure, in-line densimeters. I insist on in-line, high-pressure densimeters for calling flush volumes on treatments and also as a secondary backup to primary densimeters used by most service companies on their frac blenders.

If I lose all densimeters on a treatment, and I have secondary personnel present on the treatment, I will move to the frac blender and visually monitor sand as it is going into the blender. By monitoring the volumes coming out of the sand storage unit one can approximate the amount of proppant pumped. If it is early in the treatment and you lose your densimeters, you should abort the treatment.

Frac vans, a bane or a blessing?

In the early days of acidizing and fracturing, and today in cementing, a pressure recording device and sometimes a visual readout from a flowmeter were the only measurements made on the job site. Many times these devices were left on the tailgate of a pickup, and the oil company man and service company supervisor monitored them during the treatment. When the first frac vans were instituted, experienced stimulation management personnel thought these were an indication that the industry was getting soft. The vans were unnecessary equipment that would isolate the customer from the treatment. In reality, the frac vans have been a god. send, allowing calm, controlled decision-making away from the noise and sometimes, rain, snow, sleet, and other diversions that occur close to the high-pressure pumping equipment. The frac van does somewhat isolate the oil company man from the treatment. This is why more than one oil company person should supervise treatments of any size. It is easy to seclude yourself in the monitoring van, assume the digital readouts are accurate, and never really know what is going on in the treatment.

One very common item that I have to request the service companies to do on many treatments is to move the treatment van so they can have clear, open observation of the tree, the frac blender, and treating iron. It is not uncommon for the frac van to be parked in an area where there is not access to ready observation of the major happenings on the treatment.

One of the very first of the complex computer vans built by the service companies was a device that I accused the service company of malice aforethought. The van was built in such a manner as to be primarily a dungeon. The only means of observation out of this van was in the very front. There were no windows on either side. The company person sat in the back and observed the TV screen, which, by the way, on the early vans was delayed from real time by a matter of about 5 seconds. In my first experience in this van, I very quickly learned to move to the front and stand by the actual treater. Since that time, I have never worked in the back end of this particular type of unit.

Pay attention

Electronic monitoring equipment has distinct deficiencies. First of all, like any mechanical or electrical equipment, it can be wrong. Service companies constantly are calibrating and checking this equipment.

Second, there is always the possibility that the readouts do not necessarily reflect what is actually occurring. This may be due to the service company's dampening – adjusting equipment to read occasionally – the readouts or, in rare cases, displaying data that are not representative of what, is taking place. An example of this sort of thing is the slip factor coefficient on most flowmeters. As I mentioned, turbine flowmeters are not linear when monitoring fluids of differing viscosities. During a fracturing treatment, you might pump thin fluids for a prepad and switch to cross-linked gel for the remainder of the treatment.

Observe a treater on a frac treatment. He frequently changes the slip factor of his flowmeter to correlate with the person strapping the tanks, He is recalibrating the flowmeter during the treatment so the fluids rate shown matches actual use. This is normal and is necessary because of the shortcomings of turbine flowmeters. No one is trying to cheat you by changing slip factors during the treatment. However, these slip factors can be altered to indicate much higher pumping rates than the actual rates. This amounts to changing the efficiency of the meter (i.e., you could indicate a 12 bbl./min. rate while you are really pumping ten) by plugging a 120% slip factor into the meter. The ramifications of doing this are obvious. Unfortunately, some service companies can plug in dummy density values, indicating false sand concentrations. Minimal pre-job and post-job checking eliminates such travesties. If all the proppant necessary for a treatment is on the site before the job and all of it is used during the job, fudging is eliminated.

Dampening the pressure and density readouts can cause serious treatment problems. On one fracturing treatment, a pressure readout was dampened so it would only pick up a reading every 15 seconds, Disaster can occur in 15 seconds, Iron can be destroyed and people can be killed in milliseconds. Ask your service company personnel how often they are picking up signals to display on digitals or their strip charts. You can make a much smoother chart if you take fewer pressure signals. The same goes for a readout and strip-charting of rate. The service company should take pressure readings at least once a second. This creates charts that are not as pretty and clean as many people would like, but these will be much closer to reality and will increase safety.

If you have a strip chart full of straight lines from a service company, there is something wrong with the monitoring equipment. It is very difficult with either triplex pumping equipment or intensifier equipment to obtain a perfectly straight line, regardless of which service company you use. Rates and pressures vary during a treatment. I saw a strip chart that was, to say the least, unbelievable. The treatment started at approximately 3,000 lbs. and finished at 3,000 lbs. What made this treatment so implausible is that it was a foam frac pumping nitrogen as well as fluid and adding 1-4 lbs. sand/gal fluid. The increased hydrostatic head from the sand had to change the pressure! Additionally, this treatment down tubing should certainly have shown pressure fluctuations because of fluctuations in the gas pump rate. A too beautiful chart is an indication that the service company is dampening its monitoring equipment too much or that the data may not be credible.

Types of treatment vans

There are three basic types of monitoring vans available in the industry: minivans, large frac vans, and computer frac vans.

Minivans

Most service companies supply a minivan (see Figure 15-3). In this small van, virtually all pressures, rates, and densities can be recorded. These vans are used on small treatments, acidizing treatments, and sometimes on cementing treatments. The use of monitoring vans on cementing treatments is limited because of a lack of customer requests. Additionally, most cementing treatments are quick, and the amount of equipment on the locations is often minimal compared with large fracturing or acidizing treatments. Usually, these small vans have strip-chart recorders for densities, pressure, and rate (typically two pressures and two rates). These vans are equipped to give you a readout of how much fluid has been pumped in each stage and the total fluid pumped.

Large vans

The larger frac vans (see Figure 15-4) basically are equipped like the minivans but have a lot more room. These may have more sophisticated numbers, pressures, and rates. The ability to program in stages and to break down data and information may be superior, but basically the bread-truck-type frac vans and the minivans accomplish the same thing. The larger vans are more comfortable, so they are used on most moderate-to-large fracturing treatments.

Standby equipment

Most oil companies and service companies require or supply standby vans, in case monitoring equipment or the van goes down during a treatment. Most of these vans are powered by diesel or gasoline generators. The majority of these van failures have been due to the generators. In addition to having a generator in the van, many service companies have battery pack standby available to be sure they have power during the treatment.

Figure 15-3 Interior and exterior of small frac van. These types of vans are typically used for acid or cementing treatments.

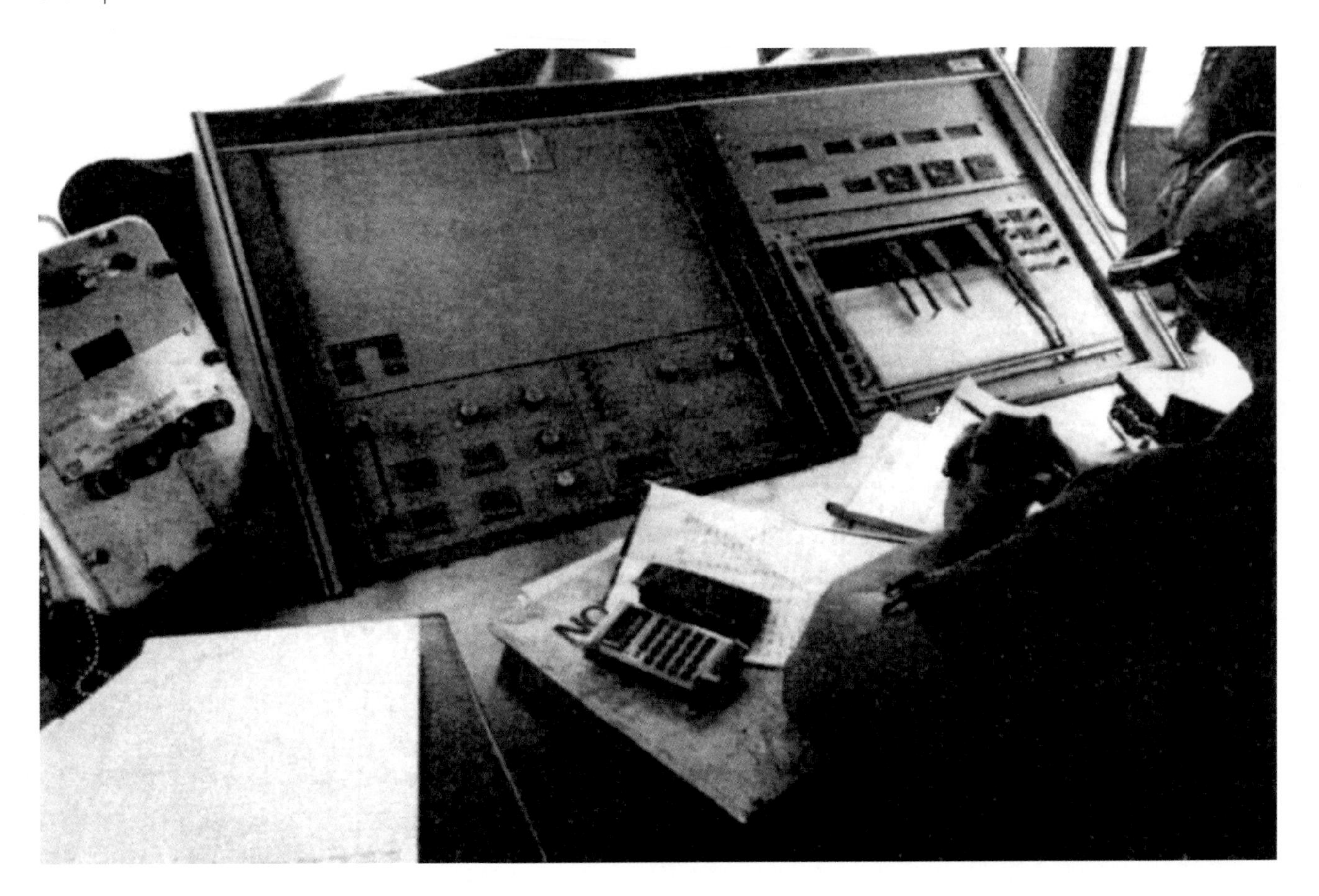

Figure 15-4 Large frac van, inside (top) and outside (bottom)

For very large treatments involving substantial capital risk, request that the service company have an electronics technician available. This person can work out the kinks, assist in calibration, and troubleshoot any problems during the treatment. Few oil service company people are able to repair electronic equipment.

Always have backup equipment for monitoring pressure, flow rate, and density. Pressure backup to compensate for electronic readout is something akin to a Martin Decker gauge or a strip-chart device. This precaution keeps you from aborting the job if your electronic equipment fails. The best backup for turbine meter readouts is having someone on the tanks, strapping them. Most service companies have standby blenders and standby equipment to use while they repair their flow metering equipment.

Calculating pump rates

For density measurement, I mentioned utilizing the service companies' ability to meter sand through their sand screws. This means it is feasible to continue a treatment if a frac van fails. Treatments are often terminated when the customer cannot see the pump rate displayed on a digital readout. Most service company treaters and operators can look at their trucks and tell you what the pump rate is at a specific pressure. They calculate hydraulic horsepower and know what gear the pumps are in. The customer ought to be aware of the rate capabilities of various pumps for such an occurrence. On an overseas high-rate operation, both the primary and standby flowmeters failed. Because we know how many pumps were running and what gear they were in, we continued the treatment, confident of the pumping rate. This is not the best situation, but it allows you to continue the treatment safely and thus save a lot of money. If you lose pressure readout and have no standby equipment, terminate the treatment. If you have both the electronic transducer and a typical Martin Decker hydraulic readout, losing both of these during a single treatment would be highly unusual.

Computer vans

A third type of monitoring van in the industry is the computer frac van. (See Figures 15-5 to 15-15.) Most service companies have spent vast sums installing very expensive computers to print out for the customer such things as Nolte plots (pressure vs. log of time) and to give pre- and post-fracture buildup data. Real time [10]Nolte data is a valuable tool that can be used in a stimulation treatment. But for a Nolte plot to be useful, you must have accurate bottom-hole treating pressure data. If you are not utilizing a monitor string, [11]then measurements of bottom-hole treating pressure are certainly questionable.

---

10 Take the bottom-hole treating pressure while the job is under way.

11 Method for taking real time data. By pumping down the annular space, you can monitor the real time bottom-hole treating pressure by subtracting the hydrostatic weight of the fluid in the tubular goods, yielding accurate bottom-hole treating pressure data.

Figure 15-5 Interior view of The Western Company computer van.

Figure 15-6 Interior view of The Western Company quality control van.

Figure 15-7 Exterior view of The Western Company's newest computer van.

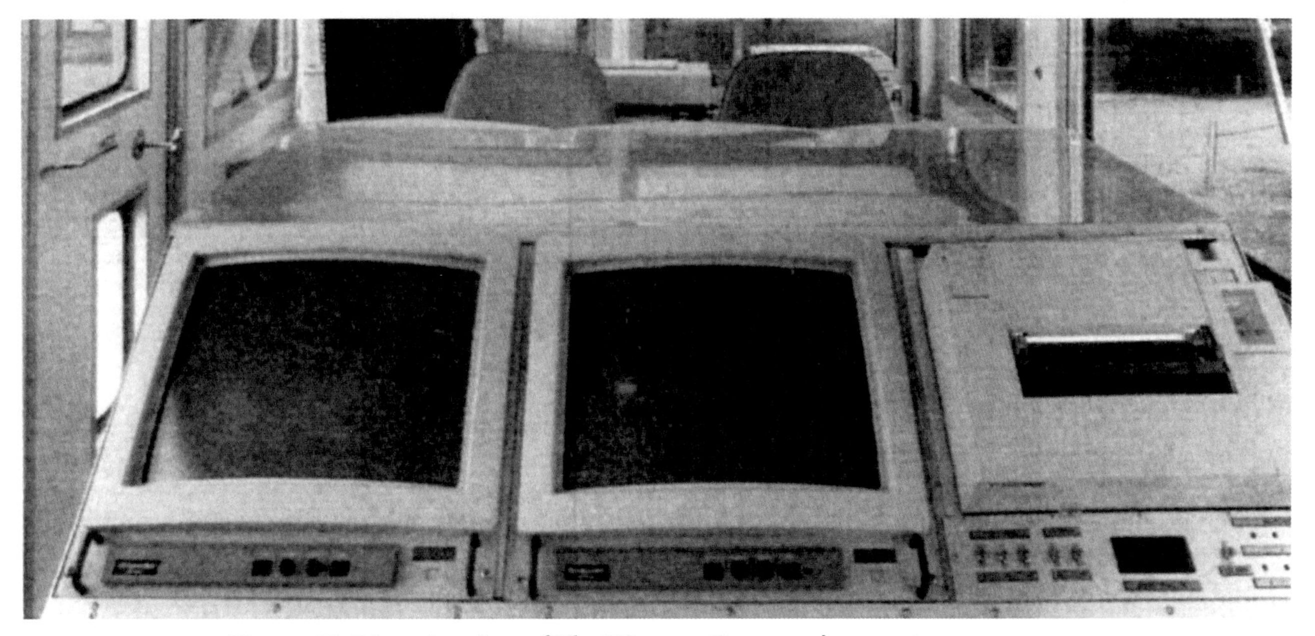

Figure 15-8 Interior view of The Western Company's newest computer van.

Figure 15-9 Exterior of Halliburton Energy Services' treatment command center.

Figure 15-10 Interior of Halliburton Energy Services' treatment command center.

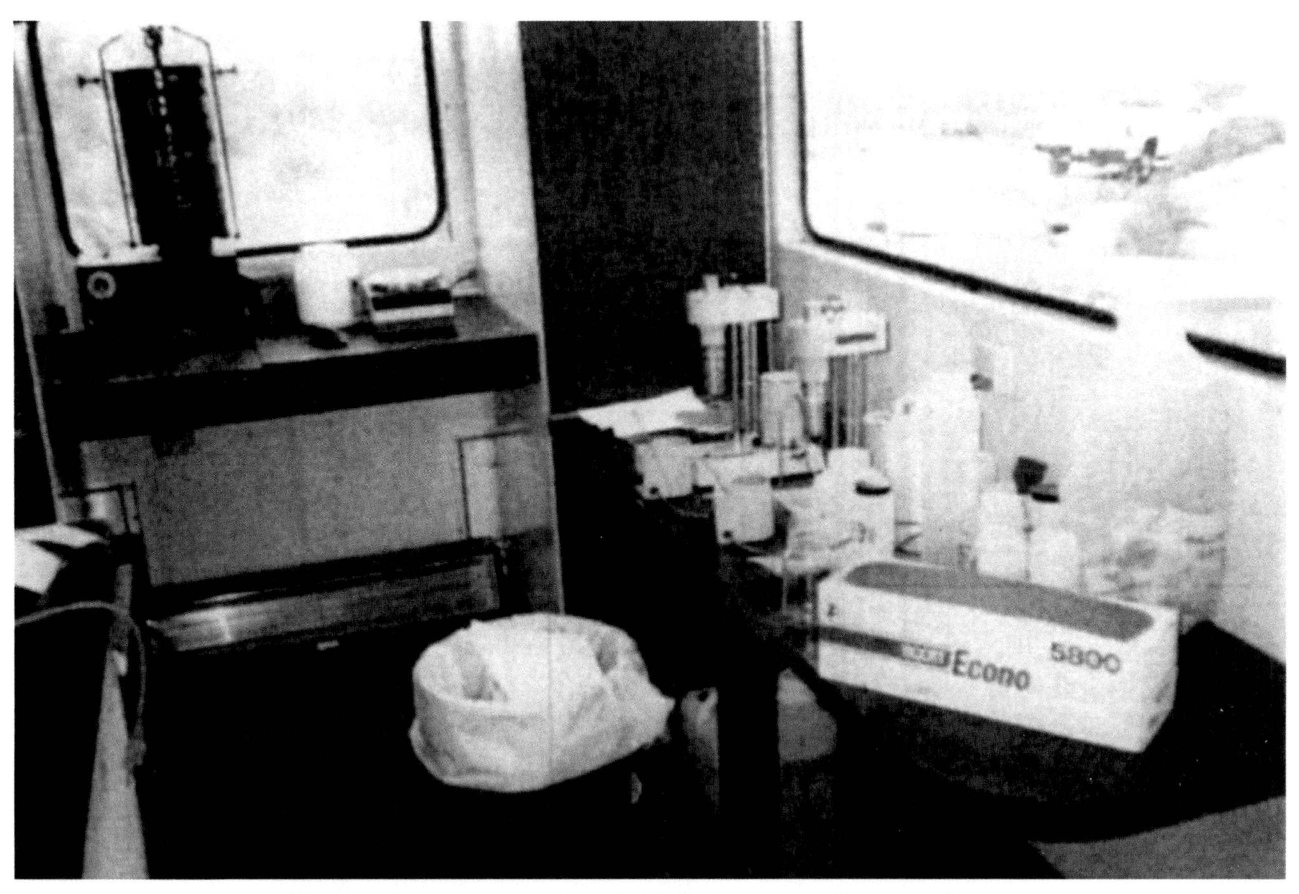

Figure 15-11 Interior view of a Halliburton quality control van.

Figure 15-12 Exterior view of Dowell's treatment monitoring vehicle.

Figure 15-13 Interior view of Dowell's treatment monitoring vehicle.

In the future, we hope to on a regular basis give you a real time picture of the growth and shape of the hydraulic fracture during a treatment. Presently, a fairly substantial number of treatments are being conducted where service companies and/or consultants are attempting to do real time analysis of fracturing treatments. Some successful work has been conducted by personnel utilizing the GRI sponsored frac program where real time analysis, including history matching, has yielded some interesting data about fracture geometry. This type of analysis requires very precise information about the formation, including stress profiles, permeability, and of course, also includes data that has been obtained through previous stress tests and mini-fracs on the same well. The problem that we continue to have concerning real time analysis relates to a total dearth of real data concerning the vast majority of wells that we treat. When we are able to have large quantities of data, then the computer vans will be truly worthwhile. They are, of course, very beneficial when properly programmed to yield pre- and post-fracture buildup data. In reality, during the treatment, computer vans without downhole data and precise formulation variable accomplish little more than the standard frac van. Many service companies use these to strip-chart such things as crosslinker concentrations and dry additive addition.

Figure 15-14 Another view inside the Dowell treatment monitoring vehicle.

Our industry has a long way to go before it fully utilizes the available computer capabilities. Several jobs have been terminated prematurely because Nolte plot data were misinterpreted. Personnel saw pressure changes that they took to be increasing bottom-hole treating pressures. The pressure increase may have resulted from a thick gel coming off the bottom of the tank, a rapid increase in cross-link time, a decrease in cross-link time, or other changes. These phenomena can easily be mistaken for bottom hole pressure rather than friction pressure. Without good monitoring and communications equipment to show what is actually happening, plotting or interpreting net pressure data correctly is very difficult.

Figures 15-5 to 15-15 show various frac vans and computer vans from the service companies. The tendency today is to move toward a combination treatment computer van. This allows all operations to be conducted in one single unit. I favor a single van, but one of my major considerations is that the company person stands or sit directly behind the person in charge for the service company and that both of them be looking at real time data and decisions be made only by them. Suggestions can be made by other technical personnel, but the final decisions need to be made only by those two people.

Figure 15-15 Dowell's newest "FRAC CAT" computer van.

Monitoring equipment recommendations

Some general rules that apply to monitoring equipment are:

1. Make certain the service company has calibrated all equipment that monitors pressure, rate, and density before the treatment.
2. Discuss with your service company representative backup capabilities such as secondary vans, alternative pressure readouts, rate readouts using a strap, standby flowmeters, and density measuring using sand screws. See that there is a plan if primary instruments fail. If you attend enough treatments, you will see all of them fail.
3. During the treatment, closely observe fluid volume readouts and correlate those with actual measurements to check the flowmeter readings. Occasionally, calculate hydraulic horsepower with rate and pressure, and watch the trucks in operation to be sure that everything fits. Service companies have run out of ·fluid halfway through a treatment because the wrong meter factor was put into the monitoring equipment. I have been on locations where the pump rate x the treating pressure indicated that some of the trucks on the location were putting out more than 2,000 hhp when the brake horsepower on the truck was less than 1,000. People make mistakes. Making spot observations during a treatment sometimes alleviates these errors.
4. Be reasonable. Sophisticated electronics fail, particularly in an oilfield environment. Have standby or alternative monitoring equipment and, on larger treatments, maintenance personnel available.

Good communication is essential

The communication network between the monitoring vans or monitoring area and the rest of the operating personnel on the location is very important. Most service companies utilize walkie-talkie radios with headsets. Equipment is susceptible to wear and tear and simple malfunction. Many times the most dangerous situation that can occur on a fracturing treatment is loss of communication between the treater and his pump operators. Some companies keep the pump operators out of the weather in enclosed vans. This can be beneficial as long as good communication exists. Check continually to be sure that everyone hears and responds throughout the treatment. The operator's radio may malfunction early in the job or at some point in the treatment, and he may not realize it. He may think people just are not talking to him. If a screen-out or some malfunction should happen so that you need to shut down quickly and communication doesn't exist, a disaster can occur. In a screen-out situation, one triplex pump left pumping by mistake can, at the least, split the tubing and, at the worst, rupture surface pipe, resulting in lost lives. I like the treater to be able to see at least one person whom the pump operators and the blender operators can see. Even if the communication equipment failed completely, you could continue the job utilizing hand signals, the way treatments were conducted in days past. As I related earlier, this procedure can be effective and can keep people on their toes; a safe operation can exist without direct communication. Have the service company check all communications equipment before the job. From time to time, ask the treater to see that he is in direct communication with all personnel, particularly those running high-pressure pumping equipment.

Guidelines

1. Have two pressure monitors, one electrical and one hydraulic (the hydraulic is a backup).
2. Have two flow-rate systems, one a flowmeter and either another flowmeter or a stroke counter as a backup.
3. Have two density measuring devices available. The primary device should be a radioactive densimeter. The secondary device could be a mud cup (on a cement job) or counting screw revolutions (on proppant frac treatments).
4. Have more than one person on the location but only one person in the frac van; have these two communicate.
5. Periodically check actual volumes by strapping versus readouts on electronic equipment.
6. Query service company personnel as to the sample frequency on pressure and rate recordings. Sample at least once per second. Excessive dampening can be dangerous. Determine if there is any delay in pressure readout from what the pump operators see.
7. Have standby vans on moderate-to-large jobs and have standby generators.
8. On larger jobs, request that an electronic technician be available.
9. Be aware of the rate and horsepower capabilities of the service company's pumps, so the job can continue if the flow-rate indicators temporarily fail.
10. Terminate the treatment if all the pressure measuring devices fail.
11. Utilize computer vans as a supplement to standard vans. Computer vans are only as good as the information that is fed to them.
12. Constantly check communication equipment.

CHAPTER 14

MONITORING AND READOUT HARDWARE (2022)

We have, without a doubt, come a long way in our monitoring from the standpoint of both hardware and software. Today one can stand in your home looking at your computer screen and monitor a frac job in Siberia. The ability to do this compared to conducting a frac job in the middle of nowhere with the job being conducted using hand signals. Jobs were conducted with operators standing on the trucks and no one wore protective gear other than hard hats and steel toed boots. See photographs at the end of the chapter. If it was raining, we got wet and did our best to protect paper strip charts recording from flow meters and pressure gauges. Most of the early fracturing was done with oil down tubulars, which resulted in low rates and high pressure. We started out with river sand and oil or water. Like we earlier discussed, early jobs were completed with gasoline and gelled with napalm. Searching for viscosity, oil external emulsions, which also yielded high friction, was a common fluid. Materials with the consistency of liquid soap were also used. For the most part our pioneers stayed away from water because of potential damage to the reservoir. Fracturing was going on internationally with lots of heavy oil fracs conducted in Venezuela on Lake Maracaibo. It was not unusual to have a frac job run with a Martin Decker circular chart and if you were lucky, you had a gauge with flow rate or personnel would count strokes on the pumps to ascertain rate. They also had personnel on top of tanks to counter check the pump rate. Cementing trucks were initially used and world war II Allison airplane engines were utilized.

As we discussed in the acidizing chapter, fracture acidizing was equally popular as fracturing with sand. Gradually horsepower increased from around 500 to greater than 1000 HHP and better monitoring equipment particularly when the decision was made to move to protected areas for the pressure and rate to be monitored. The initial frac vans for Halliburton were converted bread trucks, and simply monitored, in a controlled environment, pressure, rate, and eventually density. It was interesting to be part of the move to the vans as we were still in a macho phase, and some management felt that the vans were a passing phase. Dowell, now Schlumberger, also used the vans to house the pump operators who ran the pumps remotely. BJ, Halliburton, and the Western Company all had personnel standing on the equipment. The jobs I was involved with in my early career were all executed outside in open air. At Halliburton we had what was designated as a fracometers, which gave us paper strip charts of the job. The paper charts were open to rain, wind, snow etc. It was not manly to not stand and be pelted by rain sleet etc. In fact, an Engineer in West Texas was fired because he watched a job with binoculars from his car.

In south Texas, there was a major problem from management when a treater rigged up a trailer for the densitometer with a canvas covering to protect the equipment and the personnel. These were different times in the oilfield as the person assigned to the pump truck was responsible for the tractor, the two power engines and had to pack the pumps. A highly rated job title was the blender operator who truly had absolute control of the treatment because all additive and sand concentration were up to him. Blender operators had to buy very little beer at the bars after jobs. Today none of these things are the responsibility of individuals but are relegated to computer control systems. Maintenance of the equipment is done by mechanics and special employees, typically at the service facility.

The highest quality service that I saw in the late 60's and early 70's was in far south Texas where quality work was performed. At that time. virtually no jobs were pumped without a true ballout the day before the frac. A true ball-out is where ball sealers are dropped sealing off all the perforations assuring that all holes will be open for the frac. As mentioned earlier almost all jobs were done down tubulars, so rate was typically low. In the 60's and 70's, the most exciting part of the job was initiation of the frac when the treater dropped his raised arm, and all the pump operators came on-line at once. In fact, there was typically an unspoken contest when bringing the trucks online and anyone who did not get online because of being slow had to buy the beer after the job for the entire crew. A scene that I wished I had photographed was driving by a roadhouse with the entire frac crew enjoying adult beverages. I had stayed for flowback of the well and traveling outside of Sheridan Texas, I saw this amazing lineup of trucks parked far closer than possible on location so that everyone could go inside.

Times have changed. The jobs described above typically were done in hours and not days like the present zipper frac operations. All the wells treated were verticals and for the most part only a single stage was completed. I remember a weeklong operation in far south Texas where we fraced multiple stages on a well. The jobs were long, approximately 6-7 hours, and the wells were opened up and flowed for a few hours. Subsequently a bridge plug was run and the next zone was perforated. This series of jobs almost killed me as I was traveling with the frac crew who after doing maintenance on the trucks and gelling up for the next day, would head south over the border to party. I saw multiple times the crews would arrive at daylight, without sleep, and would pull on their coveralls and go to work. I would be lying if I told you that everyone was sober the day of the frac. After 3-4 days I was both physically and mentally exhausted. Those "good old days" were interesting but did not bode well for family stability. Most of the guys were out of high school and were making good money but working 90-100 hours per week was not a life conducive to a good home life. I should state that this was before most of the DOT regulations were in place and was a much different world in relation to work ethic, safety, and dictated time off. From an Engineering standpoint, cementing was in many ways even tougher on home life and worst of all were the tool men who worked extremely long hours but like their frac counterparts had to dress all their own tools and truly were many tines under a great deal more pressure relating to technical aspects and performance.

Today the world has changed with much needed DOT regulations and more emphasis on safety. I would state that, during downturns, safety takes a much lesser portion of priorities. A typical worker in the oilfield works 12 hour shifts and many have two weeks on and two weeks off and gone are the days of (we hope) people are not under the influence during job operations. Fire protectant gear, hearing protection, goggles, and associated safety devices are required.

Instead of having to bear the brunt of bad weather the great majority of employees are shielded inside vans where very impressive data is presented and recorded, instead of hand signals and sound power headphones one can address the pump operator at the same time as you are instructing the person in charge of the operation. Unless there is no satellite connection, all the data is transmitted all over the world through the internet. (Note the data van photograph section which gives a historical perspective on how far we have come.)

CHAPTER 15

RE-INVENTIONS OF PAST FAILURES AND MASSIVE OVERUSE OF "INTUITIVE REASONING" (2022)

The saying that, unless we study the past, we are doomed to repeat it. This saying is painfully true in our oilfield. I am going to discuss re-invention of various processes that failed initially but were brought back, in some cases, multiple times to enjoy even more failure. The other subject of this chapter is what I term "Intuitive Reasoning" which many times simply does not work in the oilfield. I tell my peers that God is mad at the oilfield and has created, particularly in the US, very heterogeneous reservoirs which create massive problems for modelers and data search analysts who base their world on homogenous isotropic media. To give an example of how intuitive reasoning has doomed our work, one could select the Austin Chalk. Because of an initial success on a vertical well with 40-pound crosslink gel and 300,000 pounds of sand of the discovery well, literally thousands of wells were treated with the identical treatment. Those of us who have worked the chalk since 1980 now know that even for the vertical wells, this treatment was not optimum even for the first well. Hindsight is brilliant but the "dreaded successful first well in the field" has doomed many operators who overlook the fact the first well was probably on a structural geologic high and may have many reservoir properties wildly different than wells which are only 10 acres away, not to mention 1-mile offsets. I will also attempt to dispel some egregious myths which have plagued our industry since I started in 1965.

Mistaken results based on laboratory core data

We as an industry need to closely look at how particular data is run in a laboratory setting. For high permeability reservoirs the selection of core plugs is typically straight forward. One uses sidewall cores or selects a clean whole core and makes core plugs without much subjectivity. In the case of tight gas sands, i.e., .05 to .5 millidarcy rock, the technician is very careful not to core where a fracture exists. Therefore, in my opinion, what one measures with tight fractured rock with subjective core plug selection, is absolutely the worst-case permeability for oil and gas production. I should note that "producing permeability" can be quantitatively measured in the field using standard FET analysis. We have discovered that some of the nano-Darcy wells had fluid efficiencies in the 30 % range indicating effective permeability of the reservoir approaching or exceeding many conventional reservoirs. This discrepancy is due to the natural fractures present in the matrix. This data has been verified by the high productivity of ultralow matrix perm rock exceeding, in many cases, 4000 barrels per day. We will discuss more in the slick water chapter. I would also state, fearing severe punishment from Dr. John Lee, that we have experienced enhanced producing permeability due to extreme brittleness of reservoir rock when treated with water and small proppant.

This problem of misinterpreted laboratory data was brought out when the initial completions of the Cotton Valley sands were conducted using massive hydraulic fracturing and were designed around matrix permeabilities of much less than .001 millidarcy perm. A tremendous amount of reservoir work was conducted evaluating flow rates post frac and using the very low perm numbers, it was stated that propped fracture lengths of exceeding 2000 feet were achieved. This type of intense study led many to believe that the Cotton Valley sands could be depleted on, if not 320 acres but perhaps 640, acre spacing. Several years following the completion of many wells on 640 acres spacing, low tech operators drilled wells 300 feet from the original completion finding no depletion. It turns out in the area studied the producible permeability was dominated by not the matrix but the tiny fracture systems in the rock. Prior to the advent of horizontal completions, many areas were down spaced to 10-20 acre spacing with great success. The problem of bad analysis of fracture length was further exacerbated by bad assumptions on fracture containment and dependency on single planar fracture systems.

Other examples of what I term "worst case results" are capillary suction tests with very finely ground core measuring travel time across a set distance. These tests, including imbibition tests with follow up nitrogen flow are not realistic of potential damage or the wildly overplayed "Water Sensitivity", a phenomenon, which has plagued our industry for more than 50 years. When you crush the low perm rock down to 200 mesh or smaller, you create massive surface area which is not realistic based on the frac treatment conducted. We will discuss water sensitivity in a separate heading.

We as an industry need to look hard at the facts relating to severe reservoir heterogeneity. We do not frac any 8-inch cores (relating to API proppant testing cells). Early in the GRI work near wellbore drill-throughs were conducted and to my knowledge we never saw less than 5 fractures in the near wellbore vicinity. I believe we have been tied to single planar fractures because they are easy to model. I do also believe that with surgical perforating and viscous fluids we have been able to achieve something close to single fractures. This conclusion is based on many post job buildups where we had pre frac job flow properties. Tied into this thinking is the myth of contained hydraulic fractures. In my 50 + years of supervising and monitoring frac jobs I can count less than 9 treatments where from the moment we started pumping pressures were increasing i.e., we were fracture contained. These treatments were exciting to say the least. In one case we saw 7,200 pounds of net pressure from the dead string. For a period of 5 hours, we were on more than a 1 to 1 slope. This treatment, documented in a previous version of the book, placed more than 500,000 pounds of 20/40 bauxite below 20,000 feet in a high temperature Smackover reservoir. It was interesting that some on-site experts wanted us to go to flush even before we started proppant.

The ongoing myth that we need to pump hydrocarbons -not water

Several years back, I got a call from some investors who were excited that a brand-new means of stimulating oil and gas reservoirs, using propane, was being introduced into the oilfield. They got upset when I related that it was not only not new but had little chance of success. They were unaware that a mixture of propane, CO_2, and methanol had been utilized and many papers published in the late 60's and early 70's gave the process as the answer to all oilfield problems. Initially the Gas Frac Process, developed by Dowell, was advertised as a eutectic mixture of propane and CO_2 which, with varying mixes, could revert to a gaseous medium at a specific temperature. As was found later the system had used gelled methanol to yield viscosity. The fluid system also incorporated 5 % toluene to keep the alcohol viscosifier in solution when CO_2 was introduced. I was a chemist working for Halliburton at the time and once we discovered the toluene factor, I developed a fluid called Vapor Frac which was gelled alcohol mixed with CO_2. We even developed Vapor Frac II which used gelled condensate or diesel mixed with CO_2. Although it was the very early 70's we had remote blenders to keep employees off the blending equipment during the frac.

Although SPE papers were written with glowing results reported, both the Gas Frac and Vapor frac faded into obscurity, only to be reinvented 40+ years later. They were expensive and dangerous, but their failure was not primarily due to their danger or complexity. They failed for the same reason that very large slick water fracturing is working. For extremely low permeability reservoirs, a massive amount of surface area must be created. The very complex fractures, created with low viscosity fluids and fractures held open with very small sand, have shown that small fracture volumes in tight rock simply do not work. The sales pitch was, Gas Frac was non damaging, and we could resell the frac fluid. The use of 4 million gallons of propane, not only would be expensive but would create potentially the world's largest fire hazard. The process failed not because of lack of ingenuity but had no chance of creating huge transmissibility increases in very low permeability rock. It failed in the seventies and failed again in the 2000's.

Much of the confusion about hydraulic fracturing relates to the fact that the very first treatment as well as the vast majority of fracs through most of the 60's were only damage removal fracs, not effectively stimulating production, only bypassing damage. Much of the early advertising was done by those who had little or no knowledge of the basics of fracture stimulation. There were hundreds of advertisements illustrating a poor well making only 10 barrels per day that was stimulated to many hundreds of barrels per day by a 1,000-barrel frac. When in fact the well probably had a huge skin factor relating to early damage during drilling or completion and similar results could have been achieved with a small acid breakdown or very small proppant skin frac.

One type of fluid has been re-invented multiple times during my career, methanol fracs. This is a crossover from the dreaded intuitive reasoning phenomenon. Even today I find young bright Engineers who know about the characteristics of methanol and consider it to be a great candidate for a fracturing fluid. Over the years hundreds of jobs have been done with linear, crosslinked, and foamed methanol fracs. Most of the jobs were done with some percentage of water but actual treatments have been conducted using pure gelled methanol and or methanol combined with nitrogen or CO_2. The ongoing myth of water sensitivity should have been put to rest by the thousands of slick water fracs which, as most know, are not conducted only in source rock shale but have seen success in so called water sensitive rock if very low permeabilities exist. The high cost and danger are not the only problems with methanol fracs. Alcohol systems require massive amounts of breaker to degrade the polymers because of the alcohol functioning as a reducing agent and degrading agents for enzymes. Please do not re-invent alcohol- based fluid.

One field experience was an actual 100% methanol foam treatment conducted in South Texas. After the frac we were flowing the well back and started producing fresh water, I was not brave enough to ask the customer why he had spent enormous amount of money using a methanol frac when the formation produced fresh water. I may be biased toward methanol because it almost cost my job with Halliburton, I, along with other chemists was charged with developing a high temperature fracturing fluid for Shell. With a lot of work my technician and I developed a very successful frac fluid for Halliburton termed *HyFrac*. The system was a mixture of fast hydrating HEC for batch mixing and a controlled hydration HEC to be added on the fly. This combination of linear gels was a staple product for many years.

Back to the story. While I was developing n' and K' (viscosity input for the models), I turned over this work to other technicians while my technician and I spent a couple of weeks in the field testing the product. Upon returning, I was called on the carpet and accused of dry labing data (creating false data to make the product look good). None of the technicians at the lab could duplicate our work. It was a very serious matter with my job at risk.

My group leader and the laboratory manager had my technician, and the one who could not duplicate the work, prepared samples for testing. Both technicians carefully weighed out dry samples of the gel to undergo hydration. The gentlemen who did not work directly for me carefully sifted the powdered gel into a waring blender taking almost 30 seconds to add the polymer. While that gel was hydrating, my technician, who had weighed his sample in a small plastic cup, reached behind his back and squirted methanol into the cup, swirled the cup and added the gel in a few seconds. The sample we prepared followed our data track and the other non-slurried sample failed.

The success of our test was due to the reducing capability of alcohol in stabilizing the fluid. 5 % Methanol became the stabilizer of choice for high temperature fracturing for many years only to later be replaced with sodium thiosulfate or other reducing agents negating oxygen degradation at high temperatures. Once again as I have stated before, a great number of discoveries were a result of events with little to do with technology. In this case my technician could mix the gel much faster by slurrying in alcohol with no regard to the effects on fluid stability. Another of my exploits related to trying to develop a breaker for an early gelled oil and continued to end up with higher viscosity rather than degradation. The late Dr. Bob Tiner and I have a patent on viscosity enhancement in gelled oils while trying to develop a breaker with benzoic acid.

Already covered in the previous edition of the book are my feelings about pumping either crude oil or diesel instead of water. This problem exists due to "intuitive reasoning" and misguided micro scale lab testing has led us astray. In the early days of hydraulic fracturing, there was a continuum of back-and-forth dominance of either water or oil-based fluids. Based on all of my experience there was only one formation that could not be treated with water. A relatively deep, relatively high permeability, high clay content Morrow formation in Southeastern New Mexico was truly sensitive to even KCL treated water. Halliburton was very successful with a 70 % methanol foam frac. I admit I tried to utilize KCL fluids, but the 30 + % swelling clay combined with 100 % water dramatically reduced the permeability of the formation. The Halliburton frac worked very well in this relatively permeable reservoir until the government banned their Fluorocarbon foamer. There was no other foaming agent that would work in the alcohol mix so then the formation was treated with 70 quality nitrogen foam which seemed to be functional combined with rapid flowback.

In south Texas, the poster child for water sensitive formations is the Olmos, a formation with high clay content which led the industry to frac with high gravity oil or gelled diesel for many years. Gelled hydrocarbons are included in my category of banking fluids and some of the largest hydrocarbon fracs ever conducted were in the Olmos formation. These dangerous fracs were conducted for many years until a non-technical small operator fracture treated a well in the middle of an Olmos field with crosslinked gelled water using 3 % KCL and created the best well in the field. The damage, if present, was mitigated by the much more efficient fracturing fluid. From that time on the Olmos was stimulated with water containing clay control.

The myth of clean frac fluids yielding increased production

Other phenomena which, as long as I have been in the oilfield, has created massive profitability for manufacturers of man-made gelling agents, is the concern about gel residue. I would suspect there has been more than 200 papers presented proving that low residue gels are much better than the "evil guar Gum". Guar supposedly plugged up proppant packs and drastically reduced hydrocarbon flow compared to low or non- residual fluids, which after degrading, left little or no residue.

Guar is a simple bean, grown typically in India or Pakistan, which is simply ground to a fine powder, this powder creates a viscous linear fluid that has been in use since before I began work in Halliburton research in 1968. Laboratory personnel noticed on degradation that a residue was left, and great concern was had by all. Because of these concerns, the frac industry moved quickly to derivatized guar, initially Hydroxypropyl guar and then later moved on the carboxymethyl hydroxypropyl guar as alternatives to the evil guar gum. As was discussed in the earlier version of my book cleaner fluids have less fluid loss control and can create very damaging filter cakes negating any low residue component benefit. Additionally, the manufacturers of guar have reduced the quote "residue' to less than 5 % with more filtering. There was a step change in guar when it was found that simply going through an additional grinding step not only reduced residue but yielded higher viscosity.

There is a larger problem with so much emphasis on components of the gelling system where you have 3 or 400 pounds of residue in a 1000-barrel frac. The investigators forget that the 500,000 pounds of high-quality sand you use has a minimum of 14 % fines at 4,000 pounds closure i.e. comparing 500 pounds of residue to 70,000 pounds of fines may be non-sensical. We should also realize that just simply breaking the brittle rock creates massive fines. Our hydrocarbon reservoirs during the frac process do not have smooth single planer fractures. I believe many of our misconceptions along this line relate to gravel pack work. It is critical that we have low residue and fines free sand in a gravel pack as the pack is a transmission path from a high permeability reservoir to the perforations. As time went forward and common sense entered the picture, the industry moved to guar only for temperatures approaching 300 F and only utilizing the double derivatized CMHPG for higher temperature applications in high temperature conventional wells.

The myth of massive near wellbore cooldown occurring during fracs

In the early days of utilization of the Forced Closure technique, it was observed that the fluid samples caught during flowback were broken precisely as if there was no cooldown. When one truly understands the folly of modeling single planar smooth wall fractures, a simple model of cooldown becomes unthinkable. In addition, when evaluating extremely high magnification pictures of the fracture face, we see something simulating micro scale mountain ranges. One can understand the lack of cooldown in the reservoir when massive surface area is available. The cooldown equations in fracture models that I am aware of are smooth wall plates, not the true scenario of massive rock surface with near infinite heat sink. Hundreds of thousands of fracture treatments have been conducted with the breaker schedule utilizing no cooldown. Using smooth wall cool down curves will invariably lead to either over-breaking conventional crosslink fluids or in some cases destroying the breaker. It should be noted we have seen service company recommendations where the breaker recommended was used at bottom hole temperatures 100 degrees above the dissociation temperature of the chemical used for degradation. In my opinion, you are creating the possibility of plugging reservoirs by using cooldown calculations. We have always tapered breaker loadings from low to higher numbers and then utilizing a hot breaker for the last hopper of sand where we were implementing forced closure on completion of conventional crosslinked waterfracs.

The myth of needing large numbers of perforations to produce hydrocarbons in tight reservoirs

Similar confusion reigns in perforation schemes where certain Engineers shoot hundreds of perforations in tight reservoirs where they would be better served with minimal perforations particularly when trying to treat multiple intervals with a single treatment. I have had customers shoot 600 plus holes over 4 sections of a 400-foot interval with 100 perforations in the upper zone. They are confused when all or the vast majority of their frac treatment goes into the upper interval, even when treating at 30-50 barrels per minute.

In my early career I worked on wells that had a total of 9 .32" perforations over 2000 feet. Because of the depth difference, they were able to stage the zones by dropping RCP ball sealers while pumping 12 barrels per minute.

We design treatments for up to 500-foot sections with 3-15 intervals and treating rate is set at 2 barrels per minute per perf. With .32" perforations that translates into 1000 psi of perf friction. A typical treatment would be 3 intervals at 7 perforations of .32" and a pump rate of 40 barrels per minute. For those who are worried about restriction to flow, there is none when you investigate what kind of drawdown you have with 21 .32" chokes.

In relation to perforations, we have always believed in minimizing the number of perforations in conventional reservoirs where we attempt to have a single planar fracture. I had a customer who trusted me and allowed us to treat 300' plus intervals with a 10-foot section of perforations. This was the cotton Valley Taylor formation in far northeastern Texas. Memory serves that we treated around 25 wells, each with around 1 million pounds of sand. The geologists were aghast at the perforation scheme and post frac perforated around 10 wells top to bottom before they realized that they achieved no additional production by adding 1800 .37" perforations. One might be able with deep penetrating charges to bypass some near wellbore damage, but the surface area of multiple perforations is minimal.

Perhaps the major problem in understanding how fracturing works is that we must compartmentalize permeability ranges. I will generalize by stating that high permeability reservoirs that can produce on their own i.e., 50 millidarcy + reservoirs can, unless damaged, produce on their own and should be highly perforated, perhaps with deep penetrating guns, to account for perforation tunnel effects. When you approach single digit perm then hydraulic fracturing shines. Combined with geo-pressure, such as exists in South Texas, fracture treatments achieve success by enhanced surface area but many times the success is achieved by accelerated production i.e. achieving 10 BCF in one year rather than 10 years.

For what has been termed tight conventional sands, i.e., .005 to .1 millidarcy, everything relates to creation of massive surface area.

At a frac pack conference in the early days of frac packs, a discussion was held regarding a screen out of a large interval that had been perforated with 3,600 1" holes. The low rate frac treatment screened out very early and the audience was asked for suggestions to negate the screen out. Incidentally, this was and is standard procedure for high perm Miocene completions in the Gulf of Mexico.

Being a tight sand guy, I suggested that the operator could shoot a minimal 10-foot section and minimize the probability of multiple fractures occurring and suggested 20 perforations. There was a loud commotion in the room as a gentleman stood up and said every D*** fool knows you cannot produce 10 million cubic feet of gas through 20 perforations. He also quickly stated that he had access to Nodal analysis, and I needed to be careful with recommendations. I suggested he turn off the perf tunnel part of Nodal analysis as there is no perf tunnel restriction after a frac. Literally thousands of wells have been completed with minimal perforations with post frac skins in the -3 -5 range.

The conclusion by many that we have piston like displacement of fluids and proppant regardless of viscosity.

Our industry has been utilizing the concept of staging our proppants such that the smaller proppant is the earliest pumped followed by 1 or more stages of larger proppant. This has held true for every type of fluid. In my opinion, backed by stokes law, unless you have a perfect proppant transport fluid, that the last proppant pumped is not at the wellbore. By my definition, a perfect proppant fluid would have between 500 and 1000 centipoise at bottom hole temperature and a shear rate of 37 reciprocal seconds. This type of viscosity is easily achievable with conventional crosslinked gels and delayed breaker systems for temperatures from 70 degrees up to 350 degrees Fahrenheit. For those new to the industry, refer to pictures in the original chapter 5 on page 65. When the first crosslinked gel was introduced to the industry it was a tremendously sellable commodity.

A perfect proppant transport fluid cannot be achieved with 10-to 15 pounds per 1000-gallon gels which appear viscous but simply do not yield a fluid which will suspend sand. We will discuss this more in the next myth relating to crosslink fluids. Any fluid such as foam, linear gel, gelled oil, or low concentration crosslinked gels are simply banking fluids where the initial proppant settles out near wellbore. This creates a phenomenon called equilibrium bed height illustrated in image 1 from a 50 + year old paper from the early days of fracturing. Once the bed is formed, based on viscosity, rate, and proppant size, the remaining proppant is transported over the bed and any late proppant is not at the wellbore.

Typically, the major denial relating to this subject is that the job was put away even with high concentrations at the end. We found success as a company designing perfect proppant transport fluids and achieving better productivity than offset wells which had poorer fluids. I might digress and define a successful frac treatment. In my service company experience, a successful frac was "Dirt below the surface" or simply placing all the chemicals and proppant outside the perforations. An early or even late screen out was very problematic for both the service company and the oil producer. I do hope we have moved forward with a successful frac is one that optimizes profitability for both the service company and the operator.

The concept of last proppant pumped remaining at the wellbore is carried to the extreme when one considers the proppant carrying capacity of water. We have now completed hundreds of fracs with slick water with a relatively simple process termed *"Counterprop"*. In the treatments we place the larger proppants first followed by the smaller proppant. An example of a Counterprop design is shown in image 2. The basic thinking is that the larger proppant will remain at the wellbore and the smaller proppant will travel further out due to its smaller size in the Newtonian low viscosity environment. It will also be noted that this particular job is designed with large pad and numerous, non-proppant laden sweeps. We term this type of treatment as a *"Waterfrac Sweep"*. This type of design, with and without the Counterprop angle, has been executed successfully in many thousands of fracs for at least 20 years throughout the world (see image 3 for an example). We believe this process allows for creation of more surface area, better dispersion of proppant, and time after time has outperformed similar sized jobs containing much more proppant. I have, since the beginning of the waterfrac explosion, stated that volume of the treatment was more critical than the volume of proppant.

An area of some controversy has come up regarding resin coated sand. The use of resin coated sand for sand control has been a very successful process when using viscous perfect proppant transport fluids. Tailing in with curable resin was extremely successful when used with piston like displacement fluids. The use of curable resin sand being used at the end of slick water treatments has not worked well. The reason relates to the discussion in the paragraphs describing early proppant settling at the wellbore and tail in proppant being placed away from the perforations. The solution has been, where proppant production if anticipated is to over displace the well at least 200 barrels.

This technique works very well with no damage to connectivity of the proppant pack. You do not move a settled equilibrium proppant pack from the wellbore once it has settled. We have customers who regularly over displace waterfracs with more than 600 barrels of water at max rate. It should also be noted that one does not call flush on slick water treatments when the last concentration is dropping at the inline densitometer, but displacement is called at zero concentration at the wellhead

For conventional high viscosity frac theorists, over flushing is disastrous. Where piston like displacement occurs, over-flushing can remove proppant from the near wellbore area. For perfect proppant transport treatments, we recommend under-flushing at least 2 barrels and calling flush when the inline densitometer starts dropping below max proppant concentrations.

It should be noted that there is some small percentage of reservoirs that simply produce proppant even with over flushing in slick water treatments. From my experience, normally but not totally, these wells are hard rock and relatively shallow, in these cases one should initiate the job using a curable resin coated sand as the first proppant pumped allowing the cured pack to negate proppant production at the perforations. We discussed resin coated sand in some detail in chapter 11.

Confusion about what is broken gel.

In the early days of fracturing with viscous fluids there was a great deal of confusion about the definition of when a gel was totally broken. Publications from major service companies stated that any gel with a dial reading of 10 centipoises measured with a shear rate of 511 reciprocal seconds was broken. The fallacy relates to measuring at high shear rate i.e., 300 rpm with a B1 bob, when in reality the shear rate for the Non-Newtonian gel in a fracture system would be closer to 10-30 reciprocal seconds yielding a down hole viscosity of 100s of centipoise plugging pore throats and fracture systems permanently. One very interesting piece of data came from one of the early horizontal wells, where horizontal cores were taken, we recovered stable crosslink gel that had been in the formation for more than 9 years at 240 degrees Fahrenheit. Early on with crosslinked gels, no one wanted to run breaker because of fear of the dreaded screen out. Many wells were, instead of stimulated, plugged with gel. The problem was exacerbated with the development of delayed crosslink crosslinked gels, which negated the inherent shear problem with metallic crosslinks. A large majority of early crosslinked gels utilized metallic crosslinkers which were very shear sensitive. As early treatments were typically executed down tubing and everyone seemed to want instantaneous crosslink gels, these early treatments resulted in the fracture being propagated with linear gel or in some cases less than even linear viscosity due to tremendous shear down the tubulars.

When delayed crosslink gels were first utilized there was a plethora of plugged wells due to a very stable crosslink gel in the formation without proper breaker.

The "old persons tale" of gels will eventually break was just a tale. Without the presence of oxygen or other degrading mechanisms typical gelling agents will remain stable at temperatures well over 250 degrees Fahrenheit. The true definition of a broken gel is one that yields less than 2 centipoise or less at 511 reciprocal seconds shear at ambient temperature. There is a plethora of breakers that can be delayed or only function at high temperatures allowing for stable gel systems for long times at elevated temperature exceeding 300 degrees F. These breakers are described in the original chapter 8 with some updates in the new chapter.

The confusion on what is the definition of a crosslink gel

I believe that there are many in our industry that believe that if a gel is crosslinked, i.e., the linear fluid turns into a Jell-o-like consistency, that a prefect proppant transport fluid is created. For those new to the industry, refer to pictures in the original chapter 5 on page 65. I try to distinguish between a stable crosslink, (perfect proppant transport at BHT) and imitators that fall apart with increase in temperature and shear. Any water base fluid with less than 20 pounds per 1000 cannot achieve perfect proppant transport viscosity even at low temperatures. To put in simpler terms, any crosslink gel with concentrations much below 20 pounds per 1000 gallons is essentially a banking fluid that allows for proppant settling near wellbore. The ability to create perfect proppant transport fluids, particularly at higher temperatures (> 180 F) is non-existent at 12-15 pounds per 1000 gel concentration. Perhaps with technological breakthroughs this will be achieved in the future. The confusion in understanding the meaning of crosslink gel has led to the early growth and use of "Hybrid Fracs" where slick water is followed either with low concentration crosslink systems or higher gel loading systems with massive breaker loadings. The failure of hybrid systems achieving superior results has led to most of the slick water stimulation to be water and FR only.

I should state that we were forced by a competitor to execute a true Hybrid system in the early days of a slick water Granite Wash play in the Texas panhandle. We executed 10,000+ barrel stages followed by high proppant concentrations of resin coated sand with a perfect proppant transport system. Multiple stages were pumped on the vertical well. The well was not only the most expensive in the field, but it was also the poorest performer. I believe that when we use stable crosslink gel, we are driving toward a dominate fracture system which is the exact opposite of what is required in ultralow permeability reservoirs. A vast majority of so-called hybrids are simply slick water followed by higher viscosity slick water. I believe it is a mistake to mix the two approaches. My belief has been shown to be correct by the movement toward all slick water. I should also note that wellbore pathway problems, tortuosity, can create the need for viscosity to pull the proppant through the near wellbore area. Once we improved on our ability to be selective in picking landing points for our laterals and have ceased porpoising while drilling, the need for viscosity to bypass tortuosity has been dramatically decreased.

As is the case in linear gels, foams and emulsions, these low concentration crosslink systems may look good at ambient temperatures but fail to transport proppant perfectly. They are therefore banking fluids creating an equilibrium bed height at the wellbore as we described for slick water. I believe the vast majority of fracs pumped since the inception of fracturing have been banking fluids, primarily because of the difficulty of creating perfect proppant transport fluid.

There is a belief that crosslink viscosity is required to place proppant through perforations. Almost universally everyone felt it was necessary to add a rapid crosslinker to delayed crosslink systems to be sure that sand is transported through the perforations.

I believe that design engineers, because of concerns for "gel damage" or simply wanting to reduce cost strive to minimize gel concentration and are thrilled when the job is successfully completed. In conventional treatments, simply putting sand outside perforations does not ensure that the proppant is placed across the formation. Many "Successful" fracturing treatments failed to properly stimulate the interval because all the proppant settled out of zone. (See chapter 9, 1994) Both of these assumptions are wrong. One of the most successful fluids that had a very low screen out ratio was a linear gel which typically consisted of 50-60 # base gel and 20-90 pounds linear secondary gel. This fluid was the first high temperature fracturing fluid used in the industry termed HyFrac by Halliburton. Very large fracture treatments were conducted at temperatures from 200-375 degrees F carrying sand, or bauxite to great depths at concentrations exceeding 7 pounds per gallon. In later years 1000's of polyemulsion fracs were conducted carrying concentrations up to 13 pounds per gallon in the US from the late 60s until slick water became the vogue in the industry.

The only real exception of where viscous crosslink fluids or linear fluids such as High Vis polyacrylamide or simple guar gels are required to carry proppant through the near wellbore area is in formations where severe near wellbore tortuosity exists. Tortuosity and diagnostics was discussed in the fluids chapter.

Image 3

Ely Design Waterfrac sweep
WELL / LOCATION: Nadine 22 Paradise
FORMATION: wildcat
PERFS: 7 stages with 7 clusters
with 42 .37" holes 60 degree phasing

Stage	Fluid Type	Fluid Volume (gals)	Proppant Concentration (ppg)	Stage Proppant (lbs)	Injection Rate (bpm)
1.	1 % NaCL + FR	70,000	-	-	80
2.	"	10,000	1 # 100 mesh	10,000	80
3.	"	60,000	-		80
4.	"	10,000	1 # 100 mesh	10,000	80
5.	"	60,000	-		80
6.	"	8,000	.1 40/70	800	80
7.	"	8,000	-	-	80
8.	"	8,000	.25 40/70	2,000	80
9.	"	8,000	-	-	80
10.	"	8,000	.5 40/70	4,000	80
11.	"	8,000	-	-	80
12.	"	8,000	.75 40/70	6,000	80
13.	"	8,000	-	-	80
14.	"	8,000	1 40/70	8,000	80
15.	"	8,000	-	-	80
16.	"	8,000	1 40/70	8,000	80
17.	"	8,000	-	-	80
18.	"	8,000	1 40/70	8,000	80
19.	"	8,000	-	-	80
20.	"	8,000	1 40/70	8,000	80
21.	"	8,000	-	-	80
22.	"	8,000	1 40/70	8,000	80
23.	"	8,000	-	-	80
24.	"	8,000	1 40/70	8,000	80
25.	"	8,000	-	-	80
26.	"	8,000	1 40/70	8,000	80
27.	"	8,000	-	-	80
28.	"	8,000	1 40/70	8,000	80

29.	"	8,000	-	-	80
30.	"	8,000	1 40/70	8,000	80
31.	"	8,000	-	-	80
32.	"	8,000	1 40/70	8,000	80
33.	"	8,000	-	-	80
34.	"	8,000	1 40/70	8,000	80
35.	"	8,000	-	-	80
36.	"	8,000	1 40/70	8,000	80
37.	"	8,000	-	-	80
38.	"	8,000	1.25 40/70	10,000	80
39.	"	8,000	-	-	80
40.	"	8,000	1.25 40/70	10,000	80
41.	"	8,000	-	-	80
42.	"	8,000	1.5 40/70	12,000	80
43.	"	8,000	-	-	80
44.	"	8,000	1.75 40/70	14,000	80
45.	"	8,000	-	-	80
46.	"	8,000	2.0 40/70	16,000	80
47.	"	10,300	-	-	80

TOTALS: 1% NaCl Water - 590,000 gals. Plus tank bottoms*
40/70 Ottawa sand - 170,800 lbs
100 mesh sand - 20,000 lbs
The additional water is for extending flush stages where we are seeing banking and or excess pressure.

- DESIGN/FLUID CRITERIA:

1. Design Pump rate - 80 bpm.
2. Maximum Pump rate - 90 bpm.
3. Maximum treating pressure –6,200 psi.
4. HHP required 12,000 HHP plus 50% standby. Fluid –NaCl water with friction reducer
5. Based on pumping down 11,288' of 4 1/2 "11.6 # N-80 casing.

ADDITIVES:

1. Water based friction reducer.
2. Biocide if required—common bleach will suffice.

ADDITIONAL EQUIPMENT:

1. 2 in-line densiometer(s).
2. Sand sieves, and associated equipment to perform QC on location. Sand sieves on all compartments and water analysis on pit water.
3. 50% Standby horsepower.
4. Pressure relief valve on the casing and kickouts or popoff required on downhole pumps.

Completed ?

Image 2

Rocky Mountain Energy, Inc.
WELL / LOCATION: Counterprop 2-10 paradise
FORMATION: tight sand Stage 2
PERFS: 8,293-95, 8,311-13, 8,330-32, 8,305-08, 8,384-86, 8,451-53, 8,472-76 8,501-03, 8,535-37, 8,554-58
2 spf 50 .40" holes 60 degree phasing

Stage	Fluid Type	Fluid Volume (gals)	Proppant Concentration (ppg)	Stage Proppant (lbs)	Injection Rate (bpm)
1.	15 % HCL	1,500	-	-	10
2.	Brine Water + F.R.	80,000	-	-	100
3.	"	5,000	.1 # 40/70	500	100
4.	"	80,000	-	-	100
5.	"	5,000	.25 # 40/70	1,250	100
6.		80,000	-	-	100
7.	"	5,000	.5 # 40/70	2,500	100
8.		80,000	-	-	100
9.	"	5,000	.75 # 40/70	3,750	100
10.		80,000	-	-	100
11.	"	10,000	1 40/70	10,000	100
12.	"	10,000	-	-	100
13.	"	10,000	1 40/70	10,000	100
14.	"	10,000	-	-	100
15.	"	10,000	1 40/70	10,000	100
16.	"	10,000	-	-	100
17.	"	10,000	1 40/70	10,000	100
18.	"	10,000	-	-	100
19.	"	10,000	1 40/70	10,000	100
20.	"	10,000	-	-	100
21.	"	10,000	1 40/70	10,000	100
22.	"	10,000	-	-	100
23.	"	10,000	1 40/70	10,000	100
24.	"	10,000	-	-	100
25.	"	10,000	1 40/70	10,000	100
26.	"	10,000	-	-	100
27.	"	10,000	1 40/70	10,000	100

28.	”	10,000	-	-	100
29.	”	10,000	1 40/70	10,000	100
30.	”	10,000	-	-	100
31.	”	10,000	1 40/70	10,000	100
32.	”	10,000	-	-	100
33.	”	10,000	1 40/70	10,000	100
34.	”	10,000	-	-	100
35.	”	10,000	1 40/70	10,000	100
36.	”	10,000	-	-	100
37.	”	10,000	1 40/70	10,000	100
38.	”	10,000	-	-	100
39.	”	10,000	1 40/70	10,000	100
40.	”	10,000	-	-	100
41.	”	10,000	1 40/70	10,000	100
42.	”	10,000	-	-	100
43.	”	10,000	1 70/140	10,000	100
44.	”	10,000	-	-	100
45.	”	10,000	1 70/140	10,000	100
46.	”	10,000	-	-	100
47.	”	10,000	1 70/140	10,000	100
48.	”	10,000	-	-	100
49.	”	10,000	1 70/140	10,000	100
50.	”	10,000	-	-	100
51.	”	10,000	1 70/140	10,000	100
52.	”	10,000	-	-	100
53.	”	10,000	1 70/140	10,000	100
54.	”	10,000	-	-	100
55.	”	10,000	1 70/140	10,000	100
56.	”	10,000	-	-	100
57.	”	10,000	1 70/140	10,000	100
58.	”	10,000	-	-	100
59.	”	10,000	1 70/140	10,000	100
60.	”	10,000	-	-	100
61.	”	10,000	1 70/140	10,000	100
62.	”	10,000	-	-	100
63.	”	10,000	1 70/140	10,000	100
64.	”	10,000	-	-	100

65.	"	10,000	1.25 70/140	12,500	100
66.	"	10,000	-	-	100
67.	"	10,000	1.5 70/140	15,000	100
68.	"	10,000	-	-	100
69.	"	10,000	1.75 70/140	17,500	100
70.	"	10,000	-	-	100
71.	"	10,000	2.0 70/140	20,000	100
72.	"	15,000	-	-	100

1

TOTALS: Brine water - 1,080,000 gals. Plus pit bottoms*
40/70 Ottawa - 168,000 lbs.
100 Mesh - 175,000 lbs.
15 % HCL - 1,500 gallons

The additional water is for extending flush stages where we are seeing banking and or excess pressure.

- DESIGN/FLUID CRITERIA:

1. Design Pump rate - 100 bpm.
2. Maximum Pump rate – 110 bpm.
3. Maximum treating pressure –10,000 psi. Anticipated pressure at 100 bpm 6,330 psi.
4. HHP required 15,600 HHP plus 100% standby. Standby for rate.
5. Fluid Brine water with friction reducer.
6. Based on pumping down 8,293' of 5" 18 # casing.

ADDITIVES:

1. Water based friction reducer. Cationic or copolymer for brine use.
2. Biocide –common bleach will suffice.
3. Scale inhibitor

ADDITIONAL EQUIPMENT:

1. 2 in-line densiometer(s).
2. Sand sieves, and associated equipment to perform QC on location. Sand sieves on all compartments and water analysis on pit water.
3. 100% Standby horsepower.
4. Pressure relief valve on the casing and kickouts or popoff required on downhole pumps.
5. Have ball gun on site and 75 mid-range bioballs if required.

Completed ?

Image A

Pictures of my work at Haliburton research; the middle picture includes the late Dr. Jiten Chatterji and Dr. Marlin Holtmeyer.

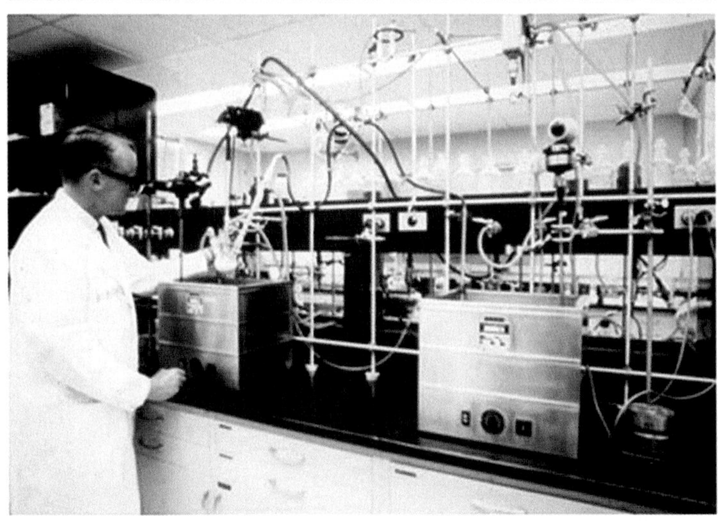

Image 1

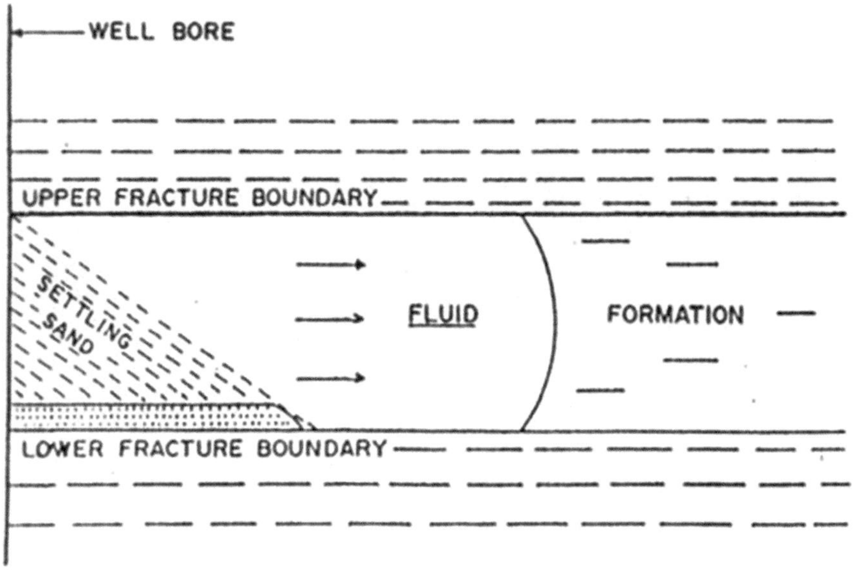

Early buildup of settled sand bank.

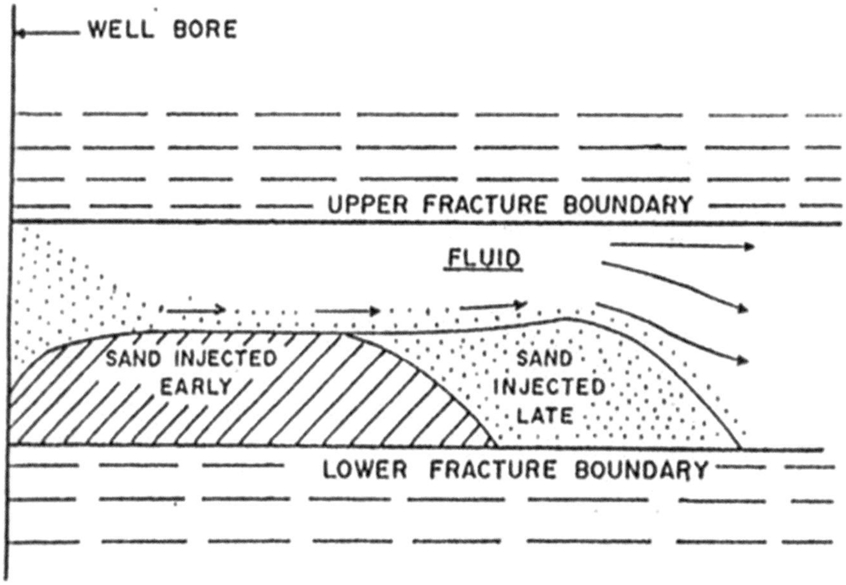

Probable final position of sand injected late in treatment.

Fifty-plus year-old slides illustrating the effects of Stoke's Law in Newtonian fluids.

CHAPTER 16

THE PHILOSOPHY OF QUALITY CONTROL (1994)

There are few endeavors, such as the space program and nuclear war, where there are more unknowns than in the stimulation of oil and gas wells. A fracturing treatment is always exciting because of its aura of mystery. This is especially true if the well is a little deeper or a little hotter than usual or if you are attempting a new procedure, pumping a new fluid, or trying out a new system.

In my early travels in the oilfield, I was terribly naive. I had a basic knowledge of what was going on downhole. I looked at computer designs and assumed that those programs were derived from real tests and were real data. Like Mark Twain, the older and more informed I became, the more frightened and confused I was about fracture design theory. I discovered we knew little about what was going on downhole and about what made up the formation.

Fracturing

Just before the mid-1960s, fracturing industry scholars were certain that all hydraulic fractures were horizontal. It made sense that fractures would propagate at the bedding planes. We have learned since that they don't. Rock mechanics research and field proof show that the majority of fracture systems at any depth are vertical. Frequently, I stood before crowds and said that the company I was working for knew everything there was to know about fracturing except the length, width, and height of the fractures. If you give me the height, I can calculate the length and width of the fracture system. Customers are sometimes incredulous when they hear this-the simple truth.

Recently in south Texas, I met with various oil company representatives. I was put on the spot and asked to evaluate the fracture height. Three of us estimated the fracture height from logs. We knowledgeable people gave the fracture height as 60 ft, 250ft and 350ft. The oil company man gave the smaller fracture height because his decision was based on economics. The other two figures were based on the fact that there were minimal barriers to fracture growth. The company man's decision was valid economically. Using a 350 ft high frac to treat a 50 ft net interval wouldn't make much sense. He, of course, wanted us to stay in the pay zone.

We are moving toward pinning down fracture height determinations. We are learning more as we go along about heterogeneous formations that we are trying to stimulate, but we are a long way from being scientific. There are many variables in the actual treatment design.

I often tell customers that most of our stimulation computer designs have a minimum of 30 variables. Usually, we can pin down only two or three of these; and, in many cases, one of these two or three is the name of the company. Other known variables may be (if we are lucky) temperature and depth. We assume Young's Modulus and Poisson's Ratio in the formation.* These assumptions may be made from a sonic log or from an offset well. We approximate fracture height. Fracture width estimations are based on the equation we use. The equation chosen is often the whim of the supplier, the service company, or the oil company. What I am saying is that there are numerous variables in choosing fracture designs and materials. With experience, we get pretty good at estimating. When we get real post-frac data, many times we find our assumed figures were fairly close to the actual numbers.

Carry out treatments as designed

When we don't know what chemicals were used on a stimulation treatment, evaluating data is difficult. Many years ago several customers and I attempted to 'write some SPE papers. I could not utilize wells from various parts of the country because the treatments were not carried out as designed. Over the years, I have seen many instances when the fluid that was designed was not used, when the proppant concentration or amount was changed, when the designated acid strength was varied, or when the additives were not put into the well. These alterations may have been mistakes, or they may have been larceny.

For many years, my personal goal and a philosophy of the company I own has been to do what we promised; we take a lot of pride in that. There is no reason why a treatment cannot be executed as planned. Customers would very much like the jobs run as planned rather than altered to save money or to increase the service company's profits.

Oil service companies and equipment glutted the market when oil prices dropped. The 1981-1982 oil surplus put many of these companies in a precarious position. Bidding wars raged, and losses for many companies have persisted and even increased. Many service companies have folded. When you take the food out of their children's mouths, the parents will resort to anything. I am not accusing any company or anyone, but the tremendous pressure to be profitable when bidding prices exceed 50% of 1981 prices increases the possibility that jobs may be conducted differently than designed. On the other hand, the oil company must take extra care to quality control and to be sure the jobs are conducted as designed.

Regardless of pressure, bidding, or cost cutting, the oil company should always be given the job designed. Every attempt should be made to conduct a treatment correctly. But not everyone feels this way.

Many service companies through good personnel and high-quality work have established trust between themselves and various oil companies. I have been on several treatments where the customer didn't even show up. He trusted us to pump and do the job as planned. There are people who take advantage of this. With the possible demise of the company looming ahead, the pressure is sometimes overwhelming. Unfortunately, an honest service company that runs the additives paid for and conducts the treatment as planned competes with those that cut back on polymer, proppant, surfactants, acid inhibitor, or acid strength. You may say, "They'll be found out in the end." This does not always happen.

Some years ago, a moderate-sized treatment was conducted. Because of a malfunction in sand screws, no proppant was run. The entire treatment was pumped with the customer standing by the treater with the treater marking concentrations on the chart. Of course, it was an uneventful job; the well did not respond. The customer gave up the lease thinking he had given it his best shot. This sounds a bit unreal, something that couldn't happen. It did. Numerous treatments have been run when one-third, or certainly less than one-half of the planned proppant was run. I've been on a location with the customer and counted the sand storage a competitor was using on a treatment. There wasn't enough sand storage on the location to hold even one-half of that for the designed treatment.

Finding out someone has short-changed you on sand or proppant is easy. People who don't run surfactants, who run very low concentrations of acid inhibitor, and who cut back gel strength are harder to detect. But with a little diligence, these deceptions can be unveiled. Everybody should compete straight up, on the same basis. We are all trying to stimulate oil and gas wells. Unless treatments are conducted as designed, we have no way of knowing if we can achieve what we are attempting. We've come a long way in gathering pre-frac data, in running treatments as designed, and in evaluating them afterward. With proper quality control, with diligent oil company men overseeing conscientious service companies, every treatment can be conducted as planned. The success ratio of treatments will increase accordingly.

I have been accused (probably because of comments such as those in the previous paragraphs) of being hard and critical of various service companies. These comments, which were part of the original edition of this book, were written at a time when I was personally aware of larcenous behavior going on in the oilfield. I want to state unequivocally that there is no service company today, to my knowledge, cheating anyone on fracture treatments. What occurred in the 1960s, 1970s, and to some extent in the 1980s, was some individuals who work for those companies took advantage of specific events to short customers for their own personal gain. As has been discussed in previous chapters, large lawsuits such as those that went on with Parker and Parsley in west Texas have, I believe, once and for all, caused the demise of any of this sort of thing going on in the industry. We still have equipment that fails and we still have human beings on locations. These human beings make mistakes. It is my conviction that intense quality control must be done on every job, not to catch people and not to harass individuals, but simply to be assured because of the complexity of the systems and the equipment that we use that jobs are conducted as planned.

As I was editing and going through my previous book, I changed most of the chapters tremendously and in fact have changed my mind and changed some theoretical basis. Other than events and occurrences that have happened since I wrote that book, I have changed very little about what was said.

The discounts continue. Service companies are continually bought out. We are seeing a tremendous amount of aging in the pumping equipment within our industry, but I also see some very bright things on the horizon. I recently did some work in the Rocky Mountains where one service company was building three quality control vans for one location. Quite frankly, I hadn't seen more than one quality control van in any station anywhere in the United States prior to this. In-fact, prior to 1992 a quality control van was a very unusual piece of equipment. It only existed in the eyes of consultants and perhaps the Gas Research Institute. Service companies taking quality control and job execution very seriously will be the ones that will survive. I am continually amazed, as I go out on treatments day after day, to find lackadaisical and in many cases a very negative attitude toward doing specific quality control measures. The basic attitude of some service companies is still, "Leave us alone and we will do a good job. And trust us; we always do the very best." It is not a question of trust. It is a very difficult problem to execute a treatment such as a hydraulic fracture treatment with cross-link fracturing fluid as planned. It is far more difficult than I dreamed, even when I wrote the original edition of this book. We are still seeing at least 70% of the time potential job failure situations where we do intense quality control. As the service companies start to see the importance of this, as they start to build the quality control vans, and as they start to train their own people, then I think this situation will turn around. I can state that I never dreamed that the problem of execution and quality control on fracturing treatments was as serious as it truly is. A great deal of the problem has been due to the lack of training and due to the severe economic problems that have existed in the oilfield service industry. Secondly, it has been because of the lack of concerted effort by upper management to take quality assurance seriously. What I am most impressed about lately is that this attitude is starting to change and the results are forthcoming.

Summary

- Stimulation: Many Unknowns.
- More Information = Better Design = Improved Results.
- Treatments can be executed as designed.
- Good quality control by the oil company and the service company will make a successful treatment.

JOHN W. ELY

Mr. Ely started his career with Halliburton Co., in 1965, working as a technician for the Analytical group while completing his college work. He graduated from Oklahoma State University in 1968 with a Batcheler's in chemistry. On returning to Halliburton, Mr. Ely served as chemist and senior chemist in fracturing research before transferring to International Operations in 1973. While in fracturing research, he was instrumental in the development of the first high temperature fracturing fluids. He also developed nonaqueous energized systems. In 1973 he transferred to his first assignment overseas was in south Iran as a district engineer. In 1975, he was promoted to technical adviser, Eastern Hemisphere, and transferred to Bahrain. He traveled and worked in eleven Mideast countries in this position. In 1976, John transferred to Dubai, U.A.E. The following year, he was promoted to technical adviser, International Operations, and was based in Duncan, Oklahoma where his primary duty was to coordinate all phases of research with international field operations. Ely joined Nowsco Services in 1980 as Engineering Manager. His responsibilities included overseeing chemical and mechanical research and coordinating training for field engineers.

In 1985, John joined S.A. Holditch and Associates as Vice-President of Stimulation Technology. In addition to designing and supervising hundreds of stimulation treatments, he was involved in research on fracturing fluids under the auspices of the Gas Research Institute. Additionally, he served as an expert witness in areas involving completion and stimulation of oil and gas wells.

In May 1991, John with three partners founded Ely & Associates Corp. This company provides a blend of practical and technical expertise on well completion, stimulation fluids and equipment, and reservoir analysis. John holds several patents and has numerous publications, including the first and second edition of books titled "Stimulation Treatment Handbook/ An Engineer's Guide to Quality

Control". He is also a contributing author to the 1989 and 2020 issue of the S.P.E. monograph on hydraulic fracturing (Recent Advances in Hydraulic Fracturing), writing chapters on hydraulic fracturing fluids. Presently John is writing the third edition of his book this time titled "Frac without a K" which updates the previous versions and goes into depth about slick water fracs. He has also published a chapter in the "Modern Fracturing" publication by BJ Services. Mr. Ely typically teaches hydraulic fracturing 3-5 times a year. Ely and Associates has has become known as the premier fracture design and oversight organization in the industry. Particularly in the last 15 years John has become recognized as the leading expert in the design and implementation of the water fracs. Ely has designed literally thousands of water fracs using technology learned while exploiting both source rock reservoirs and tight conventional rock. His "Waterfrac Sweep" process has been successful in shale, tight sands, and fractured carbonates. Ely and Associates presently is involved in more than 5000 wells per year throughout the world. Mr. Ely is a member of the American Chemical Society, The Society of Petroleum Engineers, and is a fellow of the American Institute of Chemistry. He was the recipient of the prestigious John Franklin Carll award for excellence in petroleum engineering for 2020. In 2021, he was awarded the Legend of Hydraulic Fracturing award at the SPE Hydraulic Fracturing Conference.

CHAPTER 16

PHILOSOPHY REVISITED AND ATTACK ON OUR INDUSTRY (2022)

Upon reading the 1994 chapter *"A philosophy of Quality Control"*, I was tempted to not write the 2021 version because virtually everything stated in the chapter still is pertinent. I have watched with dread the statement that fracturing is just a manufacturing process, and no real oversight is needed. These statements combined with the vicious attack on fracturing by frac scholars such as Yoko Ono and Matt Damon have put a discreet pall on our industry. These wise people who do not even know how to spell frac, combined with the pandemic, have created, without a doubt a scenario where our industry is teetering on the precipice of failure.

We are all excited to see the huge capability of analyzing data at massive speed. I am very concerned that we are conducting massive treatments where we have no idea within experimental accuracy how much fluid, proppant, or what size and or size distribution is pumped.

All of us are excited to have found, not necessarily from science, ways to produce oil and gas from nanodarcy rock. Without knowing what and how much we pumped the finest software cannot calculate a means to optimize the treatments to achieve even better success. There are straightforward diagnostics to optimize treatments in real time but have no function if we are 20-25 % off on not only rate and volume but have no idea what size distribution of proppant we have pumped.

I do not want to finish the book on a sour note so will state that I still believe in our industry and the young personnel coming forward and they will right the ship in the direction of quality and technology.

CHAPTER 17

DIRECTIONAL KILL BLOWOUTS (2022)

\What does directional kill blowouts have to do with hydraulic fracturing. In truth very little, as one of the criteria in that type of operation is not to reach fracture pressure during kill operations. For those not familiar with directional kill operations, it amounts to taking control of an uncontrolled well from the bottom up. What I am describing has nothing in common with what occurred in Kuwait or all the pictures of kill teams setting off explosives to put the fire out of an uncontrolled well to allow personnel to go to the wellhead to make repairs. Directional control killing is required typically when there is nothing left on the surface to repair. Another example is pictures I will show of wells in Sumatra where the gas contained hydrogen sulfide and the producing temperature was over 300 degrees F. In the case of Sumatra, we had to keep the well burning to protect the nearby area from H2S, a deadly gas. See images a-1.

That still does not answer why I am including this in the book. The absolute truth is that these types of operations were a major part of my life and I believe very interesting, and I do not plan on writing another book but believe most of the oilfield people who read this book for fracturing would be interested in my exploits in directional kill wells. If you are not interested, then cease and desist and move back to Steven King etc.

My first experience with directional kill attempts was a huge blowout off Dubai, UAE in 1975 and 1976. Dubai Petroleum Company was drilling a well planned for approximately 9,000 feet when the drill bit encountered a high-pressure zone around 3000 feet and for whatever reason the drilling personnel lost control and the rig caught on fire later to disappear below the seabed in about 200 feet of water. If you are interested, I presented an SPE paper covering the blowout in the national SPE meeting in 1977. The paper is SPE 6835. In 1975 the ability to drill directly into a blowout had not been perfected but standard drilling practices could allow us to drill close enough to hopefully be able to kill the well from below. Two drilling rigs started drilling and quickly got close to the producing blowout well. The problem was that there existed huge salt cavities that the injection of sea water only enlarged and the well continued producing unabated. I was the Halliburton technical rep and formulated an idea to pump controlled hydration polymer to enhance the ability to kill the well directionally. See images J-O.

Eventually the idea became a Halliburton patent, and the product was used on the Ixtoc-I blowout and was planned to be used on an uncontrolled well in Syria. I made trips to Conoco's headquarters in the US explaining the concept of the slow hydration polymer and proposed use of a guar crosslinked polymer to plug off the salt cavities. Obviously, they bought my sales pitch and Halliburton eventually shipped more than 2 million pounds of cellulose polymer and several hundred thousand pounds of HPG and the other components of a crosslinked gel on leased 747 airliners. The effort put forth by Halliburton and Dowell would have made a good John Wayne movie.

The first attempt appeared to be working with the well dying when a decision was made to switch polymers, resulting in the fire roaring back to life. We pumped all our polymer, all our mud, all our cement and watched the fire grow back to its original intensity. The decision was made to reorder everything and start again. We got everything loaded and were headed out when a very strong storm came into the Gulf shutting operations down. The storm moved the drillship into the primary jack up, sinking the jack up, and killing the tool pusher. The next morning the fire was out and the explanation was the salt cavities had collapsed. Haliburton went to one of the injection wells and pumped 30,000 sacks of radiated cement. We never located the cement after logging all injection wells. Divers went down in the cavity where the blowout occurred and found no evidence of the lost rig 200 feet below the surface and abandoned all further work.

I was deeply involved with 2 high pressure sour gas wells in Sumatra. One of the blowouts was due to drilling problems and the other was due to an earthquake allowing for the first use of dynamic kill procedures and the Magrange tool, allowing for direct intercept downhole. These wells encompassed more than 7 months of my life. The location was just south of Vietnam and had very similar topography. Halliburton made a major effort in responding to Mobil's need shutting down manufacturing to only build skids for the blowout. These 2 blowouts alone would make a decent book.

The next blowout occurred in the bay of the Campeche Sound and the well was known as Ixtoc-I. See images P-W. On this well I made 19 trips to Mexico and on both this blowout and the Sumatra kill operations, I was fortunate enough to get well acquainted with Red Adair. He was a very dynamic person with no resemblance to John Wayne in the Hell Fighters. Directional kill was not Red Adair's forte, but he depended on Halliburton to do the technical and grunt work. I consider myself privileged to have been around such a prominent individual. It was, and remains, my singular opinion that the Ixtoc-I oil spill was as big, if not bigger, than that which we mostly recently witnessed during the Macondo oil well blowout in the Gulf of Mexico. We did pump polymer and had pumping barges and boats for the duration of the blowout. The well effectively died on its own. In addition to the polymer used, a very large conical sombrero was put over the fire but had been under designed. Also, very dense balls were pumped into the wellhead without any effect.

I participated in another blowout in Argentina, where the wellbore tubulars were blown completely out of the well. I do not have pictures of this event, but others do. The blowout was near the Colorado river in Argentina.

Image A

Pictures in timed sequence of occurrences with the well. Since this was the second occurrence, we were able to kill the well in less than 50 days, utilizing dynamic kill operations. Mr. Elmo Blunt was in charge of the dynamic kill operation.

Image B

Image C

Image D

No. 13
Monitors installed to cool C-III-8 wellhead to minimize valve and flange failure - 21 April 1980

Image E

No. 14
Monitor coverage to cool C-III-8 wellhead - 21/4/80

Image F

Image G

Image H

Image I

Image J

Image K

Top - Picture of fire at night, through 200-feet of water.

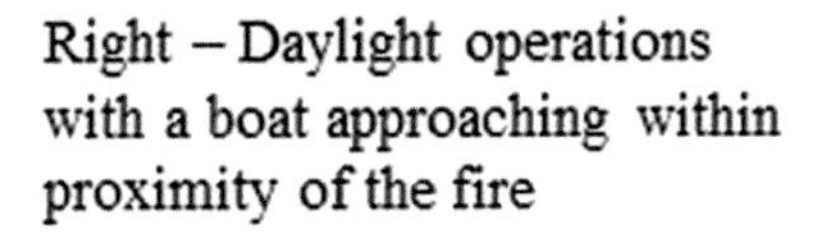

Right – Daylight operations with a boat approaching within proximity of the fire

Image L

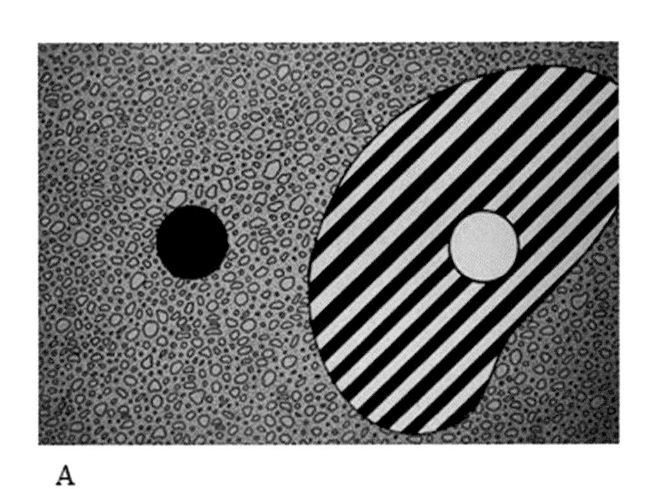

A

B

C

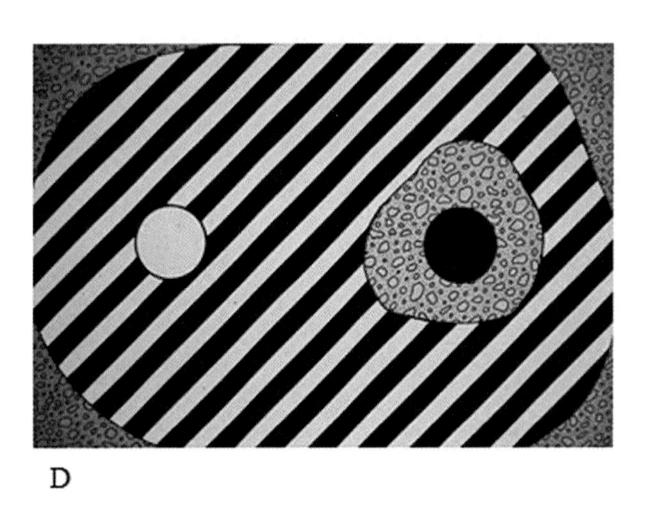

D

Illustrations of how the polymer is used to control the flow of kill fluid into the well.

Image M

Distant picture of the drill ship that eventually, during a very large storm, drifted into the jack-up. This turned the jack-up over and into the water.

Image N

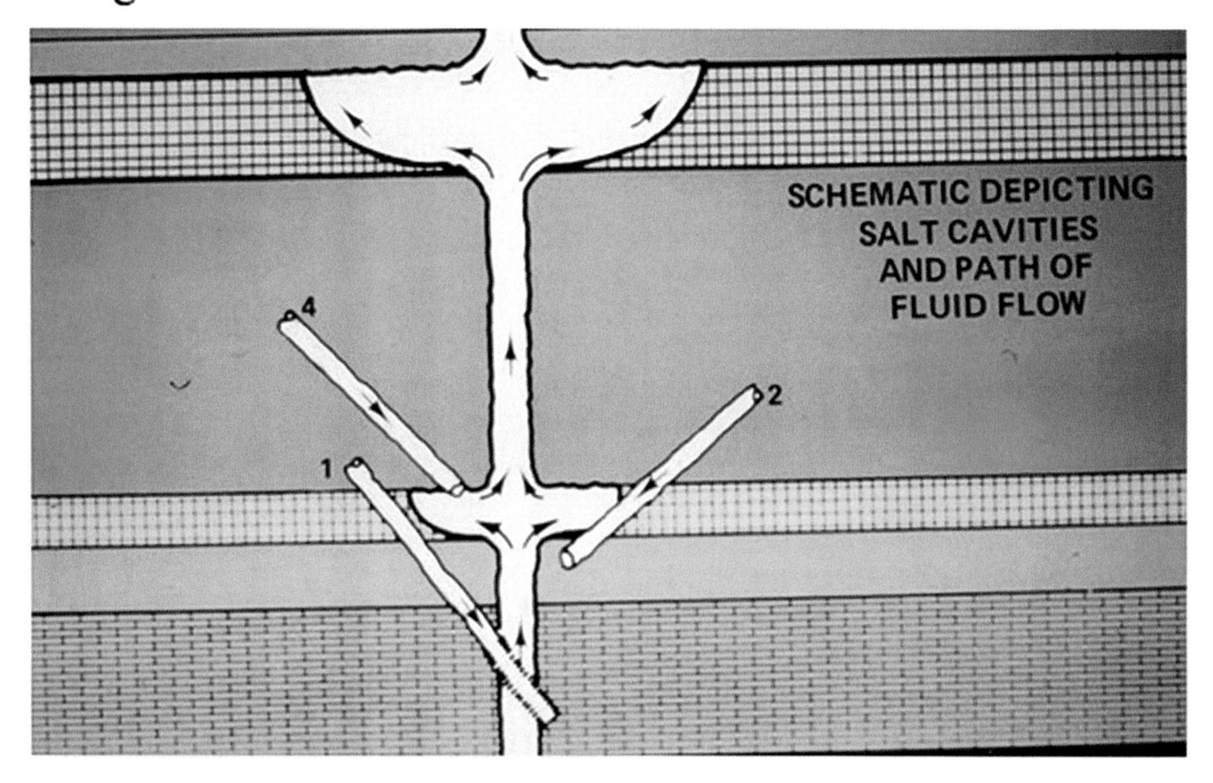

Image O

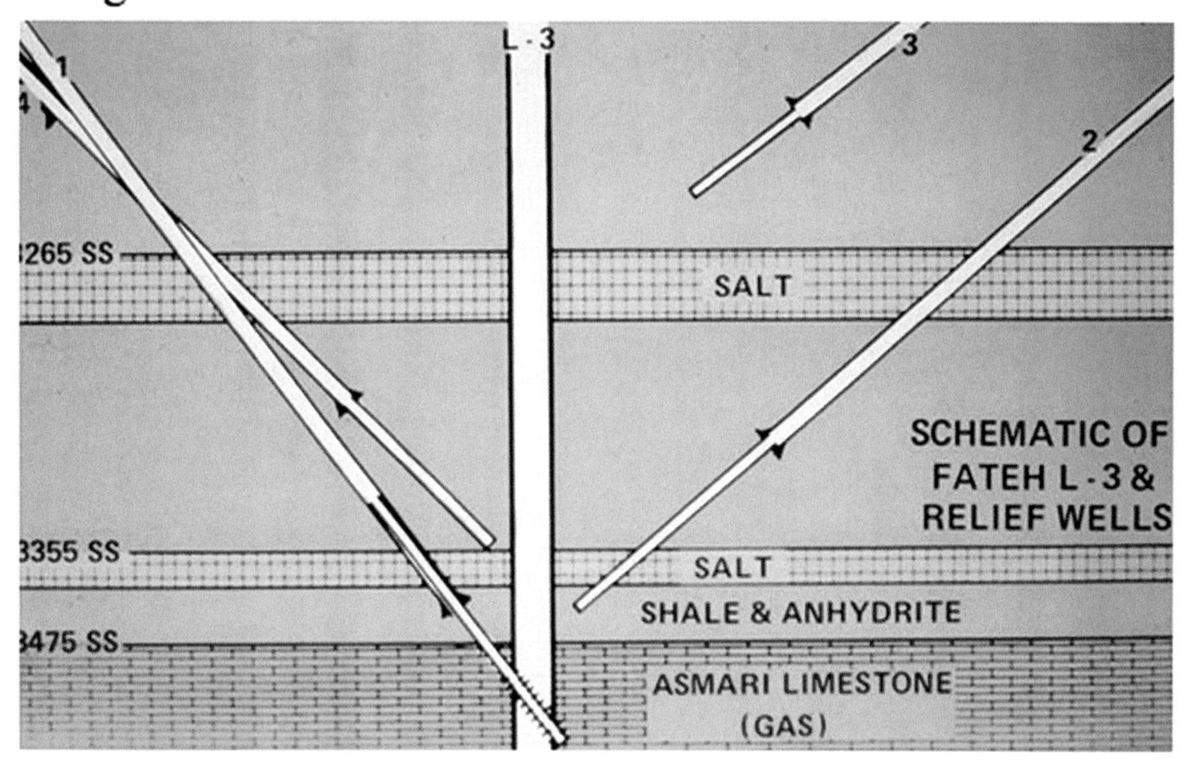

Image P

First failed attempt with the polymer made us flow the well back and ended up with 'noodles' of HEC polymer on the surface of the water. Approximately 10,000-feet of these noodle segments, by my estimation.

Image Q

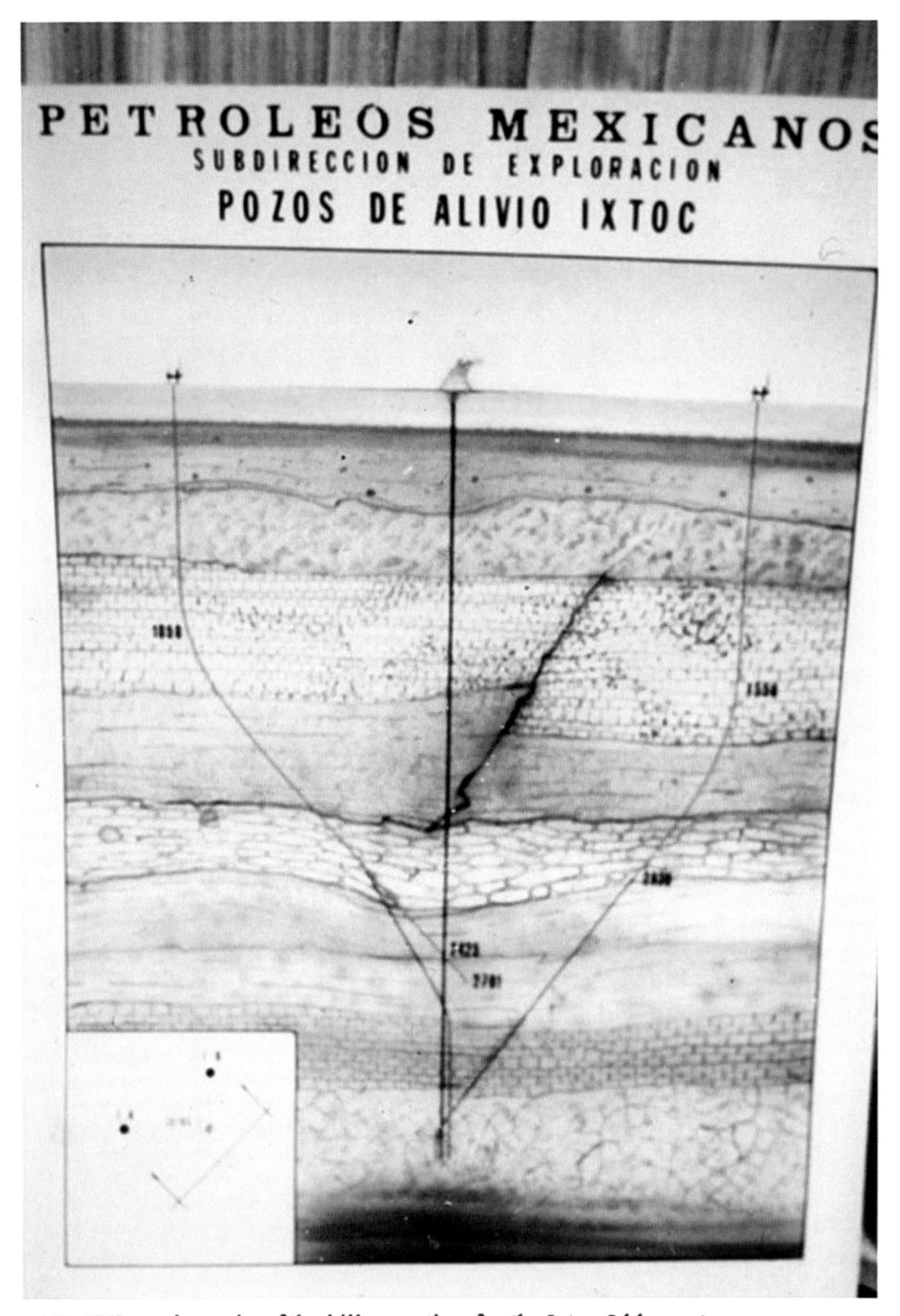

PEMEX – schematic of the kill operation for the Ixtoc-I blowout.

Image R

Top – Red Adair walking with Haliburton-Mexico Division Engineer

Right – Manifold utilized for pumping the infamous weighted balls.

Image S

Image T

Haliburton – pumping vessel, loaded with pumping units.

Image U

Viewpoint from the jack-up drilling rig of the blowout.

Image V

Image W

Two illustrations of oil migrating North, from the Campeche Sound of Mexico.

Printed in the United States
by Baker & Taylor Publisher Services